Progress in Probability
Volume 20

Series Editors
Loren Pitt
Thomas Liggett
Charles Newman

Probability in Banach Spaces 6

Proceedings of the Sixth International Conference, Sandbjerg, Denmark 1986

U. Haagerup
J. Hoffmann-Jørgensen
N.J. Nielsen
Editors

1990

Birkhäuser
Boston · Basel · Berlin

U. Haagerup
Matematisk Institut
Odense Universitet
DK-5230 Odense M
Denmark

J. Hoffmann-Jørgensen
Matematisk Institut
Aarhus Universitet
DK-8000 Aarhus C
Denmark

N.J. Nielsen
Matematisk Institut
Odense Universitet
DK-5230 Odense M
Denmark

With 10 Illustrations.

ISSN: 0892-063X

Library of Congress Cataloging-in-Publication Data
Probability in Banach spaces 6 : proceedings of the sixth
  international conference, Sandbjerg, Denmark 1986 / U.
  Haagerup, J. Hoffmann-Jørgensen, N.J. Nielsen, editors.
       p.   cm.—(Progress in probability, ISSN 0892-063X ; v. 20)
    "Selection of papers . . . 6. International Conference on
  Probability in Banach Spaces, Sandbjerg, Denmark, June 16–21, 1986"-
  -Pref.
    Includes bibliographical references.
    ISBN-13: 978-1-4684-6783-3
    1. Probabilities—Congresses.   2. Banach spaces—Congresses.
  I. Haagerup, U.   II. Hoffmann-Jørgensen, J.   III. Nielsen, N. J.
  IV. International Conference on Probability in Banach Spaces (6th :
  1986 : Sandbjerg, Denmark)   V. Series.
  QA273.43.P77   1990
  519.2—dc20   89-18363

Printed on acid-free paper.

ISBN-13: 978-1-4684-6783-3   e-ISBN-13: 978-1-4684-6781-9
DOI: 10.1007/978-1-4684-6781-9

Camera-ready copy prepared by the editors using TeX.

9 8 7 6 5 4 3 2 1

Preface

This volume contains a selection of papers by the participants of the *6. International Conference on Probability in Banach Spaces, Sandbjerg, Denmark, June 16–21, 1986.* The conference was attended by 45 participants from several countries. One thing makes this conference completely different from the previous five ones, namely that it was arranged jointly in Probability in Banach spaces and Banach space theory with almost equal representation of scientists in the two fields. Though these fields are closely related it seems that direct collaboration between researchers in the two groups has been seldom. It is our feeling that the conference, where the participants were together for five days taking part in lectures and intense discussions of mutual problems, has contributed to a better understanding and closer collaboration in the two fields.

The papers in the present volume do not cover all the material presented in the lectures; several results covered have been published elsewhere.

The sponsors of the conference are:

> The Carlsberg Foundation,
>
> The Danish Natural Science Research Council,
>
> The Danish Department of Education,
>
> The Department of Mathematics, Odense University,
>
> The Department of Mathematics, Aarhus University,
>
> The Knudsen Foundation, Odense,
>
> Odense University,
>
> The Research Foundation of Aarhus University,
>
> The Tuborg Foundation.

The participants and the organizers would like to thank these institutions for their support.

The Organizers.

Contents

Participants

A. de Acosta, *Case Western Reserve University, Cleveland, Ohio, USA.*

K. Alexander, *University of Washington, Seattle, Washington, USA.*

N.T. Andersen, *Texas A & M University, College Station, Texas, USA.*

C. Borell, *Chalmers Tekniska Högskola, Göteborg, Sweden.*

J. Bourgain, *Institut des Hautes Études Scientifiques, Bures-sur-Yvette, France.*

P. Casazza, *University of Missouri, Columbia, Missouri, USA.*

J.P.R. Christensen, *Københavns Universitet, København, Denmark.*

V. Dobric, *Gradevinski Institut, Zagreb, Yugoslavia.*

R. Dudley, *Massachusetts Institute of Technology, Cambridge, Massachusetts, USA.*

E. Eberlein, *Universität Freiburg, Freiburg, Germany.*

X. Fernique, *Université Louis Pasteur, Strasbourg, France.*

T. Figiel, *Polska Akademia Nauk, Sopot, Poland.*

P. Gaensler, *Universität München, München, Germany.*

D.J.H. Garling, *University of Cambridge, Cambridge, England.*

N. Ghoussoub, *University of British Columbia, Vancouver, Canada.*

E. Giné, *Texas A & M University, College Station, Texas, USA.*

V. Goodman, *Indiana University, Bloomington, Indiana, USA.*

Y. Gordon, *Technion, Haifa, Israel.*

S.E. Graversen, *Aarhus University, Aarhus, Denmark.*

S. Guerre, *Université de Paris VI, Paris, France.*

M. Hahn, *Tufts University, Medford, Massachusetts, USA.*

B. Heinkel, *Université Louis Pasteur, Strasbourg, France.*

J. Kuelbs, *University of Wisconsin, Madison, Wisconsin, USA.*

H. König, *Universität Kiel, Kiel, Germany.*

M. Ledoux, *Université Louis Pasteur, Strasbourg, France.*

W. Linde, *Friedrich-Schiller-Universität, Jena, DDR.*

J. Lindenstrauss, *The Hebrew University, Jerusalem, Israel.*

P. Mankiewicz, *Polska Akademia Nauk, Warszawa, Poland.*

M. B. Markus, *University of Texas, Austin, Texas, USA.*

V. D. Milman, *Tel Aviv University, Tel Aviv, Israel.*

A. Pajor, *Université de Lille I, Villeneuve d'Ascq, France.*

S. Schechtman, *Weizmann Institute, Rehovot, Israel.*

C. Schütt, *Universität Kiel, Kiel, Germany.*

A. Szankowski, *Hebrew University, Jerusalem, Israel.*

M. Talagrand, *Université de Paris VI, Paris, France.*

N. Tomczak-Jaegermann, *University of Alberta, Edmonton, Canada.*

L. Tzafriri, *Hebrew University, Jerusalem, Israel.*

G. Wittstock, *Universität des Saarlandes, Saarbrücken, Germany.*

W.A. Woyczynski, *Case Western Reserve University, Cleveland, Ohio, USA.*

A. Wojtaszyk, *Polska Akademia Nauk, Warszawa, Poland.*

J.E. Yukich, *Lehigh University, Bethlehem, Pennsylvania, USA.*

J. Zinn, *Texas A & M University, College Station, Texas, USA.*

Organizers.

U. Haagerup, *Odense University, Denmark.*

J. Hoffmann-Jørgensen, *Aarhus University, Denmark.*

N.J. Nielsen, *Odense University, Denmark.*

On the identification of the limits in the law of the iterated logarithm in Banach spaces

A. DE ACOSTA[1] AND M. LEDOUX

1. Introduction.

In the very recent paper [7], Ledoux and Talagrand have character-ized the Banach space valued random vectors which satisfy the bounded law of the iterated logarithm (BLIL), as well as those which satisfy the compact law of the iterated logarithm (CLIL) (for a description of these questions, see [2], [3], [7]).

In the latter situation, it is well known that

$$\text{cluster set } (\{S_n/a_n\}) = K \qquad \text{a.s.,} \qquad (1.1)$$

$$\lim_n d(S_n/a_n, K) = 0 \qquad \text{a.s.,} \qquad (1.2)$$

where $a_n = (2n \log \log n)^{\frac{1}{2}}$, S_n is the sum of n independent copies of the random vector X and K is the unit ball of the reproducing kernel Hilbert space of $\mathcal{L}(X)$; K is compact in this case.

However, when X satisfies the BLIL but not the CLIL, (1.1) and (1.2) need not be true (see [5], [8]). A general result of the form (1.1), in which K may not be compact, was proved in de Acosta and Kuelbs [1], but far less is known about (1.2). In fact, only in the Hilbert space case is it known that if X satisfies the BLIL then (1.2) (and (1.1)) hold ([1], Corollary 4); a weaker result in more general spaces is proved in [1], Theorem 2.

The aim of the present paper is to determine, or, at least, provide non-trivial bounds for, the quantities

$$\overline{\lim}_n p(S_n/a_n),$$

[1] This author's research was partially supported by an NSF grant.

where p is any continuous seminorm (Theorem 2.1 and Corollary 2.2), and derive from these bounds statements which are close to (1.2) (Corollary 2.3). As a consequence of the method of proof, which follows closely the Gaussian randomization approach and the arguments in [7], our results are sharp when the random vector is of the form gX, where g is a standard Gaussian r.v. independent of X, somewhat less accurate when X is symmetric and still less accurate for general X. The problem of the exact identification of the limits for general X appears to be a difficult one.

Notation. For $n \geq 1$, $a_n = (2n \, LLn)^{\frac{1}{2}}$, where $LLn = \log\log n$ for $n \geq 3$, $LLn = 1$ for $n = 1, 2$. S_n or $S_n(X)$ will denote the n-th partial sum of independent copies of a random vector X taking values in a separable Banach space B. p will denote a continuous seminorm on B; for simplicity, we will assume that p is equivalent to the norm $\|\cdot\|$ on B, although the results are clearly true without this assumption. p^* is the dual norm on the dual space B^*. g will denote a standard Gaussian r.v., independent of X.

2. Inequalities for the limits.

We introduce the following quantities:

$$\Lambda_p(X) = \overline{\lim}_n p(S_n/a_n)$$

(let us recall that by the zero-one law, $\Lambda_p(X)$ is a.s. constant);

$$\sigma_p(X) = \sup\{(Ef^2(X))^{\frac{1}{2}} : f \in B^*, \, p^*(f) \leq 1\};$$
$$\Gamma_p(X) = \overline{\lim}_n E \, p(S_n/a_n).$$

It is proved in [7] that the conditions
(i) $Ef^2(X) < \infty$ and $Ef(X) = 0$ for all $f \in B^*$,
(ii) $E(\|X\|^2/LL\|X\|) < \infty$,
(iii) $\{S_n/a_n\}$ is stochastically bounded

are equivalent to

$$\Lambda_p(X) < \infty. \tag{2.1}$$

The following result, formulated along the lines of Th. 7 of [2], makes (2.1) more precise. In the statement we use the notation

$$c_g = \int_0^\infty (2P(\{|g| > t\}))^{\frac{1}{2}} dt;$$

an easily obtained but not sharp estimate is $c_g \leq 4$.

Let us remark that clearly $\sigma_p(gX) = \sigma_p(X)$.

Theorem 2.1. *Assume that X satisfies (i)–(iii). Then*

(1) $\max\{\sigma_p(gX), \Gamma_p(gX)\} \leq \Lambda_p(gX) \leq \sigma_p(gX) + 2\Gamma_p(gX)$.

(2) *If X is symmetric,*

$$\max\{\sigma_p(X), \Gamma_p(X)\} \leq \Lambda_p(X) \leq (\pi/2)^{\frac{1}{2}}(\sigma_p(X) + 2c_g\Gamma_p(X)).$$

(3) *For general X,*

$$\max\{\sigma_p(X), \Gamma_p(X)\} \leq \Lambda_p(X) \leq \pi^{\frac{1}{2}}\sigma_p(X) + (2\pi)^{\frac{1}{2}}2c_g\Gamma_p(X).$$

We shall need the following lemma. Let us recall that the Lorentz space $L_{2,1}$ is defined as the class of real valued r.v.'s ξ such that

$$\|\xi\|_{2,1} = \int_0^\infty (P(\{|\xi| > t\}))^{\frac{1}{2}} dt < \infty.$$

It will be useful to introduce the quantity

$$\widetilde{\Gamma}_p(X) = \sup_n Ep\left(\frac{S_n(X)}{a_n}\right).$$

Statement (2) of the Lemma is not needed in the rest of the paper, but it is used in the proof of (1).

Lemma.   *For any random vector X and any symmetricl r.v. $\xi \in L_{2,1}$, independent of X,*

(1) $\Gamma_p(\xi X) \leq 2^{\frac{1}{2}} \|\xi\|_{2,1} \Gamma_p(X),$

(2) $\tilde{\Gamma}_p(\xi X) \leq 2^{\frac{1}{2}} \|\xi\|_{2,1} \tilde{\Gamma}_p(X).$

Proof: The argument is similar to that in [6], pp. 363–364. Obviously, we may assume that $\Gamma_p(X)$ is finite. Also, it is enough to prove that for any positive $\xi \in L_{2,1}$,

$$\Gamma_p(\xi X) \leq \|\xi\|_{2,1} \Gamma_p(X), \qquad\qquad (2.2)$$

$$\tilde{\Gamma}_p(\xi X) \leq \|\xi\|_{2,1} \tilde{\Gamma}_p(X); \qquad\qquad (2.3)$$

in fact, (1) (resp., (2)) for symmetric ξ follows by applying (2.2) (resp., (2.3)) to ξ^+ and ξ^-.

We will establish (2.3) first. Let $\xi = I_A$, where A is a measurable set. Then $S_n(\xi X)$ and $S_{S_n(\xi)}(X)$ have the same distribution. By the independence of ξ and X, we may write $P = P_\xi \otimes P_X$ and $E = E_\xi E_X$ (the meaning of these symbols being the obvious one). We have

$$
\begin{aligned}
\tilde{\Gamma}_p(\xi X) &= \sup_n Ep\left(\frac{S_{S_n(\xi)}(X)}{a_n}\right) \\
&= \sup_n E_\xi E_X p\left(\frac{a_{S_n(\xi)}}{a_n} \frac{S_{S_n(\xi)}(X)}{a_{S_n(\xi)}}\right) \\
&\leq \sup_n E_\xi \left(\frac{a_{S_n(\xi)}}{a_n}\right) \tilde{\Gamma}_p(X) \\
&\leq (P(A))^{\frac{1}{2}} \tilde{\Gamma}_p(X),
\end{aligned}
\qquad (2.4)
$$

since $E_\xi\left(\frac{a_{S_n(\xi)}}{a_n}\right) \leq \left(E_\xi\left(\frac{a_{S_n(\xi)}}{a_n}\right)^2\right)^{\frac{1}{2}} \leq \left(E\left(\frac{S_n(\xi)}{n}\right)\right)^{\frac{1}{2}} = (P(A))^{\frac{1}{2}}.$

Next, suppose that ξ is a positive r.v. in $L_{2,1}$ and for each $\varepsilon > 0$, define $\xi_\varepsilon = \sum_{k=1}^{\infty} \varepsilon I_{\{\xi > \varepsilon k\}}$. Then $\xi_\varepsilon \leq \xi$, $\lim_{\varepsilon \to 0} \xi_\varepsilon = \xi$ and

$\lim_{\epsilon \to 0} \|\xi - \xi_\epsilon\|_{2,1} = 0$. We have, by (2.4),

$$\tilde{\Gamma}_p(\xi_\epsilon X) \leq \sup_n \sum_{k=1}^{\infty} \epsilon Ep\left(\frac{S_n(I_{\{\xi > \epsilon k\}} X)}{a_n}\right)$$

$$\leq \sum_{k=1}^{\infty} \epsilon \tilde{\Gamma}_p(I_{\{\xi > \epsilon k\}} X) \tag{2.5}$$

$$\leq \sum_{k=1}^{\infty} \epsilon (P\{\xi > \epsilon k\})^{\frac{1}{2}} \tilde{\Gamma}_p(X)$$

$$\leq \|\xi\|_{2,1} \tilde{\Gamma}_p(X),$$

where the last inequality follows easily from the definition of $\|\xi\|_{2,1}$. Since, for each $n \geq 1$, we have by (2.5)

$$Ep\left(\frac{S_n(\xi X)}{a_n}\right) < \varliminf_m Ep\left(\frac{S_n(\xi_{1/m} X)}{a_n}\right)$$

$$\leq \|\xi\|_{2,1} \tilde{\Gamma}_p(X),$$

it follows that for every positive $\xi \in L_{2,1}$, (2.3) holds.

The proof of (2.2) follows similar steps. As before, we write

$$Ep\left(\frac{S_n(\xi X)}{a_n}\right) = E_\xi\left\{\left(\frac{a_{S_n(\xi)}}{a_n}\right) Exp\left(\frac{S_{S_n(\xi)}(X)}{a_{S_n(\xi)}}\right)\right\}.$$

By Fatou's lemma, which is applicable in view of the inequality

$$\sup_n \frac{a_{S_n(\xi)}}{a_n} Exp\left(\frac{S_{S_n(\xi)}(X)}{a_{S_n(\xi)}}\right) \leq \tilde{\Gamma}_p(X) < \infty,$$

and the strong law of large numbers, we have

$$\Gamma_p(\xi X) \leq E_\xi \varlimsup_n \left\{\left(\frac{a_{S_n(\xi)}}{a_n}\right) Exp\left(\frac{S_{S_n(\xi)}(X)}{a_{S_n(\xi)}}\right)\right\}$$

$$\leq (P(A))^{\frac{1}{2}} \Gamma_p(X).$$

Next,

$$\Gamma_p(\xi_\varepsilon X) \leq \overline{\lim}_n \sum_{k=1}^{\infty} \varepsilon E p \left(\frac{S_n(I_{\{\xi > \varepsilon k\}} X)}{a_n} \right)$$

$$\leq \sum_{k=1}^{\infty} \varepsilon \Gamma_p(I_{\{\xi > \varepsilon k\}} X)$$

$$\leq \|\xi\|_{2,1} \Gamma_p(X).$$

Here the second inequality is justified by the steps leading to (2.5). Finally, for positive $\xi \in L_{2,1}$,

$$\Gamma_p(\xi X) \leq \Gamma_p(\xi_\varepsilon X) + \Gamma_p((\xi - \xi_\varepsilon)X)$$

$$\leq \|\xi\|_{2,1} \Gamma_p(X) + \tilde{\Gamma}_p((\xi - \xi_\varepsilon)X)$$

and since $\tilde{\Gamma}_p((\xi - \xi_\varepsilon)X) \leq \|\xi - \xi_\varepsilon\|_{2,1} \tilde{\Gamma}_p(X)$ by (2.3), letting $\varepsilon \to 0$ we obtain (2.2).

Proof of Theorem 2.1: The left inequalities in (1)–(3) were already proved in [2] (notice that the assumption $E\|X\|^2 < \infty$ was not used there). In order to establish the right inequality in (1), by the Borel-Cantelli lemma it is enough to prove: for every $\varepsilon > 0$, there exists an increasing sequence of integers $\{n_k\}$ with $\lim_k n_k = \infty$, such that

$$\sum_k P\{ \max_{n_k < n \leq n_{k+1}} p(S_n(gX)/a_n) > \sigma_p(gX) + 2\Gamma_p(gX) + \varepsilon \} < \infty,$$

or, by the symmetry of gX and Levy's inequality,

$$\sum_k P\{ p(S_{n_{k+1}}(gX)) > (\sigma_p(gX) + 2\Gamma_p(gX) + \varepsilon)a_{n_k} \} < \infty.$$

It is easily seen that it is actually enough to prove that for every $\varepsilon > 0$ and $\beta > 1$, setting $n_k = [\beta^k]$,

$$\sum_k P\{ p(S_{n_k}(gX) > (\sigma_p(gX) + 2\Gamma_p(gX) + 8\varepsilon)a_{n_k} \} < \infty. \tag{2.6}$$

Let $U_i^{(k)} = X_i I(p(X_i) \le \varepsilon a_{n_k})$, $V_i^{(k)} = X_i - U_i^{(k)}$, for $i = 1, \ldots, n_k$. Since, by assumption (ii),

$$\sum_k P\left\{ p\left(\sum_{i=1}^{n_k} g_i V_i^{(k)} \right) > 0 \right\} \le \sum_k n_k P\{p(X) > \varepsilon a_{n_k}\} < \infty,$$

in order for (2.6) to hold it suffices that

$$\sum_k P\left\{ p\left(\sum_{i=1}^{n_k} g_i U_i^{(k)} \right) > (\sigma_p(gX) + 2\Gamma_p(gX) + 7\varepsilon)a_{n_k} \right\} < \infty. \quad (2.7)$$

To prove (2.7), we will show that

$$\sum_k P\left\{ p\left(\sum_{i=1}^{n_k} g_i U_i^{(k)} \right) \right.$$
$$\left. - E_g p\left(\sum_{i=1}^{n_k} g_i U_i^{(k)} \right) > (\sigma_p(gX) + 2\varepsilon)a_{n_k} \right\} < \infty \quad (2.8)$$

and

$$\sum_k P\left\{ E_g p\left(\sum_{i=1}^{n_k} g_i U_i^{(k)} \right) > (2\Gamma_p(gX) + 5\varepsilon)a_{n_k} \right\} < \infty, \quad (2.9)$$

where E_g denotes integration with respect to the sequence $\{g_i\}$. Statement (2.9) follows from the corresponding arguments in [7]. In fact, by the proof of Hoffmann-Jørgensen's inequality ((3.3), p. 364 of [4]), and using the fact that $p(U_i^{(k)}) \le \varepsilon a_{n_k}$, one obtains as in [7], Lemma 3.4,

$$P\left\{ E_g p\left(\sum_{i=1}^{n_k} g_i U_i^{(k)} \right) > (2\Gamma_p(gX) + 5\varepsilon)a_{n_k} \right\}$$
$$\le \left(P\left\{ E_g p\left(\sum_{i=1}^{n_k} g_i U_i^{(k)} \right) > (\Gamma_p(gX) + 2\varepsilon)a_{n_k} \right\} \right)^2.$$

By the definition of $\Gamma_p(gX)$ and a simple contraction argument, for all k large enough

$$Ep\left(\sum_{i=1}^{n_k} g_i U_i^{(k)}\right) \leq Ep(S_{n_k}(gX)) \leq (\Gamma_p(gX) + \varepsilon)a_{n_k};$$

hence (2.9) will hold once it is proved that

$$\sum_k \left(P\left\{E_g p\left(\sum_{i=1}^{n_k} g_i U_i^{(k)}\right) - Ep\left(\sum_{i=1}^{n_k} g_i U_i^{(k)}\right) > \varepsilon a_{n_k}\right\}\right)^2 < \infty,$$

which follows from the arguments in the proof of Lemma 3.4 of [7].

It remains to prove (2.8). The proof is based on the exponential inequality for Gaussian tails proved in [9]: for any centered B-valued Gaussian vector G, any $t > 0$,

$$P\{|p(G) - Ep(G)| > t\} \leq 2\exp\{-t^2/2(\sigma_p(G))^2\}, \qquad (2.10)$$

and on certain bounds obtained in [7]. It will be convenient to use the following notation:

$$T_k = \sum_{i=1}^{n_k} g_i U_i^{(k)}, \qquad \rho_k(\omega) = \{\sup_{p^*(f)\leq 1} \sum_{i=1}^{n_k} f^2(U_i^{(k)}(\omega))\}^{\frac{1}{2}},$$

$$\eta_k(\omega) = \left(\sup_{p^*(f)\leq 1} |\sum_{i=1}^{n_k}\{f^2(U_i^{(k)}(\omega)) - Ef^2(U_i^{(k)})\}|\right)^{\frac{1}{2}}, \qquad \sigma = \sigma_p(gX).$$

We may assume that $\sigma > 0$ (if $\sigma = 0$, then $\mathcal{L}(X) = \delta_0$ and (2.8) is trivial); also, let us observe that $\rho_k \leq \eta_k + n_k^{\frac{1}{2}}\sigma$.

By the independence of g and X we may write $P = P_X \otimes P_g$ (the meaning of these symbols being obvious). Assume first that $\rho_k(\omega) = 0$. Then

$$P_g\{\omega': |p(T_k(\omega,\omega')) - E_g p(T_k(\omega,\cdot))| > (\sigma + 2\varepsilon)a_{n_k}\} = 0, \qquad (2.11)$$

since $\sigma_p(T_k(\omega,\cdot)) = \rho_k(\omega)$. On the other hand, if $\rho_k(\omega) > 0$, we write

$$P_g\{\omega' : |p(T_k(\omega,\omega')) - E_g p(T_k(\omega,\cdot))| > (\sigma + 2\varepsilon)a_{n_k}\}$$
$$\leq P_g\{\omega' : \eta_k(\omega)|p(T_k(\omega,\omega')) - E_g p(T_k(\omega,\cdot))| > \varepsilon\rho_k(\omega)a_{n_k}\}$$
$$+ P_g\{\omega' : n_k^{\frac{1}{2}}\sigma|p(T_k(\omega,\omega')) - E_g p(T_k(\omega,\cdot))| > (\sigma + \varepsilon)\rho_k(\omega)a_{n_k}\}.$$
$$(2.12)$$

To bound the first term in the right hand side of (2.12) we use (2.10):

$$E_g\left\{\left|\frac{p(T_k(\omega,\cdot)) - E_g p(T_k(\omega,\cdot))}{\rho_k(\omega)}\right|^6\right\} \leq \int_0^\infty (6t^5)2e^{-t^2/2}dt = C < \infty.$$

Therefore the first term is bounded by

$$C(\varepsilon a_{n_k})^{-6}(\eta_k(\omega))^6. \qquad (2.13)$$

Using (2.10) again, we find that the second term in the right hand side of (2.12) is bounded by

$$2\exp\{-(\sigma + \varepsilon)^2 LLn_k/\sigma^2\}. \qquad (2.14)$$

Now (2.11)–(2.14) yield

$$\sum_k (P_X \otimes P_g)\{|p(T_k) - E_g p(T_k)| > (\sigma + 2\varepsilon)a_{n_k}\}$$
$$\leq C\varepsilon^{-6}\sum_k a_{n_k}^{-6} E_X \eta_k^6 + 2\sum_k \exp\{-(\sigma + \varepsilon)^2 LLn_k/\sigma^2\}.$$

The second sum is clearly finite, while the finiteness of the first sum follows from [7] (see the argument for the control of (II) and Lemma 3.6 in [7]). This completes the proof of (2.8), and hence that of (1).

Assertion (2) follows from the Lemma and the inequality

$$\Lambda_p(X) \leq (\pi/2)^{\frac{1}{2}}\Lambda_p(gX), \qquad (2.15)$$

valid for symmetric X; (2.15) is proved by writing $\Lambda_p(X) = \lim_n E\{\sup_{m \geq n} p(S_m/a_m)\}$, using a standard contraction argument and finally recalling $E|g| = (2/\pi)^{\frac{1}{2}}$.

A similar argument, using $EX = 0$, shows that for general X,

$$\Lambda_p(X) \leq \Lambda_p(X - X'), \tag{2.16}$$

where X' is an independent copy of X. Also, by (2),

$$\begin{aligned}
\Lambda_p'(X - X') &\leq (\pi/2)^{\frac{1}{2}}\{\sigma_p(X - X') + 2c_g\Gamma_p(X - X')\} \\
&= (\pi/2)^{\frac{1}{2}}\{2^{\frac{1}{2}}\sigma_p(X) + 4c_g\Gamma_p(X)\} \\
&= \pi^{\frac{1}{2}}\sigma_p(X) + (2\pi)^{\frac{1}{2}}2c_g\Gamma_p(X),
\end{aligned}$$

which together with (2.16) proves (3). This ends the proof of Theorem 2.1.

It is well known (see e.g. [7], Prop. 2.3) that under the assumption (iii)' $S_n/a_n \xrightarrow{p} 0$,
we have $\Gamma_p(X) = 0$ (and also $\Gamma_p(gX) = 0$ by the Lemma). This yields

Corollary 2.2. *Assume that X satisfies (i), (ii) and (iii)'. Then*

(1) $\Lambda_p(gX) = \sigma_p(gX)$.

(2) *If X is symmetric, $\sigma_p(X) \leq \Lambda_p(X) \leq (\pi/2)^{\frac{1}{2}}\sigma_p(X)$.*

(3) *For general X, $\sigma_p(X) \leq \Lambda_p(X) \leq \pi^{\frac{1}{2}}\sigma_p(X)$.*

In the next corollary, K is the unit ball of the reproducing kernel Hilbert space of $\mathcal{L}(X)$ (see [3]); we shall use the well-known fact that

$$\sigma_p(X) = \sup_{x \in K} p(x). \tag{2.17}$$

Corollary 2.3. *Assume that X satisfies (i), (ii) and (iii)'. Then*

(1) $\lim_n d(S_n(gX)/a_n, K) = 0$ *a.s..*

(2) *If X is symmetric, $\lim_n d(S_n(X)/a_n, (\pi/2)^{\frac{1}{2}}K) = 0$ a.s..*

(3) *For general X, $\lim_n d(S_n(X)/a_n, \pi^{\frac{1}{2}}K) = 0$ a.s. .*

Proof: Let U be the closed unit ball of B. For $\varepsilon > 0$, define p_ε by

$$p_\varepsilon(x) = \inf\{\lambda > 0 : x \in \lambda(K + \varepsilon U)\}.$$

By (2.17), $\sigma_{p_\varepsilon}(gX) = \sigma_{p_\varepsilon}(X) \leq 1$; it follows from Corollary 2.2 that $\Lambda_{p_\varepsilon}(gX) \leq 1$, and this easily implies: for every $\varepsilon > 0$,

$$p\{S_n(gX)/a_n \in K + \varepsilon U \quad \text{eventually}\} = 1.$$

This proves (1). Assertion (2) and (3) are proved in the same way.

Remark: It was proved in [1] that under assumptions (i) and (iii)',

cluster set $(\{S_n(X)/a_n\}) =$
$$\text{cluster set } (\{S_n(gX)/a_n\}) = K \ a.s.. \quad (2.18)$$

This result and assertion (1) of Corollary 2.3 give a complete picture of the almost sure limiting behavior of $\{S_n(gX)/a_n\}$ in general Banach spaces under assumptions (i), (ii) and (iii)'.

If the Banach space is of type 2, then (ii) and $EX = 0$ imply (iii)'. It follows that if X takes values in a space of type 2, then we have a complete picture of the limits in the LIL for gX under assumptions (i) and (ii).

References

[1] A. de Acosta and J. Kuelbs, *Some results on the cluster set* $C(\{S_n/a_n\})$ *and the LIL*, Ann. Probability 11 (1983), 102–122.

[2] A. de Acosta, J. Kuelbs and M. Ledoux, *An inequality for the law of the iterated logarithm*, pp. 1–29 in "Probability in Banach spaces IV", Lecture Notes in Math. 990. Springer-Verlag, 1983.

[3] V. Goodman, J. Kuelbs and J. Zinn, *Some results on the LIL in Banach space with applications to weighted empirical processes*, Ann. Probability 9 (1981), 713–752.

[4] J. Hoffmann-Jørgensen, *Sums of independent Banach space valued random variables*, Studia Math. 52 (1974), 159–186.

[5] J. Kuelbs, *When is the cluster set of $\{S_n/a_n\}$ empty?*, Ann. Probability 9 (1981), 377–394.

[6] M. Ledoux, *The law of the iterated logarithm in uniformly convex spaces*, Trans. Amer. Math. Soc 294 (1986), 351–365.

[7] M. Ledoux and M. Talagrand, *Characterization of the law of the iterated logarithm in Banach spaces*, Ann. Probability 16 (1988), 1242–1264.

[8] G. Pisier, *Le théorème de la limite centrale et la loi du logarithme iteré dans les espaces de Banach*, Séminaire Maurey-Schwartz 1975–76, Exposés 3 et 4. Ecole Polytechnique, Paris (1976).

[9] G. Pisier, *Probabilistic methods in the geometry of Banach spaces*, pp. 167–241 in "Probability and Analysis", Lecture Notes in Mathematics 1206, Springer-Verlag, 1986.

Department of Mathematics and Statistics, Case Western Reserve University, Cleveland, Ohio 44106, USA

and

Département de Mathématique, Université de Strasbourg, 7 Rue René Descartes, 67084 Strasbourg Cedex, France

Analytic and Empirical Evidences
of Isoperimetric Processes

CHRISTER BORELL

Dedicated to the memory of Antoine Ehrhard

1, Introduction.

In this paper we will tell about different isoperimetric inequalities of either empirical or theoretic nature. The underlying models come from economy or game theory.

First recall the ideal picture of stock price fluctuations as Brownian motion (see e.g. Black and Scoles [2], Cootner [6], Granger and Morgenstern [11], Harrison and Pliska [13], and Samuelson [19]). Actually, the true mathematical equations of the so-called Einstein-Wiener Brownian motion go back to Bachelier in his now famous work on French stock price dynamics [1].

Our own stock price information is restricted to weekly returns of fifteen Danish stocks during the period from January 1th, 1971 to the middle of September 1985. Below we will mainly scrutinize the Brownian motion picture in a way of giving the stationary Gaussian Markov process or the so-called Ornstein-Uhlenbeck process as the hypothetic equilibrium. The connection between isoperimetry and the Ornstein-Uhlenbeck operator $-\Delta + x \cdot \nabla$ in $\mathbf{R}^n$ was initiated by my late dear friend Antione Ehrhard [7] to whom this paper is dedicated. The same context was later extended to the time dependent situation by the author in [3]. As will be seen below, there is a very significant interplay between isoperimetric inequalities and the number of so-called runs of a sequence. The latter concept, is, of course, very well-established in the analysis of stock price fluctuations (see e.g. [11]).

Second recall the ideal picture of the net outcome of a fair game as Brownian motion. Also, taking care of the interest rate the resulting isoperimetry is analogous to the one met with in hyperbolic geometry. Due to the lack of a finite invariant measure the property is mainly of so-called pure interest.

Acknowledgement: We are very much indebted to Professor Peter Jenenrgren, Odense, and Lektor Bjarne G. Sørensen, Odense for their kind generosity to bring us the information about the Danish stock prices.

2. Isoperimetric processes

Consider an Abelian group T and measurable space $(E, \mathcal{E})$. Let $\mathcal{E}_0 \subseteq \mathcal{E}$ and suppose $(\ )^*: \mathcal{E}_0 \to \mathcal{E}_0$ is a mapping. A stationary E-valued stochastic process $X = (X(t); \ t \in T)$ possessing the invariant measure $\theta = \mathcal{L}X(t)$ is said to be *-isoperimetric if

(1) $\theta(A^*) = \theta(A) > 0$, all $A \in \mathcal{E}_0$
   and

(2) $\mathbf{P}[X(t + \Delta t) \in B | X(t) \in A] \leq \mathbf{P}[X(t + \Delta t) \in B^* | X(t) \in A^*]$, all $t, \Delta t \in T$ and all $A, B \in \mathcal{E}_0$.

Example 2.1: Put

$$\theta(dx) = e^{-|x|^2/2} dx / (2\pi)^{n/2}, \quad x \in \mathbf{R}^n,$$

and denote by $\mathcal{E}_0$ the class of all open or closed subsets of $\mathbf{R}^n$ with positive θ-measure. To any $A \in \mathcal{E}_0$, we associate an open half-space $A^* = (x; x_n > a)$ such that (i) holds. If W stands for the normalized n-dimensional Wiener process, the solution of the following Itô equation

$$\begin{cases} dX(t) = -\tfrac{1}{2}X(t)dt + dW(t) \\ \mathcal{L}X(t) = \theta \end{cases}$$

is *-isomeripetric [3]. For the foundation of diffusion theory, see e.g. Ikeda and Watanabe [14].

An affine image of the process X is called an Ornstein-Uhlenbeck process. In one dimension, an Ornstein-Uhlenbeck process is the same as a stationary Gaussian Markov process.

Example 2.2: Suppose

$$\theta(dx) = e^{2R(x)}dx, \quad x \in \mathbf{R},$$

is a probability measure with $R' \in Lip_{loc}$ and denote by $\mathcal{E}_0$ the class of all open or closed subsets of $\mathbf{R}$ of positive θ-measure. If $A \in \mathcal{E}_0$, let $A^* = \{x; x > a\}$ satisfy (i). Now, given $\sigma > 0$, the solution of the Itô equation

$$\begin{cases} dX(t) = \sigma^2 R'(X(t))dt + \sigma dW(t) \\ \mathcal{L}X(t) = \theta \end{cases} \tag{2.1}$$

is isoperimetric if and only if

$$\theta([a,b]) = \theta([r,+\infty[) \Rightarrow \theta([a-\varepsilon, b+\varepsilon]) \geq \theta([r+\varepsilon,+\infty[), \quad \varepsilon > 0, \ (2.2)$$

(Borell [4]). The condition (2.2) implies the existence of a point of symmetry of θ. Conversely, if θ possesses a point of symmetry and if the distribution function $F: x \curvearrowright \theta(]-\infty, x])$ is log concave, then (2.2) holds. Recall that F is log concave if R is concave. ∎

For short, if (2.2) is true, the solution of (2.1) will be called a one-dimensional isoperimetric diffusion process.

Next suppose $X = (X(t); t \in T)$ is a real-valued stationary process with the invariant measure θ. Furthermore, suppose it is possible to pick a sequence

$$-\infty = a_0 < a_1 < \cdots < a_{m-1} < a_m = +\infty$$

such that $\theta(]a_{k-1}, a_k]) = 1/m$, $k = 1, \ldots, m$. Put $e_k =]a_{k-1}, a_k]$ and $E = \{e_1, \ldots, e_m\}$. Define $X_{(m)}(t) = e_k$ on $\{X(t) \in e_k\}$, the process $X_{(m)} = (X_{(m)}(t); t \in T)$ is stationary with the following transition probability matrix

$$\begin{aligned}[p_m(\Delta t; X)]_{ij} &= \mathbf{P}[X_{(m)}(t + \Delta t) = e_j | X_{(m)}(t) = e_i] \\ &= m\mathbf{P}[X(t) \in e_i, X(t + \Delta t) \in e_j].\end{aligned}$$

Note that the invariant distribution of $X_{(m)}$ is a uniform one. Note also that

$$p_m(\Delta t; X) = p_m(\Delta t; f \circ X)$$

for each measurable strictly increasing function $f : \mathbf{R} \to \mathbf{R}$. Below, the matrix $p_m(\Delta t; X)$ will be called the m-state transition matrix of X.

Theorem 2.1. *Let X be a one-dimensional isoperimetric diffusion process. Then*

a) $[p_m(\Delta t; X)]_{ij} = [p_m(\Delta t; X)]_{ji} = [p_m(\Delta t; X]_{m+1-i, m+1-j}$,

b) $[p_m(\Delta t; X]_{11} \geq [p_m(\Delta t; X)]_{ij} \geq [p_m(\Delta t; X)]_{1m}$,

c) *the map* $0 \leq \Delta t \curvearrowright [p_m(\Delta t; X)]_{ii}$ *is convex and decreasing*

d) *the matrix*

$$p_3(\Delta t; X) = \begin{bmatrix} a & b & c \\ b & d & b \\ c & b & a \end{bmatrix}$$

satisfies

$$a \geq d \geq \frac{1}{3} \geq b \geq c.$$

e) *setting* $E = \{e_1, \ldots, e_m\}$, $\mathcal{E}_0 = 2^E$, *and*

$$A^* = \{e_1, \ldots, e_{\text{card}A}\}, \quad A \in \mathcal{E}_0,$$

the process $X_{(m)}$ *is* *-isoperimetric.

Proof: Introducing, $H = -\frac{1}{2}(d/dx)^2 - R'd/dx$,

$$E_x f(X(T)) = [e^{-tH} f](x),$$

$$Ef(X(0))g(X(t)) = \langle f, e^{-tH} g \rangle_\theta = \langle e^{-tH} f, g \rangle_\theta,$$

and

$$\langle Hf, f \rangle_\theta = \frac{1}{2} \| \nabla f \|_{\theta, 2}^2,$$

for smooth functions. Now viewing X as an isoperimetric diffusion process, Theorem 2.1 is immediate. ∎

Below we will represent an m by m matrix A graphically as the graph of the sequence $A(1,1), \ldots, A(1,m), A(2,m), \ldots, A(m,m)$ (the first row comes first, then the second row and so on). The matrix $\begin{bmatrix} a & 1-a \\ 1-a & a \end{bmatrix} = A_a$ will be called flat, if $a = \frac{1}{2}$, and U-shaped, if $a > \frac{1}{2}$; to expand further on this, the matrix in Theorem 2.1 d) is said to be W-shaped. Moreover, if $\frac{1}{2} < a_1 > a_2$, A_{a_1} is said to be more U-shaped than A_{a_2}.

The m-state transition matrix of a stationary process $X = (X(n))_{n=1}^N$ is related to so-called runs as follows. Suppose $E = (e_i)_1^m$ is any partition of the range of X and represent each $X^{-1}(e_i)$ as a mutually disjoint union of intervals (= a set of consecutive integers) $\bigcup_\alpha I_{i\alpha}$. In the stock literature the $I_{i\alpha}$ are called runs (see e.g. [11]). Let R denote the total number of runs for all i. In particular, if $e_i = (\zeta(i-1), \zeta(i))$, where $\zeta(m-i)$ is the i/m-quantile of the sequence $X(1), \ldots, X(N)$, the estimator

$$q_m(k; X) = m\{(N-k)^{-1} \sum_{n=1}^{N-k} 1_{[X(n) \in e_i, X(n+k) \in e_j]}\}_{1 \le i, j \le m}$$

of $p_m(k; X)$ satisfies the equation

$$R = 1 + \frac{N-1}{m} \sum_{i \ne j} [q_m(1; X)]_{ij}$$

or

$$R = 1 + (N-1)(1 - \frac{1}{m} \text{ Trace } q_m(1; X)).$$

To avoid too much cluttering with letters, write $\hat{p}_m(k, X) \stackrel{\triangle}{=} q_m(k; X(\omega))$. To say that $\hat{p}_2(1; X)$ is U-shaped thus implies that $R < (N+1)/2$.

Needless to say, the above notation also makes sense if X is nonstationary.

3. Renaming the Danish stocks

To begin with we rename the Danish stocks as follows:

ST1 ← Privatbanken

ST2 ← Den Danske Bank

ST3 ← Handelsbanken

ST4 ← Provinsbanken

ST5 ← Codan Forsikring

ST6 ← Det Danske Luftfartsselskab, A.

ST7 ← Det Danske Trælastkompagni

ST8 ← Wessel & Vett, præf.

ST9 ← Østasiatisk Kompagni

ST10 ← DFDS

ST11 ← De Danske Sukkerfabrikker

ST12 ← De Forenede Bryggerier, ord. (A)

ST13 ← Nordisk Kabel- og Trådfabrikker

ST14 ← F.L. Schmidt & Co., B

ST15 ← Superfos, ord.

Throughout, $P(n)$ denotes the price of a fixed stock at the time point $n(= 0, \ldots, N = 765;$ weeks). Moreover, we put $Q(n) = ln\ P(n)$ and $\xi(n) = Q(n) - Q(n-1)$. Of course, the prices are adjusted to splits and dividends, etc.

4. The random walk model

The random walk hypothesis or model (RW) states that the returns $\xi = (\xi(n))$ constitute a sequence of stochastically independent and identically distributed random variables (abbreviated $\xi \in i.i.d.$). In [15], among others exploiting run tests, Jennergren and Toft-Nielsen reject the RW for Danish stocks during the period 1973–1975 (using daily returns). For other interesting information about the Danish stock market, see Sørensen's articles [22] and [23].

To test the RW using the m-state transition matrices for the returns may cause some minor troubles depending on sharp boundaries. Henceforth, let $\gamma \in i.i.d.(\gamma(n) \in N(0;1))$ be stochastically independent of ξ and put $\xi_\varepsilon = \xi + \varepsilon\gamma.\mathrm{stand}\xi$, where $\varepsilon = 10^{-10}$.

Figure 4.1 graphs the $\hat{p}_2(1,\xi_\varepsilon)$. Viewing Figure 4.1 as a 5×3-matrix, the common entry of the i'th row (from the top) and the j'th column (from the left) corresponds to the $(3(i-1)+j)$'th stock. The same convention will be used throughout the paper. As the reader may see, the estimated 2-state transition matrices of the $|\xi_\varepsilon|$ (Figure 4.2) are more U-shaped than the corresponding matrices for the ξ_ε.

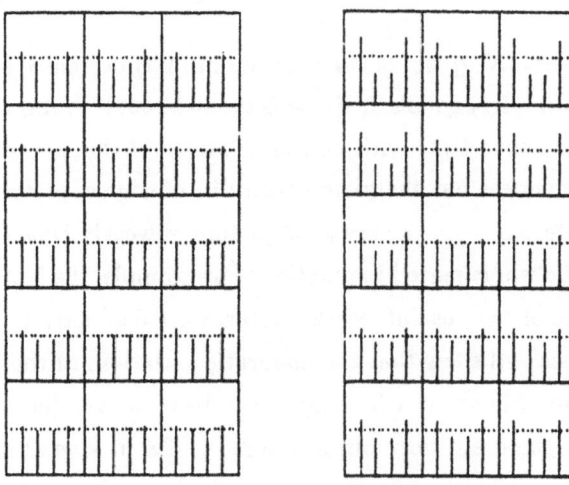

Figure 4.1. $\hat{p}_2(1;\xi_\varepsilon)$ Figure 4.2. $\hat{p}_2(1;|\xi_\varepsilon|)$
The line $p = 1/2$ is dotted.

Recall that if $X \in i.i.d.$ possesses marginal distributions of continuous type then the probability distribution of $q_2(1;X)$ is uniquely determined. Let $z_N(\alpha)$ denote the α-quantile of $q_{22}(1;X)$. Now if $\xi_{k,\varepsilon}$ refers to ξ_ε for the k'th stock, we have

$$\#\{k; [\hat{p}_2(1;\xi_{k,\varepsilon})]_{22} > Z_N(\alpha)\} = \begin{cases} 11, & \text{if } \alpha = 0.10 \\ 8, & \text{if } \alpha = 0.05 \end{cases}$$

and

$$\#\{k; [\hat{p}_2(1; |\xi_{k,\epsilon}|)]_{22} > Z_N(\alpha)\} = 15, \quad \text{if} \quad \alpha = 0.05.$$

Performing a similar test over the period $n = 1, \ldots, [N/2]$ yields

$$\#\{k; [\hat{p}_2(1; \xi_{k,\epsilon})]_{22} > Z_{[N/2]}(\alpha)\} = \begin{cases} 8, & \text{if} \quad \alpha = 0.10 \\ 4, & \text{if} \quad \alpha = 0.05 \end{cases} \tag{4.1}$$

$$\#\{k; [\hat{p}_2(1; |\xi_{k,\epsilon}|)]_{22} > Z_{[N/2]}(\alpha)\} = \begin{cases} 14, & \text{if} \quad \alpha = 0.10 \\ 13, & \text{if} \quad \alpha = 0.05 \end{cases} \tag{4.2}$$

Over the last $[N/2]$ weeks the function in (4.2) decreases by one and the function in (4.1) equals 9, if $\alpha = 0.10$, and equals 7, if $\alpha = 0.05$. By iterating the above tests with new simulations of the γ, there may be some small changes but the main pattern is, of course, preserved.

The RW does not give a very good picture of weekly Danish stock returns over the whole period in question. For example, the 2-state transition matrices of the absolute weekly returns are distinctly U-shaped. A closer analysis will show that the quadratic variations of the log-prices, in general, are bigger in a bull than in a bear market (cf. Section 7). In addition, there are very obvious trend shifts in several time series considered here (Figure 4.3). To simplify figures and formulas, we from now on replace Q by $Q - Q(0)$.

The empirical distributions of the $\xi(n)$ are very very far from Gaussian in the present case. In studies of other markets, it has often been stressed that the returns may be mixings of a Gaussian i.i.d. (see e.g. Fama [8], Granger [12], Granger and Morgenstern [11], Leitch and Paulson [16], and Praetz [18]). Moreover, the first series correlation coefficient of the returns is most often positive. Under these circumstances the $p_2(1; \xi)$ should be U-shaped. Indeed, suppose $\eta(n) = \text{const.} + \sigma(n)\gamma(n)$, where $\sigma \geq 0$ and γ are stochastically independent, $\sigma, \gamma \in i.i.d.$ and $\gamma(n) \in N(0; 1)$. Put $\xi(1) = \eta(1)$ and $\xi(n+1) = \rho\xi(n) + \eta(n+1)$, $n \geq 1$,

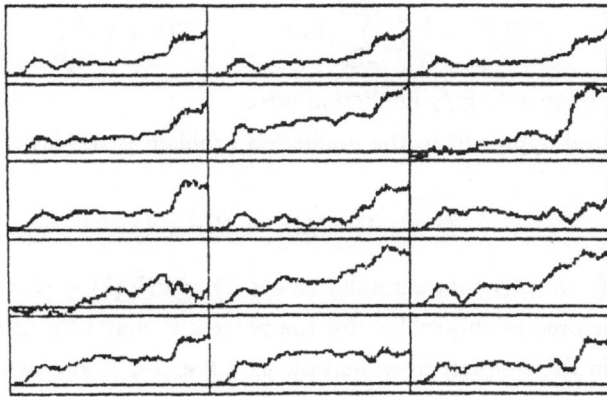

Figure 4.3. Q

where $\rho > 0$ is very small. By conditioning on σ, we conclude that $p_2(k; \xi)$ is U-shaped as the Ornstein-Uhlenbeck process is isoperimetric (a direct proof is, of course, simple). Note that the sequence $(\xi(n))$ converges in law iff $E ln^+ |\eta(n)| < +\infty$ (see e.g. Wolfe [24]). Summing up, the presence of isoperimetric inequalities as in Danish stock returns could eventually be universal. Very likely, the whole effect should not be caused by non-stationarities.

For the trouble with the fluctuating quadratic variation of stock log prices in connection with option prices, see the review article by Smith [21].

5. The prediction problem

Put $\mathcal{A}(n) = \sigma(Q(k); k \leq n)$ and

$$\Gamma(n+1) = E[Q(n+1)|\mathcal{A}(n)].$$

If Q is a random walk, $\Gamma(n+1) = Q(n) + \mu$, where μ is the mean of $Y(n)$. To any hypothesis H fixing $\Gamma(n+1)$, let

$$PE[H] = \left(\frac{1}{N} \sum_{n=0}^{N-1} (\Gamma(n+1) - Q(n+1))^2 \right)^{\frac{1}{2}}$$

be the corresponding ℓ_2 prediction error.

A natural approach to the prediction problem is to seek a decomposition

$$Q(n+1) = L(n) + V(n+1)$$

where $L(n)$ is $\mathcal{A}(n)$-measurable, $\mathcal{A}(n+1) = \sigma(V(k); k \leq n+1)$, and where the prediction problem for the process V may be mastered. Proceeding in the simplest possible way, fix a $0 < \rho < 1$, and define

$$\begin{cases} V(n+1) = \rho V(n) + \xi(n+1) \\ V(1) = \xi(1) \end{cases} \tag{5.1}$$

so that

$$Q(n) = (1-\rho)V(1) + \cdots + V(n-1)) + V(n) = T(n) + V(n), \quad n \geq 1.$$

Here it is constructuve to think of $V(n)$ as an overreaction of the next week's contribution to the long random trend $T(n)$.

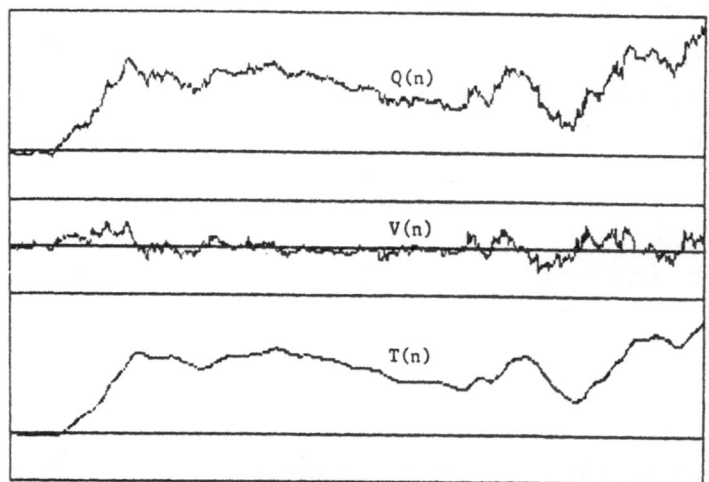

Figure 5.1. ST9

We define the OU hypothesis or model of stock prices by stating that V is an Ornstein-Uhlenbeck process. Actually under the OU it will often only be used that V is stationary, ergodic, and Gaussian. The Brownian motion model of stock prices implies the OU in equilibrium. The converse is true if $\rho = \text{corr}[V(n), V(n+1)]$ and only then.

Of course, there is a variety of equilibrium candidates for the equation (6.1) and they are all well-known if ξ is an i.i.d. (see e.g. [24]).

To examine the OU it could be natural to delete a few terms in the very beginning of the series but we will not do that here. For convenience, however, if $E\xi(n) = \mu$, $n \geq 2$, we redefine $\Gamma(1) = \mu$.

Under the OU, the $\xi(n)$, $n \geq 2$, are identically distributed with the mean μ, say, and

$$\Gamma(n+1) = \begin{cases} \mu & , \quad n = 0 \\ (1-\rho)(V(1) + \cdots + V(n)) + C_V V(n) + (1 - C_V)\mu_V & , \quad n \geq 1 \end{cases}$$

where

$$V(n) \in N(\mu_V; \sigma_V^2)$$

and

$$c_V = \sigma_V^{-2} E[(V(n) - \mu_V)(V(n+1) - \mu_V)] = \text{corr}[V(n), V(n+1)].$$

Note that under the RW

$$\Gamma(n+1) = \begin{cases} \mu & , \quad n = 0 \\ (1-\rho)(V(1) + \cdots + V(n)) + \rho V(n) + \mu & , \quad n \geq 1. \end{cases}$$

Choosing $\rho = 0^+, 0.1, \ldots, 0.9$,

$$PE[OU] < PE[RW], \quad \text{all stocks.} \tag{5.2}$$

If we rank the smallest prediction error as number one, the next smallest as number two and so on, the OU gives the following average ranks $\bar{r}$

ρ	0^+	0.1	0.2	0.3	0.4	0.5	0.6	0.7	0.8	0.9
$\bar{r}$	7.40	7.13	6.87	6.33	5.67	5.33	5.07	4.33	3.87	3.00

Looking closer at big values of ρ, we have

ρ	0.84	0.86	0.88	0.90	0.92	0.94	0.96	0.98
$\bar{r}$	5.73	4.87	3.67	2.73	2.47	3.60	5.47	7.47

Performing the same test over the period $n = 1, \ldots, [N/2]$, 149 of the 150 inequalities in (6.2) are true. Moreover,

ρ	0^+	0.1	0.2	0.3	0.4	0.5	0.6	0.7	0.8	0.9
$\bar{r}$	7.67	7.33	6.87	6.47	6.47	5.67	4.53	3.60	3.33	3.07

and

ρ	0.84	0.86	0.88	0.90	0.92	0.94	0.96	0.98
$\bar{r}$	4.40	3.80	3.00	2.80	3.13	4.47	6.40	8.00

In the following we set $\rho = 0.92$ if not otherwise stated.

6. The absence of martingale structures

Assume $\xi = (\xi(n))$ consists of random variables with the same expectation μ. Put

$$\eta(a) = E[\xi(n+1)|V(n) > a].$$

If the process $(Q(n) - n\mu)$ is a martingale, $\eta = \mu$. In contrast to this, an estimation of η gives a strictly increasing function up to a natural upper bound of the argument.

Example 6.1

a	$-\infty$	-0.05	0	0.05	0.1
$10^3_{\hat{\eta}}$					
$ST3$	3.00	3.23	4.97	5.31	9.54
$ST6$	4.82	5.99	7.62	11.23	11.05
$ST13$	3.38	4.01	5.77	6.12	8.68
$ST15$	2.96	4.69	7.26	8.91	10.74

Under the OU, this context can be understood from a parameter estimation. Indeed, using the notation $E[X; A] = E[X1_A]$,

$$E[\xi(n+1); V(n) > a] = E[V(n+1) - \rho V(n); V(n) > a] =$$
$$(c_V - \rho)E[V(n) - \mu_V; V(n) > a] + (1 - \rho)\mu_V P[V(n) > a].$$

Recall that $\mu = (1 - \rho)\mu_V$. Now setting $\emptyset = c_V - \rho$,

$$\eta(a) = \mu + \emptyset E[V(n) - \mu_V | V(n) > a]$$

where the non-negative function

$$a \curvearrowright E[V(n) - \mu_V | V(n) > a]$$

increases in the strict sense.

By estimating $\emptyset$ we get strictly positive values for all stocks. Under the OU, the $\hat{\eta}$ therefore increase strictly. The reader should note that $\emptyset$ is strictly positive only for thirteen stocks during the periods $n = 1, \ldots, [N/2]$ and $n = [N/2] + 1, \ldots, N$.

Once the parameter ρ is fixed, a very simple strategy, based on the above experiences, reads as follows:

hold the stock during the weeks $1 + \{V(\{1, \ldots, N-1\}) > a\}.$

The net return $s(a)$ per unit time may be computed from the equation

$$\mathcal{N}(a)\ell n\ 0.9925 + \eta(a)P[V(n) > a] = s(a)P[V(n) > a]$$

where $\mathcal{N}(a)$ equals the number of "cuttings" of the level $V(n) = a$ per unit time (here we assume $N = +\infty$; the transaction cost on the Danish stock market is 0.75%). In particular, choosing $a = m_V =$ median $V(n)$ as we may here

$$\hat{s}(m_V) = \hat{\eta}(m_V) + ([\hat{p}_2(1; V)]_{12} + [\hat{p}_2(1; V)]_{21})\ell n\ 0.9925.$$

In the present situation, we have

	ST1	ST2	ST3	ST4	ST5	ST6	ST7	ST8
$10^3 \hat{\mu}$	3.20	3.31	3.00	3.63	5.10	4.78	3.39	3.28
$10^3 \hat{s}(m_V)$	4.88	3.84	4.62	5.83	8.01	8.52	6.55	5.35

	ST9	ST10	ST11	ST12	ST13	ST14	ST15
$10^3 \hat{\mu}$	2.19	1.70	3.99	3.99	3.83	2.36	2.94
$10^3 \hat{s}(m_V)$	4.20	4.95	8.15	9.02	4.50	2.46	6.57

Evidently, reasonable martingale structures are very far off for individual log prices; among others, the transition matrices $p_2(1; V)$ must be too U-shaped relative to the size of the corresponding $\eta(m_V)$.

Next we want to compare the ρ making the return per unit time, $s(m_V) = s(m_V; \rho)$, maximal and the ρ giving the minimal prediction error. Ranking the highest return per unit time as number one and so on we have the following average ranks $\bar{r}$

ρ	0.84	0.86	0.88	0.90	0.92	0.94	0.96	0.98
$\bar{r}$	5.60	5.27	4.20	3.87	3.87	2.80	4.60	5.80

Note that the minimum is attained at a slightly greater value than for the prediction problem.

Over the period $n = 1, \ldots, [N/2]$, we have

ρ	0.84	0.86	0.88	0.90	0.92	0.94	0.96	0.98
$\bar{r}$	4.60	4.93	3.60	2.87	3.27	3.60	5.53	7.60

so that the least average rank is attained at exactly the same value as for the prediction problem.

7. Series correlations

Put

$$c(k) = c_V(k) = corr[V(n), V(n + k)]$$

and denote by $\ell n\, c_*$ the least square solution of the equation

$$kx = \ell n\, \hat{c}(k), \quad k \in I = \{5, 10, 15, 20, 25, 30\}.$$

Figure 7.1 gives a bar chart of the $\hat{c}(k)$, $k \in I$, as well as the points (k, c_*^k), $(k+1, c_*^{k+1})$, $k \in \{0\} \cup I$, connected by line segments. Of course, the resolution could be better. Recall that if V is an Ornstein-Uhlenbeck process, then $c(k) = c(1)^k$.

For the sake of comparison, in some special cases, we next plot the $\hat{p}_2(k; V)$ and the exact value of the $p_2(k; U)$ if U is an Ornstein-Uhlenbeck process with the first series correlation coefficient c_*. The values of the $p_2(k; U)$ are indicated by short horizontal like segments (Figures 7.2 and 7.3).

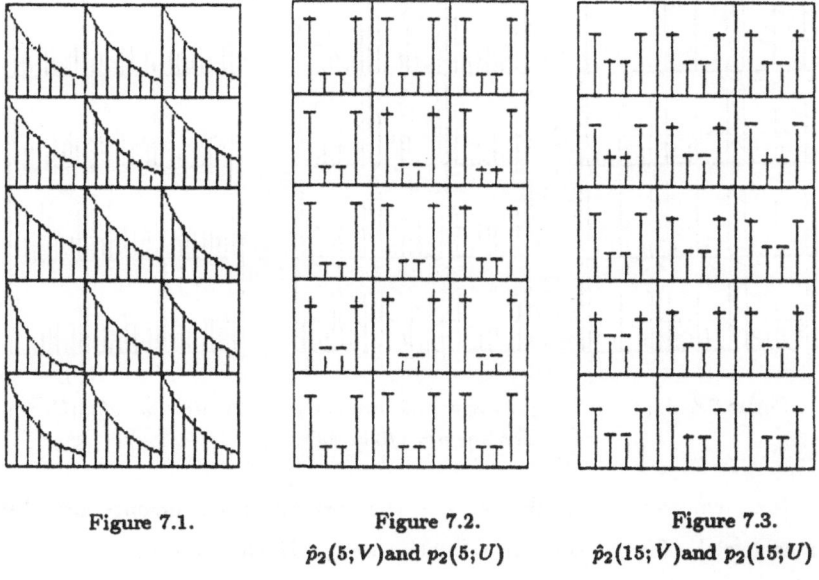

Figure 7.1.	Figure 7.2.	Figure 7.3.
	$\hat{p}_2(5; V)$ and $p_2(5; U)$	$\hat{p}_2(15; V)$ and $p_2(15; U)$

Recall that

$$[p_2(1; U)]_{11} = \frac{1}{2} + \frac{1}{\pi} \arctan c_*(1 - c_*)^{-\frac{1}{2}}, \quad c_* > 0.$$

The transition matrices indicate a slightly stronger correlation of the V than of the corresponding U. However, the pattern is not quite unique. We will not enter a statistical test, which in any case is simple to perform by simulation.

Looking at the 3-state transition matrices $\hat{p}_3(3; V)$ they are all typically W-shaped and agree well with the matrix in Theorem 2.1 d) (Figure 7.4). After a time delay of twelve weeks, for example, the W-shapes are somewhat disturbed. Of course, the same thing happens to a simulated Ornstein-Uhlenbeck process $U^{(12)}$ of the length N and the first series correlation coefficient $\hat{c}(12)^{\frac{1}{12}}$ (Figures 7.5 and 7.6)

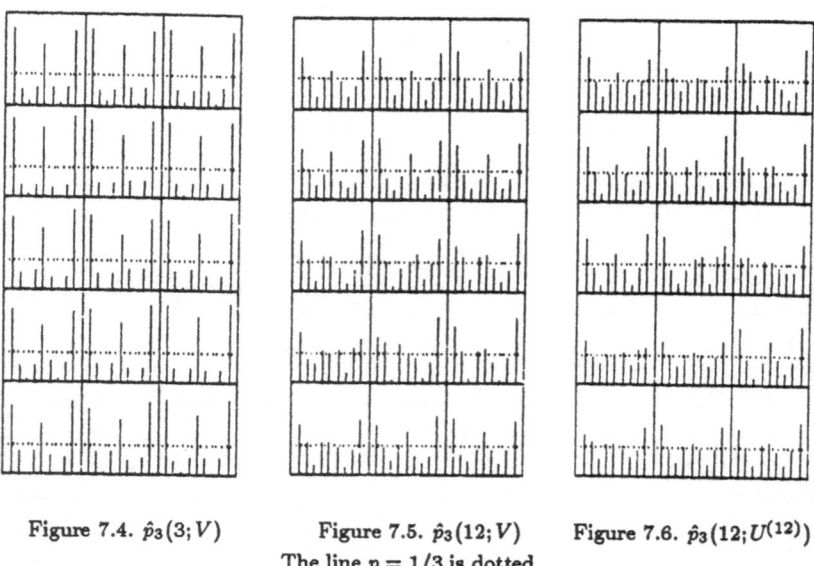

Figure 7.4. $\hat{p}_3(3; V)$ Figure 7.5. $\hat{p}_3(12; V)$ Figure 7.6. $\hat{p}_3(12; U^{(12)})$
The line $p = 1/3$ is dotted.

Next we want to check how well the correlation coefficients actually measure the fluctuations on a relative scale. To this end, put

$$Z = (V - \mu_V)/\sigma_V$$

and

$$R^{(d)} = E[\max_{0 \leq k \leq d} Z(n + k) - \min_{0 \leq k \leq d} Z(n + k)].$$

If V is stationary Gaussian, then by Slepian [20] (see also Fernique [9])

$$(c_i(k) \leq c_j(k), \quad k = 1, \ldots, d, \quad i \neq j \Rightarrow R_i^{(d)} \geq R_j^{(d)}. \tag{7.1}$$

Denoting by $\psi(d)$ the percentage of the cases in which the right-hand side of (7.1) is true when the left hand side is true,

$$\psi(2) = 81/99, \quad \psi(5) = 73/85, \quad \text{and } \psi(9) = 61/76.$$

By simulations, one easily finds out that these fractions are significantly higher for Ornstein-Uhlenbeck processes with estimated parameters.

Actually, only a few of the V possess a distribution close to the Gaussian one. If X is a real random variable, put

$$F(X) = (\mathbf{P}[-\frac{k}{2} < (X - EX)/\text{stand } X \leq \frac{k}{2}])_{k=1}^{7}.$$

Recall that

$$F(N(0;1)) = 0.383, 0.683, 0.866, 0.954, 0.988, 0.997, 1.000.$$

The arithmetic mean $\bar{F}(V)$ of the $F(V)$ equals

$$\bar{F}(V) = 0.454, 0.747, 0.881, 0.944, 0.974, 0.989, 0.994.$$

Eventually, the deviation from the Gaussian distribution here depends on trend shifts. In fact, putting

$$T_k = (k-1)255 + \{1, \ldots, 255\}, \quad k = 1, 2, 3,$$

we have

$$\bar{F}(V(T_1)) = 0.406, 0.711, 0.864, 0.950, 0.981, 0.995, 0.999$$
$$\bar{F}(V(T_2)) = 0.397, 0.694, 0.874, 0.955, 0.986, 0.995, 0.998$$

and

$$\bar{F}(V(T_3)) = 0.404, 0.710, 0.875, 0.952, 0.979, 0.992, 0.997.$$

However, we find the time series $V(T_k)$ too short for a further analysis.

Another non-stationary, certainly interacting with trend shifts, is represented by the fluctuating variance. As a matter of fact, in thirteen cases, the variance of $\xi(n)$ (or of $V(n) - c(1)V(n-1)$) is larger in the set $[V \geq \text{median } V(n)]$ than in the set $[V < \text{median } V(n)]$ and, for several stocks, the difference is significant. We find this problem very basic and would very much like to return to it in a future paper, also considering different markets.

8. Relative strength and non-diversification

The vector process of all stock log prices is, indeed, a very complicated one. Thus, here we will only make a few remarks.

Suppose a given utility function at each instant is able to rank the stocks in question. Put $M = 15$ and let $1 \leq a \leq b \leq M$ be fixed. In [17] Levy examined the following strategy based on relative strength:

  i) Set $n = 1$ and fix an initial capital K_0(crowns).
 ii) At the time point n,
    (α) Sell all stocks not ranking amongst the b best
    (β) Use all the available capital to buy the stocks ranking amongst the a best; the capital is distributed uniformly
iii) If $n \leq N - 2$, let $n \leftarrow n + 1$ and go to (ii)
 iv) Sell all stocks at $n = N$. The final capital equals K_1.

The transaction cost corresponds to 0.75%. Moreover, we assume it is possible to trade in fractions of shares. Let

$$h = h(a, b) = (K_1/K_0)^{52/(N-1)} - 1.$$

In the present situation the buy and hold strategy yields

$$h(M; M) = 0.210.$$

For the sake of uniformity, let us say that the V ranks the stocks; stock i is better than stock j if $V_i > V_j$. The reader should note that

$\rho = 0.92$. Of course, this must be improved upon in a proper situation. Actually, Levy measured relative strength by slightly different means. Table 8.1 lists h for some choices of a and b.

Table 8.1

b	4	5	6	7	8	9	10
a							
1	0.312	0.293	0.315	0.342	0.417	0.336	0.299
2	0.329	0.335	0.338	0.312	0.303	0.299	0.277
3	0.282	0.289	0.297	0.295	0.287	0.299	0.269
4	0.262	0.267	0.291	0.288	0.281	0.288	0.263

Thus $h_{\max} = h(1,8)$. For the periods $n = 1, \ldots, [N/2]$ and $n = [N/2]+1, \ldots, N$, we have $h_{\max} = h(1,8) = 0.187$, $h(M,M) = 0.163$ and $h_{\max} = h(1,8) = 0.683$, $h(M,M) = 0.275$, respectively (in the second case, the V are computed from the very beginning of the whole period). Investigating a veriety of diferent relative strengths we have noticed a very interesting common feature: the portfolio should contain (at most) a single stock (it may be a good idea to leave the market during suitable periods). This should be a mathematical theorem but we cannot prove it. Note here that we have not even tried to define risk, which is another aspect.

To throw some light on the above problem, put $Z = (Z_1, \ldots, Z_M)$, where

$$Z_m = (V_m - \mu_{V_m})/\sigma_{V_m}.$$

Given a fixed $0 < p < 1$ and a variable set A with $\mathbf{P}[Z(n) \in A] = p$, the set A is said to be p-stable if the conditional probability $\mathbf{P}[Z(n+1) \in A|Z(n) \in A]$ is maximal. Denoting by $\mathcal{R}(A)$ the expected total number of runs of Z corresponding to the partition $\{A, -A\}$ of $\mathbf{R}^M$ and assuming Z stationary,

$$\mathcal{R}(A) = 1 + \frac{N-1}{2}(\mathbf{P}[Z(n+1) \notin A|Z(n) \in A] + \mathbf{P}[Z(n+1) \in A|Z(n) \notin A]),$$

so that, if $\mathcal{L}Z = \mathcal{L}Z(N - \cdot)$,

$$\mathcal{R}(A) = 1 + (N - 1)\mathbf{P}[Z(n+1) \notin A | Z(n) \in A].$$

Hence, conditioned on $p = \mathbf{P}[Z(n) \in A]$, a p-stable set will minimize $\mathcal{R}(A)$ and is thereby interesting to reduce the transaction costs for a variety of different strategies (cf. Section 6).

If Z is an Ornstein-Uhlenbeck process in $\mathbf{R}^M$, affine half-spaces of probability p are p-stable (Example 2.1). In one dimension, the empirical results above fully support the same. The multi-dimensional situation is, however, far more involved and we just give one example. To this end let $1 \le i \ne j \le M$, let $A \subseteq \mathbf{R}^2$ be open, and let a_i, b_j be reals satisfying

$$p = \mathbf{P}[(Z_i(n), z_j(n)) \in A] = \mathbf{P}[Z_i(n) > a_i] = \mathbf{P}[Z_j(n) > b_j]$$

and let us consider the following inequality

$$\mathbf{P}[(Z_i(n+k), Z_j(n+k)) \in A | (Z_i(n), Z_j(n)) \in A] \le$$
$$\mathbf{P}[Z_i(n+k) > a_i | Z_i(n) > a_i] \wedge \mathbf{P}[Z_j(n+k) > b_j | Z_j(n) > b_j]. \quad (8.1)$$

For fixed $\alpha = (p, k, A)$ with $A = \{(y, x) | (x, y) \in A\}$, there are only 105 interesting choices of (i, j). Table 8.2 lists the number of these cases in which (8.1) is true choosing α as specified.

Table 8.2

	$p = 0.5$ $k = 1$	$p = 0.5$ $k = 4$	$p = 0.4$ $k = 4$	$p = 0.6$ $k = 4$
A centered ball	105	104	105	101
($\sim$A) centered ball	104	104	102	102

As the reader understands the vector process of prices has been very far from a reasonable equilibrium during the second half of the period and we find it difficult to go further here. Nevertheless, the geometry of relative stock prices should be a nice field to explore further for Banach space probabilists. In fact, even under circumstances like here, relative strength may lead to fairly clean processes. As an example, put $\xi_i(n) = \xi_{\tilde{\nu}_i(n)}$, where $\tilde{\xi}_\bullet(n)$ is a rearrangement of $\xi_\bullet(n)$ in decreasing order. The processes $\nu = \nu_i = (\nu_i(n))_{n=1}^N$ are not too far from i.i.d.'s with uniformly distributed marginals, a statement we do not make any more precise here.

An ideal game and isoperimetry.

Let $dW(t) = \dot{W}(t)dt$ represent an individual's net gain from a fair game during the time interval $(t, t + dt)$. If $P(t)$ denotes the gambler's capital and ρ the rate of interest, the simplest circumstances give the equation

$$dP(t) = \rho P(t)dt + dW(t).$$

Recall that $P(t) \to \pm\infty$ with probability one as $t \to +\infty$ and that the measure

$$e^{\rho|p|^2} dp$$

is an invariant distribution of the capital (for the mathematics, see [14]). In this section, we will study the interplay between the above equation and isoperimetry. The limit case $\rho = 0^+$ is the classical case (for the definite results, see Brascamp et alias [5]).

Proceeding in a slightly more general setting, consider the Itô equation

$$\begin{cases} dX(t) = R'(X(t))dt + dW(t) \\ \mathcal{L}X(t) = \theta, \quad \theta(dx) = e^{2R(x)}dx, \end{cases}$$

where

(1) $R' \in Lip$

(2) R is even

and

(3) R is convex.

Set $\mathcal{E}_0 = \{A \subseteq \mathbf{R}, A \text{ open or closed and } 0 < \theta(A) < +\infty\}$. If $A \in \mathcal{E}_0$, suppose $a > 0$ is such that $\theta(A) = \theta(A^*)$, where $A^* = [-a, a] = \bar{B}(a)$.

Theorem 9.1. *X is *-isoperimetric.*

To prove theorem 9.1, we need

Lemma 9.1. *If $A \in \mathcal{E}_0$,*

$$\theta(A) = \theta(\bar{B}(a)) \Rightarrow \theta(A_\varepsilon) \geq \theta(\bar{B}(a+\varepsilon)), \quad \varepsilon > 0 \tag{9.1}$$

where $A_\varepsilon = A + \bar{B}(\varepsilon)$.

Proof: Without loss of generality we may assume R' is strictly increasing.

First suppose $a > 0$ is fixed, $\mu(A) = \mu(\bar{B}(a))$, and $A = [\alpha, \beta]$, $\alpha < \beta$. Set $h(\alpha, \beta) = \mu(A_\varepsilon)$, where $\varepsilon > 0$ is fixed. Of course, $h(\alpha, \beta) \to +\infty$ if $\alpha^2 + \beta^2 \to +\infty$. Therefore, if $\min h = h(\alpha, \beta)$, by the Lagrange multiplier method

$$R(\beta + \varepsilon) - R(\beta) = R(\alpha - \varepsilon) - R(\alpha)$$

forcing $\alpha = -\beta$. Accordingly, (9.1) is true if $A \in \mathcal{E}_0$ is an interval.

Next let $A = \overset{m}{\underset{0}{\cup}}[\alpha_i, \beta_i]$, where $\alpha_0 < \beta_0 < \cdots < \alpha_m < \beta_m$ and $m \geq 1$. Put $\delta_A = \frac{1}{2}\underset{i=0,\ldots,m-1}{\min}(\alpha_{i+1} - \beta_i)$. By induction on m, it suffices to treat the case $0 \leq \varepsilon < \delta_A$. If $\alpha_1 \leq 0$, choose $\tilde{\alpha}_1 < \alpha_1$ satisfying $\theta([\tilde{\alpha}_1, \alpha_1]) = \theta([\alpha_0, \beta_0])$ so that $\theta(\tilde{A}_\varepsilon) \leq \mu(A_\varepsilon)$ for $\tilde{A} = [\tilde{\alpha}_1, \beta_1] \cup (\overset{m}{\underset{2}{\cup}}[\alpha_i, \beta_i])$. If $\beta_{m-1} \geq 0$, we may reduce the number of intervals in a similar way. Finally, if $m = 1$, $\beta_0 < 0$, and $\alpha_1 > 0$, choose $\tilde{\alpha}_1 < 0$ such that $\theta([\tilde{\alpha}_1, 0]) = \theta([\alpha_0, \beta_0])$ and $\tilde{\beta}_1 > 0$ such that $\theta([0, \tilde{\beta}_1]) = \theta([\alpha_1, \beta_1])$. Defining $\tilde{A} = [\tilde{\alpha}_1, \tilde{\beta}_1]$, $\theta(\tilde{A}_\varepsilon) \leq \theta(A)$, which completes the proof of Lemma 3.1.

Proof of Theorem 9.1: The proof is a variant of the line of reasoning in [4]. Nevertheless, we submit a rather detailed proof.

To any $f \in C_0^+(\mathbf{R})$ there exists a unique even function f^* having the same distribution as f relative to the measure θ. Introducing, $H = -\frac{1}{2}(d/dx)^2 - R'd/dx$, given $f, g \in C_0^+(\mathbf{R})\setminus\{0\}$ and $t \geq 0$, we shall prove that

$$\langle e^{-tH}f, g\rangle_\theta \leq \langle e^{-tH}f^*, g^*\rangle_\theta.$$

In doing so there is no loss of generality in assuming R is real analytic.

To begin with, choose $L > 0$ so that f vanishes off $[-L, L]$. Set $T_L = \inf\{t > 0; |X(t)| > L\}$, $u(t, x) = E_x[f(X(t)); T_L > t]$, $t \geq 0$, $x \in \mathbf{R}$, and

$$\tilde{u}(t, x) = \sup\{\int_C u(t, y)d\theta(y); \ C \in \mathcal{B}(\mathbf{R}), \ \theta(C)$$
$$= \theta(\bar{B}(x))\}, \quad t \geq 0, 0 \leq x \leq L.$$

The functions u and $\tilde{u}$ are continuous. Moreover, the function $u(t, x)$, $|x| < L$, is real analytic and strictly positive for each $t > 0$.

Defining $K = -(d/dx)^2 + R'd/dx$, we first prove that

$$(\partial_t + K)\tilde{u} \leq 0 \quad \text{in} \quad t > 0, 0 < x < L, \qquad (9.2)$$

in the weak sense using C_0^∞ as test functions.

Next let $t > 0$ and $0 < x < L$ be fixed and write

$$\tilde{u}(t, x) = \int_{C(t,x)} u(t, y)d\theta(y)$$

where $\theta(C(t, x)) = \theta(\bar{B}(x))$ and

$$C(t, x) = \overset{m}{\underset{0}{\cup}}[\alpha_i, \beta_i], \quad -L < \alpha_0 < \beta_0 < \cdots < \alpha_m < \beta_m < L.$$

By simple means,

$$]0, L[\supseteq (\cup[C(t, x): \varepsilon \leq t \leq \varepsilon^{-1}, \varepsilon \leq x \leq L - \varepsilon)^-, \quad \text{all } \varepsilon > 0.$$

From

$$\tilde{u}(t + r, x) \geq \int_{C(t,x)} u(t + r, y)d\theta(y), \quad r > -t$$

we have

$$\partial_t \tilde{u}(t, x) = \int_{C(t,x)} u'_t(t, y)d\theta(y). \qquad (9.3)$$

Now set $\varphi = e^{2R}$ and define a function δ by the equation

$$\sum_{i=0}^{m} \int_{\alpha_i - \delta(r)}^{\beta_i + \delta(r)} \varphi(y) dy = \theta(\bar{B}(x + r)).$$

Then, by differentiation with respect to r,

$$\delta'(r) \sum_{0}^{m} (\varphi(\beta_i + \delta(r)) + \varphi(\alpha_i - \delta(r))) = 2\varphi(x + r)$$

so that

$$\delta'(0) \sum_{0}^{m} (\varphi(\beta_i) + \varphi(\alpha_i)) = 2\varphi(x)$$

and

$$2\delta'(0)^2 \sum_{0}^{m} (R'(\beta_i)\varphi(\beta_i) - R'(\alpha_i)\varphi(\alpha_i))$$

$$+ \delta''(0) \sum_{0}^{m} (\varphi(\beta_i) + \varphi(\alpha_i)) = 4R'(x)\varphi(x). \quad (9.4)$$

Setting

$$u_0 = u(t, \alpha_0) = \cdots = u(t, \beta_m)$$

we thereby have

$$\partial_x \tilde{u}(t, x) = 2u_0 \varphi(x) \quad (9.5)$$

as

$$\left(\int_{C(t,x+r)} - \int_{C(t,x)} \right) u(t, y) d\theta(y) \geq \sum_{0}^{m} \left(\int_{\alpha_i - \delta(r)}^{\beta_i + \delta(r)} - \int_{\alpha_i}^{\beta_i} \right) u(t, y) d\theta(y).$$

In a similar way, the inequality

$$\left(\int_{C(t,x+r)} + \int_{C(t,x-r)} - 2 \int_{C(t,x)} \right) u(t, y) d\theta(y)$$

$$\geq \sum_{0}^{m} \left(\int_{\alpha_i - \delta(r)}^{\beta_i + \delta(r)} + \int_{\alpha_i - \delta(-r)}^{\beta_i + \delta(-r)} - 2 \int_{\alpha_i}^{\beta_i} \right) u(t, y) d\theta(y)$$

yields

$$\partial_x^2 \tilde{u}(t, x) \geq \delta'(0)^2 \sum_0^m ((\partial_y(u(t, y)\varphi(y)))_{|y=\beta_i}$$

$$- (\partial_y(u(t, y)\varphi(y)))_{|y=\alpha_i}) + \delta''(0) \sum_0^m (u(t, \beta_i)\varphi(\beta_i) + u(t, \alpha_i)\varphi(\alpha_i)).$$

By applying (9.4) it follows that

$$\partial_x^2 \tilde{u}(t, x) \geq \sum_0^m (u_x'(t, \beta_i)\varphi(\beta_i) - u_x'(t, \alpha_i)\varphi(\alpha_i)) + 4u_0 R'(x)\varphi(x) \quad (9.6)$$

since $\delta'(0) \leq 1$ in view of Lemma 9.1. We now use (9.3) and have

$$\partial_t \tilde{u}(t, x) = \sum_0^m \int_{\alpha_i}^{\beta_i} u_t'(t, y) d\theta(y)$$

$$= \sum_0^m \int_{\alpha_i}^{\beta_i} (\frac{1}{2} u_{xx}''(t, y) + R'(y) u_x'(t, y)) d\theta(y)$$

$$= \frac{1}{2} \sum_0^m (u_x'(t, \beta_i)\varphi(\beta_i) - u_x'(t, \alpha_i)\varphi(\alpha_i))$$

and (9.6) gives

$$\partial_t \tilde{u}(t, x) \leq \frac{1}{2} \partial_x^2 \tilde{u}(t, x) - 2u_0 R'(x)\varphi(x).$$

Finally, using (9.5), the inequality (9.2) follows at once.

We next introduce $v(t, x) = E_x[f^*(X(t)); T_L > t], t \geq 0, x \in \mathbf{R}$, and, for given $\varepsilon > 0$,

$$\bar{v}_\varepsilon(t, x) = \int_{-x}^x (\varepsilon + v(t, y)) d\theta(y), \quad t \geq 0, \quad 0 \leq x \leq L.$$

Then

$$(\partial_t + K)\bar{v}_\varepsilon = 0 \quad \text{in} \quad t > 0, 0 < x < L,$$

noting that $v(t, \cdot)$ is even for each $t > 0$.

To complete the proof of Theorem 9.1 let $w = \bar{v}_\varepsilon - \tilde{u}$ so that $(\partial_t + K)w \geq 0$ in $t > 0$, $0 < x < L$. Since w is continuous, the minimum principle gives

$$\min_{t \leq T} w \geq \min\{w; t \leq T \text{ and } t = 0 \text{ or } x = 0, L\} \quad (0 < T < +\infty)$$

(the trouble with smoothness is simple to avoid by convolution). Moreover, if $t > 0$, the function $w(t, \cdot)$ cannot attain a minimum at the point L. Indeed, if so

$$\int_{x \leq |y| \leq L} (\varepsilon + v(t, y)) d\theta(y) \leq \int_{x \leq |y| \leq L} u(t, \cdot)^*(y) d\theta(y), \quad 0 < x < L,$$

thereby forcing $\varepsilon \varphi(L) \leq 0$, which is a contradiction. From these facts, $w \geq 0$ and letting $\varepsilon \to 0$ we conclude that $\tilde{u} \leq \bar{v}_{0+}$. Hence

$$\langle u(t, \cdot) . g \rangle_\theta \leq \langle u(t, \cdot)^*, g^* \rangle_\theta \leq \langle v(t, \cdot), g^* \rangle_\theta.$$

Finally, letting $L \to +\infty$, Theorem 9.1 follows immediately. ∎

Next, put

$$\theta(dx) = e^{|x|^2/2} dx, \quad x \in \mathbf{R}^n$$

and denote by $\mathcal{E}_0$ the class of all open or closed subsets of $\mathbf{R}^n$ with positive θ-measure. If $A \in \mathcal{E}_0$, suppose $A^* = \{|x| \leq r\} = \bar{B}(r)$ satisfies $\theta(A) = \theta(A^*)$.

Theorem 9.2. *The solution of the Itô equation*

$$\begin{cases} dX(t) = \frac{1}{2} X(t) dt + dW(t) \\ \mathcal{L}X(t) = \theta \end{cases}$$

is *-isoperimetric. Furthermore, if $A \in \mathcal{E}_0$,

$$\theta(A + \bar{B}(\varepsilon)) \geq \theta(A^* + \bar{B}(\varepsilon)), \quad \varepsilon > 0.$$

A proof of Theorem 9.2 is obtained by iterations of so-called one-dimensional symmetrizations and we do not go into details here (cf. Figiel et alias [10]).

References

[1] L. Bachelier, *Theory of speculation*, in reference [6].

[2] F. Black and M. Scoles, *The pricing of options and corporate liabilities*, J. Polit. Econom **81** (1973), 637–659.

[3] C. Borell, *Geometric bounds on the Ornstein-Uhlenbeck velocity process*, Z. Wahrscheinlichkeitstheorie verw. Gebiete **70** (1985), 1–13.

[4] C. Borell, *Intrinsic bounds on some real-valued stationary random functions*, pp. 70–95 in "Probability in Banach Spaces V", Lecture Notes in Math. **1153**. Springer Verlag 1985.

[5] H. Brascamp, E.H. Lieb and J.M. Luttinger, *A general rearrangement inequality for multiple integrals*, J. Functional Analysis **17** (1974), 227–237.

[6] P.H. Cootner, "The Random Character of Stock Market Prices," M.I.T. Press, Cambridge, MA. 1964.

[7] A. Ehrhard, *Inégalités isopérimetriques et intégrales de Dirichlet Gaussiennes*, Ann. Sci. Éc. Norm. Sup. **17** (1984), 317–332.

[8] E.F. Fama, *The behaviour of stock-market prices*, Journal of Business **38** (1965), 34–105.

[9] X. Fernique, *Régularité des trajectoires des fonctions aléatoires gaussiennes*, pp. 1-96 in "École d'Été de Probabilités Saint-Flour IV-1974", Lecture Notes in Math. **480**. Springer Verlag 1975.

[10] T. Figiel, J. Lindenstrauss and V.D. Milman, *The dimension of almost spherical sections of convex bodies*, Acta Math. **139** (1977), 53–94.

[11] C.W.J. Granger and O. Morgenstern, "Predictability of Stock Market Prices," Heath-Lexington Book, Lexington, Mass. 1970.

[12] C.W.J. Granger, "A survey of empirical studies of capital markets," E.J. Elton and M.J. Gruber Ed. North-Holland Publishing Company, Amsterdam, Oxford 1975.

[13] J.M. Harrison and S.R. Pliska, *Martingales and stochastic integrals in the theory of continuous trading*, Stochastic Processes and their Applications **11** (1981), 215–260.

[14] N. Ikeda and S. Watanabe, "Stochastic Differential Equations and Diffusion Processes," North-Holland Publishing Company, Amsterdam, Oxford, New York 1981.

[15] P.L. Jennergren and P. Toft-Nielsen, *An investigation of random walks in the Danish stock market*, Nationaløkonomisk Tidsskrift **115** (1977), 254–269.

[16] R.A. Leitch and A.S. Paulson, *Estimation of stable law parameters; Stock price behaviour applications*, Journal of the American Statistical Association **70** (1975), 690–697.

[17] R.A. Levy, *Random walks: Reality or myth*, Financial Analysts Journal **XXIII** No. 6 (1967).

[18] P.D. Praetz, *The distribution of share price changes*, Journal of Business 45 (1972), 49–55.

[19] P.A. Samuelson, *Mathematics of speculative price*, SIAM Rev **XX** (1973), 1–42.

[20] D. Slepian, *The one-sided barrier problem for Gaussian noise*, The Bell System Technical Journal 41 (1962), 463–501.

[21] C.W. Smith Jr., *Option pricing: A review*, J. Fin. Econom. 3 (1976), 3–51.

[22] B.G. Sørensen, *En filtertest af danske aktiekurser*, Nationaløkonomisk Tidsskrift 118 (1980), 140–148.

[23] B.G. Sørensen, *Regnskabsinformation og aktiemarkedets effektivitet: En empirisk analyse*, Nationaløkonomisk Tidsskrift 120 (1982), 223–241.

[24] S.J. Wolfe, *On a continuous analogue of the stochastic difference equation $X_n = \rho X_{n-1} + B_n$*, Stochastic Processes and their Applications 12 (1982), 301–312.

Mölnesjögatan 30, S-424 50 Angered, Sweden

Department of Mathematics, Chalmers University of Technology, S-412 96 Göteborg, Sweden

Weak convergence of vector valued martingales[1]

W.J. Davis, N. Ghoussoub, W.B. Johnson[2],

S. Kwapien, B. Maurey

Abstract

If $(X_n)_n$ is a Banach space valued L_1-bounded martingale which converges to a strongly measurable function X in a very weak sense, then X_n converges to X strongly, a.s. This gives an extension of the Ito–Nisio theorem to the martingale setting. We also consider the case of submartingales valued in Banach lattices.

Introduction

In [3], Ito and Nisio show that the partial sums of Banach space valued independent symmetric random variables norm converge almost surely whenever there exists a "candidate for a limit" in a very weak sense. In this note we investigate the corresponding problem for martingales.

Let E be a Banach space and let F be a subspace of E^*. Suppose that $(X_n)_n$ is an L^1-bounded, E-valued martingale and that X is a strongly measurable random variable such that $f(X_n)$ converges a.s. to $f(X)$ for each f in F. In case F is **norming** for E (that is, there is a constant C so that $\|e\| \leq C \sup\{|f(e)|; f \in F, \|f\| = 1\}$ for each $e \in E$), it is part of the folklore that $(X_n)_n$ norm converges almost surely to X. This is in fact the content of Proposition 1 which is crucial to what follows. We were surprised to find that the conclusion remains true if we only assume that F is a **total subspace of E^*** (that is, $e \in E$ and $f(e) = 0$ for all $f \in F$ implies $e = 0$). This is the main result of this note. For the various notions of measurability and integrability we refer the reader to [2]. Martingales will always be defined on a probability space $(\Omega, \mathcal{F}, P)$ and adapted to an increasing sequence of sub-σ-fields $(\mathcal{F}_n)_n$ of $\mathcal{F}$.

---

[1] This communication is a compendium by P. Ørno of remarks made by the listed authors.

[2] Supported in part by NSF MCS-8500764.

Aside from the real-valued martingale convergence theorem we shall only use the following two facts:

Lemma A.   (Davis–Johnson [5]) *Let E be a separable subspace of the dual of a separable normed space F. Then there exists an equivalent norm $\| \ \|$ on F such that whenever $\{x_n, x\}$ is a sequence in E verifying $f(x_n) \to f(x)$ for each f in F and $\|x_n\| \to \|x\|$ then $\lim_n \|x_n - x\| = 0$.*

Lemma B.   (Neveu [6]) *Let I be a countable set and, for each $i \in I$, let $(X_n^i)_n$ be a real submartingale. If $\sup_n \int \sup_i (X_n^i)^+ < \infty$, then*

(1) *For every $i \in I$, $(X_n^i)_n$ converges a.s. to a limit X_∞^i.*

(2) *The submartingale $(\sup_i X_n^i)_n$ converges a.s. to $\sup_i X_\infty^i$.*

Proposition 1.   *Let F be a normed space and let $(X_n)_n$ be an F^*-valued, L^1-bounded martingale. Suppose there exists a strongly measurable F^*-valued random variable X such that $f(X_n) \to f(X)$ a.s. for each f in F. Then $(X_n)_n$ norm converges a.s. to X.*

Proof: Since (X_n) and X are strongly measurable, they are almost separably valued, so we may assume that they all take values in separable subspaces E of F^*. Consequently, by passing to a separable subspace of F which norms E, we may assume that F is separable. Let $\| \cdot \|$ be the norm on F associated to E by Lemma A and let I be a countable dense subset of $\{f \in F: \|f\| = 1\}$. By Lemma B, the real submartingale

$$\|X_n\| = \sup\{f(X_n): f \in I\}$$

converges a.s. to

$$\|X\| = \sup\{f(X): f \in I\},$$

hence the property of the norm gives that (X_n) converges strongly a.s. to X.   ∎

Remark: Note that if $(X_n)_n$ is an L^1-bounded martingale that is valued in a Banach space E, there always exists — by the w^*-compactness of the unit ball of E^{**} — a w^*-measurable, E^{**}-valued random variable X such that $f(X_n) \to f(X)$ a.s. for each f in E^*. Proposition 1 shows that one gets norm convergence whenever X is strongly measurable or whenever it is almost separable-valued in E^{**}.

Proposition 2. *Let E be a Banach space and let $(X_n)_n$ be an L^1-bounded, E-valued martingale. Let X be an E-valued strongly measurable random variable. Then the set $Y = \{f \in E^*; \lim_n f(X_n) = f(X) a.s.\}$ is a w^*-closed linear subspace of E^*.*

Proof: Since $\{X_n, X\}_n$ are all almost separable valued we can assume without loss of generality that E is separable. Note also that Y is a linear submanifold of E^*. We claim that the weak*-closure of Ball $(Y) = \{y \in Y; \|y\| \leq 1\}$ in E^* is actually contained in Y. Indeed, let φ be in $\overline{\text{Ball}(Y)}^{w^*}$ and, since E is separable, let $(f_n)_n$ be a sequence in Ball(Y) that w^*-converges to φ. Let F be the linear span of $\{f_n\}_n$ in E^* and let S be the canonical map from E into F^*. Note that since $(f_k(X_n))_n$ converges to $f_k(X)$ a.s. for each k, Proposition 1 applies to the martingale $\{S(X_n), S(X)\}$ and we get that $\lim_n \|S(X_n) - S(X)\|_{F^*} = 0$ a.s. and so $\lim_n \sup_k |f_k(X_n - X)| = 0$ a.s. It follows that $\varphi(X_n) \to \varphi(X)$ a.s., hence $\varphi \in Y$. But this implies that Ball(Y) is w^*-closed in E^*. Thus Y is w^*-closed in E^* by the Krein-Smulyan theorem. ∎

Theorem 3. *Let E be a Banach space and let $(X_n)_n$ be an L^1-bounded, E-valued martingale. Suppose there exists a strongly measurable random variable X and a total subspace H of E^* such that $f(X_n)_n$ converges to $f(X)$ a.s. for each f in H. Then $(X_n)_n$ norm converges a.s. to X.*

Proof: Note that $H \subseteq \{f \in E^*; \lim_n f(X_n) = f(X) a.s.\} \subseteq E^*$. Since H is total in E^*, it is w^*-dense. It follows from Proposition 2 that $\lim_n f(X_n) = f(X)$ a.s. for each f in E^*. The conclusion then follows from Proposition 1. ∎

Another proof: We shall now sketch a second proof of Theorem 3 which is less "functional analytic" and more "probabilistic".

Assume first that $\sup_n \|X_n\| \in L^1$ and define for each $A \in \cup \mathcal{F}_n$ the set function $F(A) = \lim_n \int_A X_n dP$. It is easy (and standard) to show that F extends to a σ-additive vector measure on the σ-field $\mathcal{F} = \sigma(\cup_n \mathcal{F}_n)$ in such a way that $\|F\|(A) \leq \int_A \sup_n \|X_n\| dP$ for each $A \in \mathcal{F}$ where $\|F\|$ denotes the variation of F.

Note that for each $f \in H$ and $A \in \mathcal{F}$ we have $f(F(A)) = f(\lim_n \int_A X_n dP) = \lim_n \int_A f(X_n) dP = \int_A f(X) dP$. Let $A_n = \{w; \|X\| \leq n\}$. since $X 1_{A_n}$ is Bochner integrable, we have for each $f \in H$ and each $A \in \mathcal{F}$, $f(F(A \cap A_n)) = \int_{A \cap A_n} f(X) dP = f(\int_{A \cap A_n} X dP)$. Since H is total, we get that $F(A \cap A_n) = \int_{A \cap A_n} X dP$ for each $A \in \mathcal{F}$ hence $\int_{A \cap A_n} \|X\| dP = \|F\|(A \cap A_n) \leq \int \sup \|X_n\| dP$. Since X is strongly measurable, $P(A_n) \uparrow 1$ and $\int_A \|X_n\| dP < \infty$ for each $A \in \mathcal{F}$. It follows that X is Bochner integrable and $X_n = E[X | \mathcal{F}_n]$ for each n. The norm convergence of such "closed martingales" in any Banach space is well known. It also follows immediately from Proposition 1.

To deal with the case where $\sup_n \int \|X_n\| dP < \infty$, we shall use a standard stopping time argument to reduce it to the "uniformly integrable case". Recall first Doob's inequality: For each $\lambda > 0$,

$$P(\omega; \sup_n \|X_n(\omega)\| > \lambda) \leq \frac{1}{\lambda} \sup_n \int \|X_n\| dP.$$

Consider now for each $a > 0$, the stopping time:

$$\sigma_a(\omega) = \begin{cases} \inf\{n; \|X_n(\omega)\| \geq a\} & \text{if } \sup_n \|X_n(\omega)\| \geq a \\ +\infty & \text{otherwise} \end{cases}$$

This is the usual way to construct from the L_1-bounded martingale X_n a martingale $Y_n = X_{n \wedge \sigma_a}$ for which $\sup_n \|Y_n\| \in L^1$. Moreover, $f(Y_n) \rightarrow f(X_{\sigma_a})$ a.s. for each $f \in H$ where $X_{\sigma_a} = X 1_{\{\sigma_a = +\infty\}} + \sum_n X_n 1_{\{\sigma_a = n\}}$. By the first part of this proof, $(X_{n \wedge \sigma_a})_n$ norm converges a.s. to X_{σ_a}. To finish the proof it is enough to notice that Doob's inequality guarantees that $P[\sigma_a < \infty] \rightarrow 0$ a.s. $a \rightarrow \infty$. ∎

This result allows us to prove a martingale version of thje Ito–Nisio theorem [13]. The equivalence of (1) and (2) below appeared in [4], but becomes easier to prove with the aid of Theorem 3.

Theorem 4.   *Let E be a separable Banach space and H a total subspace of E^*. Let X_n be an E-valued, L_1-bounded martingale. The following assertions are equivalent:*

(1) *X_n converges a.s.*

(2) *X_n converges in distribution.*

(3) *For almost all ω, the sequence $(X_n(\omega))_n$ has a cluster point in E in the topology $\sigma(E, H)$.*

(4) *There is a distribution μ on E such that for each $f' \in H$, $f \circ X_n$ converges in distribution to $f(\mu)$.*

Proof: (1) $\Rightarrow$ (2) is obvious.

(2) $\Rightarrow$ (3): By Prohorov's theorem (see e.g., [11]), for $\varepsilon > 0$, there exists a compact set $K \subset E$ such that $P[X_n \in K] \geq 1 - \varepsilon$ for each n. Therefore, setting

$$C_K = \{\omega | X_n(\omega) \in K \text{infinitely often}\}$$

we have

$$P(C_K) = P\left(\bigcap_{i=1}^{\infty} \bigcup_{n=i}^{\infty} [X_n \in K]\right) = \lim_i P\left(\bigcup_{n=i}^{\infty} [X_n \in K]\right)$$
$$\geq \lim P[X_i \in K] \geq 1 - \varepsilon.$$

If $\omega \in C_K$ then $X_n(\omega)$ has a (strong) cluster point, and hence a $\sigma(E, F)$ cluster point in K.

(3) $\Rightarrow$ (1). For almost all ω, let $X(\omega)$ be a $\sigma(E, H)$ cluster point of $(X_n(\omega))_n$ in E. Let S be the canonical restriction map from E to H^*. By ω^*-compactness, there exists a ω^*-measurable H^*-valued random variable φ such that $(S(X_n))_n$ ω^*-congerves to φ a.s. Since S is one-to-one we get that $S(X) = \varphi$ a.s. hence it is almost separably valued and

by a theorem of Pettis [2], it is strongly measurable in H^*. Moreover, since S is one-to-one and E is separable, a theorem of Lusin insures that S^{-1} is Borel measurable which implies that X is strongly measurable in E. Now we can apply Theorem 3 to get the claim.

(2) $\Rightarrow$ (4) is obvious. It remains to show that (4) $\Rightarrow$ (1). For that again let S be the canonical map from E into H^* and let φ be the ω^*-limit of the martingale $(S(X_n))_n$. The hypothesis implies that the distribution of φ is equal to the distribution $S(\mu)$ which is tight. Hence φ is strongly measurable and almost surely valued in $S(E)$. Again the theorem of Lusin gives that $S^{-1} \circ \varphi = X$ is strongly measurable. The hypothesis gives that $f(X_n) \to f(X)$ a.s. for each f in H. The claim follows again from Theorem 3.   ∎

Remark 5: All the examples of bounded, non-convergent Banach space valued martingales have "natural limits" in the double dual which are not strongly measurable. The above results show that this is neces- sary; that is, these limits can never be strongly measurable unless the martingales converge.

Example 6: In the Ito-Nisio theorem, no boundedness condition what- soever is imposed on the partial sums of the independent random vari- ables. This cannot be the case for martingales. Indeed, it is possible to give examples of non-L_1-bounded martingales which converge over total families to very nice limit functions, but which fail to converge to the limit strongly. Here is one: Let $F_n = \sum_{k=1}^{n} h_k$ be the double or nothing martingale. That is, $h_k = 2^k(1_{[0,2^{-k-1})} - 1_{[2^{-k-1},2^{-k})})$. Take indepen- dent copies $F_{n,k}$ of this martingale, and define $X_n = \sum_{k=1}^{n} F_{k,n-k+1}e_k$, where (e_k) denotes the standard unit vector basis in c_0. If (e_k^*) denotes the unit vector basis in ℓ_1, we have

$$e_k^* \circ X_n \to -1 \quad a.s. \quad \text{as } n \to \infty \quad \text{for each } k.$$

That is, the natural limit for this martingale is the constant function -1_N in ℓ_∞. Clearly, since $\|X_n(\omega) + 1_N\|_{\ell_\infty} \geq 1$ for all ω, we cannot have strong convergence. It follows that X_n is not L_1-bounded. Of course, that is easy to compute directly.

As mentioned above, the theorem of Ito-Nisio does not follow from the above results because of the L_1-boundedness assumption imposed on the martingale. The following lemma gives one way to avoid such a hypothesis.

Lemma 7. *Let E be a Banach space and let $(X_n)_n$ be an E-valued martingale such that $\int \sup_n \|X_{n+1} - X_n\| dP < \infty$. Let X be strongly measurable random variable such that $f(X_n) \to f(X)$ a.s. for each f in a total subspace H of E^*. Then $\lim_n \|X_n(\omega) - X(\omega)\| = 0$ for almost all ω in the set $\{\omega; \sup_n \|X_n(\omega)\| < \infty\}$.*

Proof: For each $a > 0$, again let σ_a be the stopping time:

$$\sigma_a(\omega) = \begin{cases} \inf\{n; \|X_n(\omega)\| \geq a\} & \text{if } \sup_n \|X_n(\omega)\| \geq a \\ +\infty & \text{otherwise} \end{cases}$$

Note that $\int \|X_{\sigma_a \wedge n}\| dP \leq a + \int \sup_n \|X_{n+1} - X_n\| dP$. That is, the martingale $(X_{n \wedge \sigma_a})_n$ is L^1-bounded and hence norm convergent a.s. by Theorem 3. The rest follows from the fact that when $a \to \infty$, the set $\{\sigma_a = +\infty\}$ increases to the set $\{\sup_n \|X_n\| < \infty\}$. ∎

By using this lemma, the theorem of Ito-Nisio can be proved by a standard truncation argument. We shall leave the details to the interested reader.

We now deal with the case of supermartingales or submartingales valued in Banach lattices. The situation here is different in view of the following result proved in [9]. If E is a Banach lattice that is not lattice isometric to an L^1-space, then there exists a uniformly bounded E-valued positive supermartingale (X_n) such that:

(1) $f(X_n) \to 0$ a.s. for every $f \in E^*$.

(2) $(X_n)_n$ is norm convergent.

However, we can give some positive results in the case of **positive submartingales**. We shall need the following lattice renorming lemma:

Lemma C. (Davis-Ghoussoub-Lindenstrauss [7]). *Let F be a Banach lattice and let E be an order continuous closed ideal in F^*. Then there exists an equivalent lattice norm $\| \ \|$ on F such that whenever $\{x_n, x\}$ is a sequence in E verifying $\lim_n \|x_n\| = \|x\|$ and $f(x_n) \to f(x)$ for each f in F then $\lim_n \|x_n - x\| = 0$.*

As in Proposition 1, we now show that the combination of Lemmas B and C give the following result:

Proposition 8. *Let E be an order continuous Banach lattice and let F be a sublattice of E^* that is norming for E. Suppose $(X_n)_n$ is an E-valued, L^1-bounded positive submartingale and X a strongly measurable random variable such that $f(X_n) \to f(X)$ a.s. for each f in F. Then $\lim_n \|X_n - X\| = 0$ a.s.*

Proof: Note that the canonical embedding of F into E^* is a lattice homomorphism, hence the adjoint map from E^{**} onto F^* is interval preserving and its restriction to E is an isomorphic embedding since F is norming. Hence E can be identified with an order continuous closed ideal in F^*. As in Proposition 1, we can assume that E and F are separable. Let $\| \ \|$ be the norm on F associated to E by Lemma C. Since it is a lattice norm, there exists a countable family I of positive functionals in the sphere of F such that $\|x\| = \sup\{f(x); f \in I\}$ for each x in F_+^*. the same proof as in Proposition 1 now gives the norm convergence of $(X)_n$ to X. ∎

The following example shows that the above statement does not hold without some assumption on the Banach lattice (i.e., the order continuity of the norm).

Example 9: Here is a uniformly bounded ℓ_∞-valued positive submartingale $(X_n)_n$ such that $f(X_n) \to f(1)$ a.s. for all $f \in \ell_\infty^*$ but (X_n) is not norm convergent.

For each $i \in \mathbf{N}$, let Ω_i be a copy of $[0,1]$, λ_i Lebesgue measure and $\sum_i$ the Borel σ-field on $[0,1]$. Set $\Omega = \prod_{i \in \mathbf{N}} \Omega_i$, $\mathcal{F} = \prod_{i \in \mathbf{N}} \sum_i$ and

$P = \prod_{i \in \mathbb{N}} \lambda_i$. For each k, ω_k will denote the k'th coordinate of $\omega \in \Omega$. Define the sequence of ℓ_∞-valued random variables $Z_m : \Omega \to \ell_\infty$ by

$$[Z_m(\omega)]_{2^n+k} = \begin{cases} 2^{-n} & \text{if } m < n \\ 1 & \text{if } m = n \quad \text{and } (k-1)2^{-m} \le \omega_m \le k2^{-m} \\ 0 & \text{otherwise} \end{cases}$$

for each m, n and $1 \le k \le 2^n$.

It is easy to see that $(Z_m)_m$ is a positive supermartingale such that $0 \le Z_m \le 1$ where $\mathbf{1} = (1, 1, 1, \dots)$ and $(\int Z_m dP)_m$ norm converges to zero. It follows that $f(Z_m) \to 0$ a.s. for each f in ℓ_∞^*. On the other hand, $\|Z_m(\omega)\|_\infty \ge 1$ for each ω, hence it is not norm convergent. The positive submartingale $X_n = 1 - Z_n$ obviously verifies the above claim.

∎

Remark 10: the assumption in Proposition 8 that the norming set F is a sublattice and not merely a subspace of E^* seems to be relevant. Indeed, if $(e_n)_n$ denotes the unit vector basis of c_0 and if F denotes the norming subspace of ℓ_1 consisting of all $(\alpha_n)_n$ such that $\sum_n \alpha_n = 0$, we get that the "deterministic positive submartingale" $X_n = \sum_{i=1}^n e_i$ is divergent even though it is norm bounded and $f(X_n) \to 0$ a.s. for each $f \in F$.

On the other hand, if E does not contain c_0, we can weaken considerably the assumptions on F.

Proposition 11. *Let E be a Banach lattice not containing c_0 and let H be a total subspace of E^*. Suppose $(X_n)_n$ is an E-valued, L^1-bounded positive submartingale and X is a strongly measurable random variable such that $f(X_n) \to f(X)$ a.s. for each $f \in H$. Then $\lim_n \|X_n - X\| = 0$ a.s.*

Proof: Note first that $(\|X_n\|)_n$ is an L^1-bounded real submartingale hence we have Doob's inequality $P[\sup_n \|X_n\| > a] \le \frac{1}{a} \sup_n \int \|X_n\| dP$ for each $a > 0$. The same "stopping-time argument" used in the proof of Theorem 3 reduces the problem to the case when $\sup_n \|X_n\| \in L^1$.

Note that for each $A \in \mathcal{F}'_m$ the sequence $(\int_A X_n dP)_{n \geq m}$ is increasing and norm bounded, hence it is norm convergent to, say, $F(A)$ ([8], p. 34). It is clear that F extends to a σ-additive vector measure on the σ-algebra $\mathcal{F}$ generated by $\cup_n \mathcal{F}_n$. Moreover $F(A) = \lim_n \int_A X_n dP$ for each $A \in \mathcal{F}$.

On the other hand, $f(X_n) \to f(X)$ a.s. for each $f \in H$. It follows as in Theorem 3 that for each $A \in \mathcal{F}$ and each $f \in H$, $f(F(A)) = f(\int_A X dP)$ which implies (since H is total) that $F(A) = \int_A X dP$, for each $A \in \mathcal{F}$.

Note now that for each f in E^*_+, $(f(X_n))_n$ is a submartingale that must converge to $f(X)$ a.s. Hence Proposition 8 applies and we get the norm convergence of $(X_n)_n$ to X. ∎

References

[1] P. Billingsley, "Convergence of Probability Measures," Wiley, 1968.

[2] R.D. Bourgin, "Geometric Aspects of Convex Sets with the Radon-Nikodym Property," Lecture Notes in Mathematics 993, Springer–Verlag, 1983.

[3] K. Ito and M. Nisio, *On the convergence of sums of independent Banach space valued random variables*, Osaka Math. J. 5 (1968), 25–48.

[4] A. Korzeniowski, *thesis*, Wroclaw.

[5] W.J. Davis and W.B. Johnson, *A renorming of non-reflexive Banach spaces*, Proc. AMS 37 (1973), 486–487.

[6] J. Neveu, "Discrete Parameter Martingales," North Holland, 1975.

[7] W.J. Davis, N. Ghoussoub and J. Lindenstrauss, *A lattice renorming theorem and its applications to vector-valued processes*, Trans. AMS 263 (1981), 531–540.

[8] J. Lindenstrauss and L. Tzafriri, "Classical Banach Spaces II, Function Spaces," Springer-Verlag, 1979.

[9] Y. Benyamini and N. Ghoussoub, *Une charactérization probabiliste de ℓ_1*, C.R. Acad. Sc. Paris 286 (1978), 795–797.

The Ohio State University, Columbus, Ohio, USA
University of British Columbia, Vancouver, B.C. Canada
Texas A& M University, College Station, Texas, USA
University of Warsaw, Warsaw, Poland and Case Western Reserve University, Cleveland, Ohio, USA
Université Paris VII, Paris, France

Fonctions aléatoires à valeurs vectorielles

XAVIER FERNIQUE

Summary

We intend to set up fundamental structures for the study of vector valued random functions. In the first part we define general notations and we study continuity, measurability and separability properties; in the second part we define oscillations notion and we study it in the gaussian case; the last part is devoted to the regularity of gaussian random functions on **R** with vector values and stationary increments.

0. Introduction

La connaissance des fonctions aléatoires (f.a) à valeurs réelles s'est considérablement développée depuis une trentaine d'années. Cette notion reste pourtant insuffisante. En effet les modèles probabilistes utilisés dans les situations mécaniques ou physiques manient plutôt les f.a à valeurs vectorielles dans certains espaces de fonctions. On sait bien étudier la régularité de certaines d'entre elles (cf. par exemple [2], [3]) en particulier les processus de Wiener ou d'Ornstein-Uhlenbeck qui constituent deux classes de f.a. à valeurs vectorielles, gaussiennes et à accroissements stationnaires; on voit aussi se développer l'étude de suites de vecteurs aléatoires, dans le cadre par exemple de la loi du logarithme itéré, qui sont des f.a à valeurs vectorielles sur **N** ou **N̄**. Les méthodes mises au point dans ces deux situations et aussi pour analyser les f.a. à valeurs réelles permettent les études dans le cadre général. Dans ce travail, on se propose de confirmer à la suite de résultats précédents ([5], [7]) que l'étude des f.a. à valeurs vectorielles en général, gaussiennes en particulier, est moins complexe qu'on ne le prévoit; leurs propriétés se déduisent souvent de celles de classes de f.a. à valeurs réelles comme si, à un certain niveau, le hasard devenait presque déterministe.

Dans le premier chapitre, on définit les notations générales et on étudie les propriétés de continuité, de mesurabilité et de séparabilité. Le second chapitre définit les notions d'oscillations et les analyse dans le cas gaussien. Enfin le troisième chapitre étudie la régularité des f.a. gaussiennes à valeurs vectorielles et à accroissements stationnaires.

1. Généralités

1.1. Définitions et notations.. Soient $(\Omega, \mathcal{A}, P)$ un espace d'épreuves (supposé complet), T un ensemble, $(E, \|\cdot\|_E)$ un espace de Banach séparable; *une fonction aléatoire X sur T à valeurs dans E*, ou plus simplement une *f.a. vectorielle*, est un ensemble $X = \{X(t), t \in T\}$ de vecteurs aléatoires à valeurs dans E indexés par T; E' désignera le dual topologique de E et (E_1', w) sa boule unité munie de la topologie faible qui est métrisable; on notera aussi w une distance définissant cette topologie. On notera $Z = (z_n, n \in \mathbb{N})$ une suite dense dans (E_1', w) et qui sépare donc les points de E. On munit E de sa tribu topologique $\mathcal{B}(E)$ et on note μ_X la loi de X qui est une probabilité sur E^T muni de la tribu produit; on note $\mathcal{A}(X)$ la sous-tribu de $\mathcal{A}$ engendrée par X. On pourra être amené à utiliser sur T, sur E_1', sur $T \times E_1'$, des écarts, c'est à dire des (pseudo)-métriques ne séparant pas nécessairement les points; pour alléger le langage, on omettra en général les préfixes correspondants.

Deux f.a. vectorielles X et Y sont dites *équivalentes* si:

$$\forall t \in T, \quad P\{\omega\colon X(\omega, t) = Y(\omega, t)\} = 1;$$

il faut et il suffit pour cela, puisque Z sépare les points de E, que:

$$\forall t \in T, \forall z \in Z, \quad P\{\langle X, z \rangle = \langle Y, z \rangle\} = 1;$$

on dira alors aussi que Y est une *modification vectorielle* de X.

A toute f.a. vectorielle X sur T, on peut associer la f.a. réelle $\bar{X}$ sur $T \times E_1'$ définie par:

$$\forall t \in T, \forall z \in E_1', \quad \bar{X}(t, x) = \langle X(t), z \rangle;$$

l'étude de la f.a. vectorielle X ne se réduit pas à celle de $\bar{X}$ par les méthodes classiques des f.a. à valeurs réelles. Par exemple, si T est muni d'une métrique δ pour laquelle il est séparable, alors $(T \times E_1'; \delta, w)$ est métrique séparable et il existe une f.a. réelle $\widetilde{X}$ séparable sur $T \times E_1'$ et équivalente à $\bar{X}$, donc à X, au sens suivant:

$$\forall t \in T, \forall z \in E_1', \quad P\{\widetilde{X}(t,z) = \bar{X}(t,z) = \langle X(t), z \rangle\} = 1,$$

mais le maniement de $\widetilde{X}$ met difficilement en valeur les propriétés de la f.a. vectorielle X puisqu'elle n'est en général ni linéaire, ni continue sur E_1', serait-ce presque sûrement. Le paragraphe suivant montre pourtant qu'une grande régularité de $\widetilde{X}$ permet de construire des modifications vectorielles régulières de X.

1.2. Un critère d'existence de modifications vectorielles à trajectoires continues.

Théorème 1.2. *On suppose que (T, δ) est séparable; dans ces conditions, les deux propriétés suivantes sont équivalentes:*

1.2.1 *il existe une modification réelle $\widetilde{X}$ de $\bar{X}$ sur $(T \times E_1'; \delta, w)$ ayant p.s. des trajectoires continues dans* **R**,

1.2.2 *il existe une modification vectorielle Y de X sur (T, δ) ayant p.s. des trajectoires continues dans E.*

Démonstration: l'implication $1.2.2 \Rightarrow 1.2.1$ est triviale, nous prouvons l'implication inverse. Sous l'hypothèse 1.2.1., la compacité de (E_1', w) montre qu'il existe une partie négligeable $N_0 \subset \Omega$ telle que:

$$\forall \omega \notin N_0, \forall t \in T, \forall \varepsilon > 0, \exists \eta > 0 :$$
$$\delta(s,t) \leq \eta \Rightarrow \sup_{z \in E_1'} |\widetilde{X}(\omega; t, z) - \widetilde{X}(\omega; s, z)| \leq \varepsilon;$$

soit S une suite dense dans T, il existe aussi une partie négligeable $N_1 \subset \Omega$ telle que:

$$\forall \omega \notin N_1, \forall t \in S, \forall z \in Z, \quad \widetilde{X}(\omega; t, z) = \langle X(\omega, t), z \rangle.$$

Dans ces conditions, pour tout $\omega \notin N_0 \cup N_1$ et pour tout $t \in T$, l'image par $X(\omega)$ de la trace sur S du filtre des voisinages de t est un filtre de Cauchy dans l'espace complet E et il converge vers un élément $Y(\omega, t) \in E$; les raisonnements habituels montrent que l'application $t \to Y(\omega, t)$ est continue. On pose $Y(\omega) = 0$ si $\omega \in N_0 \cup N_1$ et on vérifie suivant la routine que Y satisfait 1.2.2.

1.3. Un critère d'existence de modifications mesurables. On suppose dans ce paragraphe que T est muni d'une tribu $\mathcal{T}$; une f.a. vectorielle X est dite *mesurable* si l'application $(\omega, t) \to X(\omega, t)$ de $(\Omega \times T, \mathcal{A} \otimes \mathcal{T})$ dans $(E, \mathcal{B}(E))$ est mesurable. On peut alors appliquer à X des opérations d'intégration sur l'un des facteurs ou l'espace produit en utilisant le théorème de Fubini. On sait que toute f.a. réelle n'a pas néccessairement de modifications mesurables; la situation a la même complexité dans le cas vectoriel:

Théorème 1.3. *Les deux propriétés suivantes sont équivalentes:*

   1.3.1 *Pour tout élément z de E_1', la f.a. réelle $\langle X, z \rangle$ a une modification mesurable y_z*

   1.3.2 *La f.a. vectorielle X a une modification vectorielle mesurable Y.*

La démonstration du théorème 1.3. sera basée sur le critère partiel et les deux lemmes suivants:

Théorème 1.3.3. *On suppose X intégrable au sens suivant:*

$$\forall t \in T, \quad E\{\|X(t)\|_E\} < \infty;$$

dans ces conditions, les deux propriétés suivantes sont équivalentes:
1.3.3.1. X a une modification vectorielle mesurable Y.
1.3.3.2. Pour tout $a \in \mathcal{A}$, l'application $t \to \int_a X(t)dP$ est une application mesurable de $(T, \mathcal{T})$ dans $(E, \mathcal{B}(E))$; de plus, il existe une sous-tribu $\mathcal{C}$ de $\mathcal{A}$ à base dénombrable telle que la tribu complétée $\bar{\mathcal{C}}^P$ contienne $\mathcal{A}(X)$.

Lemme 1.3.4. *Si A est un élément de la tribu produit $\mathcal{A} \otimes \mathcal{T}$, alors la tribu $\mathcal{A}(1_A)$ engendrée dans Ω par les sections $A(t) = \{\omega : (\omega, t) \in A\}$, $t \in T$ est contenue dans une tribu à base dénombrable.*

Lemme 1.3.5. *Si Y est une application mesurable de $(\Omega \times T, \mathcal{A} \otimes \mathcal{T})$ dans $(E, \mathcal{B}(E))$, alors la tribu $\mathcal{A}(Y)$ est contenue dans une tribu à base dénombrable.*

Démonstration du lemme 1.3.4: Notons $\bar{\mathcal{A}}$ la classe des éléments A de $\mathcal{A} \otimes \mathcal{T}$ vérifiant la propriété énoncée; c'est une classe monotone qui contient l'algèbre des réunions finies $\cup_{k=1}^{n} [a_k \times b_k]$ de produits d'éléments de $\mathcal{A}$ et de $\mathcal{T}$; l'énoncé en résulte.

Démonstration du lemme 1.3.5: Si l'image de Y est dénombrable, le résultat se déduit du lemme précédent; sinon puisque E est polonais, Y est limite d'une suite d'applications mesurables dont l'image est dénombrable et le résultat général s'en déduit.

Démonstration du théorème 1.3.3: Supposons 1.3.3.1 vérifié; alors pour tout $t \in T$ et tout $a \in \mathcal{A}$, les intégrales $\int_a X(t)dP$ et $\int_a Y(t)dP$ sont égales si bien que la première partie de 1.3.3.2 résulte des arguments du théorème de Fubini appliqués à Y. Sous la même hypothèse, la tribu $\mathcal{A}(X)$ est contenue dans la tribu complétée $\overline{\mathcal{A}(Y)}^P$ de sorte que la deuxième partie de 1.3.3.2 résulte du lemme 1.3.5. Inversement, sous l'hypothèse 1.3.3.2, nous notons (c_n) une base dénombrable de $\mathcal{C}$; pour tout $n \in \mathbf{N}$, $\mathcal{C}_n$ est la tribu engendrée par $(c_k, k \leq n)$ et $\mathcal{A}_n$ est la famille finie des atomes non négligeables de $\mathcal{C}_n$. Puisque X est intégrable, on a alors:

$$\forall t \in T, \ E\{X(t)|\mathcal{C}_n\} = \sum \{(\int_a X(t)dP)1_a/P(a), \ a \in \mathcal{A}_n\} \ p.s.$$

Si on note $Y_n(t)$ la somme finie figurant au second membre, la propriété 1.3.3.2 montre que Y_n est mesurable sur l'espace produit. On définit alors une f.a. vectorielle Y mesurable en posant:

$$\forall \omega \in \Omega, \forall t \in T,$$

$$Y(\omega, t) = \lim Y_n(\omega, t) \text{ si cette limite existe,}$$

$$Y(\omega, t) = 0 \text{ sinon;}$$

puisque X est intégrable et que la suite croissante $(\mathcal{C}_n)$ engendre $\mathcal{C}$, on a:

$$\forall t \in T, \ X(t) = \lim E\{X(t)|\mathcal{C}_n\} = Y(t) \ p.s.$$

de sorte que Y est une modification vectorielle mesurable de X. Le théorème est établi.

Démonstration du théorème: L'implication 1.3.2 $\Rightarrow$ 1.3.1 est immédiate, nous démontrons l'implication inverse.

(a) Supposons pour commencer X intégrable au sens 1.3.3.; sous l'hypothèse 1.3.1., la tribu $\mathcal{A}(X)$ est contenue dans la tribu $\overline{V_{z \in Z} \mathcal{A}(y_z)}^P$; de même, pour tout $a \in \mathcal{A}$, la mesurabilité de l'application $t \to \int_a X(t) dP$ résulte de celle des applications $t \to \int_a < X(t), z > dP$, $z \in Z$ qui est effectivement réalisée; ceci permet d'appliquer à X le théorème 1.3.3 qui montre que X a une modification vectorielle mesurable; c'est le résultat partiel du théorème.

(b) Dans le case général, sous l'hypothèse 1.3.1., nous définissons une f.a. vectorielle X' en posant:

$$\forall \omega \in \Omega, \ \forall t \in T, \quad X'(\omega, t) = \frac{X(\omega, t)}{1 + \sup\{\langle X(\omega, t), z \rangle z \in Z\}},$$

si le dénominateur est fini. Alors X' est intégrable au sens 1.3.3. et vérifie la propriété 1.3.1. La démonstration partielle (a) fournit une modification vectorielle mesurable Y' de X'; on constate que pour tout $t \in T$, $\|X'(t)\|_E$ est p.s. inférieur à 1 de sorte que $Y'/(1 - \|Y'\|_E)$ est une modification vectorielle mesurable de X; le théorème est établi.

Remarque 1.3.6: Les propriétés des f.a. réelles montrent que si la tribu T sur T est associée à une métrique δ pour laquelle T est séparable et si pour tout $z \in E_1'$, la f.a. $\langle X, z \rangle$ est continue en probabilité sur (T, δ), la f.a. vectorielle X vérifie 1.3.1. Elle possédera donc une modification vectorielle mesurable.

1.4 Sur l'existence de modifications vectorielles séparables. Sous beaucoup d'aspects, la séparabilité introduite par Doob est trés utile pour le maniement des f.a. réelles: toute f.a. réelle sur un espace métrique séparable T possède une modification numérique (éventuellement à valeurs dans $\bar{\mathbf{R}}$) séparable au sens de Doob, mais non nécessairement une modification réelle séparable; la notion de séparabilité semble donc partiellement inadaptée aux f.a. vectorielles: l'existence universelle de modifications séparables exigerait un affaiblissement de la notion ([8]). Nous prendrons ici au contraire une définition forte de la séparabilité et nous énoncerons des critères d'existence de modifications vectorielles séparables. Nous dirons *qu'une f.a. vectorielle X sur l'espace métrique séparable (T, δ) est séparable s'il existe une partie dénombrable dense S de T et une partie négligeable N de Ω telles que:*

$$\forall \omega \notin N, \ \forall t \in T, \quad X(\omega, t) \in \bigcap_{\varepsilon > 0} \overline{\{X(\omega, s), s \in B(t, \varepsilon) \cap S\}}$$

où l'ensemble surbarré est l'adhérence dans E de l'ensemble indiqué. En suivant pas à pas les constructions classiques des modifications séparables des f.a. réelles, on constate alors:

Théorème 1.4. *Soit X une f.a. vectorielle sur un espace métrique séparable (T, δ); alors les deux propriétés suivantes sont équivalentes:*

1.4.1. Il existe une modification vectorielle séparable de X

1.4.2. Il existe une partie dénombrable S de T telle que:

$$P_*\{\forall t \in T, \ \bigcap_{\varepsilon > 0} \overline{\{X(s), s \in B(t, \varepsilon) \cap S\}} \neq \emptyset\} = 1.$$

On notera que si on accepte d'agrandir la tribu $\mathcal{A}$ sur Ω en prolongeant P, l'existence de modifications vectorielles séparables est déterminée par la condition:

1.4.3. *Il existe une partie dénombrable S de T telle que*

$$P^*\{\forall t \in T, \ \bigcap_{\varepsilon>0} \overline{\{X(s), s \in B(t,\varepsilon) \cap S\}} \neq \emptyset\} = 1.$$

Enfin l'existence de f.a. vectorielles séparables sur un autre espace d'épreuves ayant la même loi que X est déterminée par:

1.4.4. *Il esiste une partie dénombrable S de T telle que*

$$\mu_X^*\{x \in E^T : \forall t \in T, \ \bigcap_{\varepsilon>0} \overline{\{x(s), \ s \in B(t,\varepsilon) \cap S\}} \neq \emptyset\} = 1$$

Dans certaines situations, la propriété 1.4.2 a un caractère très naturel et peut être vérifiée en utilisant le lemme suivant:

Lemme 1.4.5. *Soient E un espace polonais et S un ensemble dénombrable; on note A l'ensemble $\{x \in E^S : \{x(t), t \in S\}$ relativement compact dans $E\}$; alors A est mesurable dans E^S; de plus pour toute probabilité μ sur E^S, on a:*

$$\mu(A) = \sup\{\mu(K^S), \ K \ compact \ dans \ E\}.$$

Démonstration: Soit (x_n) une suite dense dans E; puisque E est complet, les ensembles relativement compacts y sont les ensembles précompacts; on a donc:

$$A = \bigcap_{n \in \mathbb{N}} \bigcup_{m \in \mathbb{N}} \bigcap_{t \in S}\{x(t) \in \bigcup_{j=1}^{m} B(x_j; 4^{-n})\};$$

ceci montre que A est mesurable et que pour toute probabilité μ sur E^S, on a:

$$\mu(A) = \sup_{m \in \mathbb{N}^\mathbb{N}} \mu\left[\left\{\bigcap_{n \in \mathbb{N}} \bigcup_{j=1}^{m(n)} \overline{B(x_j; 4^{-n})}\right\}^S\right],$$

c'est le résultat de l'énoncé.

Le lemme 1.4.5 fournit alors en effet:

Corollaire 1.4.6. *Soit X une f.a. vectorielle sur un espace métrique séparable (T, δ); on suppose qu'il existe une partie dénombrable dense S de T sur laquelle les trajectoires de X sont p.s. relativement compactes; alors X possède une modification vectorielle séparable.*

Le corollaire précédent reste assez général pour fournir un nouveau critère de régularité des trajectoires:

Théorème 1.4.7. *Soit X une f.a. vectorielle sur un espace métrique compact T; pour que X ait une modification vectorielle à trajectoires continues, il faut et il suffit que les deux conditions suivantes soient réalisées:*

(a) *Il existe une partie dénombrable dense S de T sur laquelle les trajectoires de X sont p.s. relativement compactes,*

(b) *Pour tout élément z de E_1', $\langle X, z \rangle$ a une modification à trajectoires continues.*

Nous démontrons la suffisance: si (a) est réalisée, le corollaire 1.4.6 montre qu'il existe une modification vectorielle séparable Y de X; les propriétés (a) et (b), le lemme 1.4.5 et la séparabilité montrent que pour tout $\varepsilon > 0$, il existe une partie compacte K de E et une partie mesurable Ω_ε telles que:

$$P(\Omega_\varepsilon) \geq 1 - \varepsilon \text{ et } \forall \omega \in \Omega_\varepsilon,\ \forall t \in T,\ Y(\omega, t) \in K,$$

$$\forall \omega \in \Omega_\varepsilon,\ \forall z \in Z,\ \langle Y(\omega, \cdot), z \rangle \text{ continue.}$$

Puisque Z sépare les points de E et donc ceux du compact K, la topologie affaiblie définie par Z sur K coïncide avec la topologie induite par E de sorte que:

$$P\{t \to Y(t) \text{ continue}\} \geq P\{\Omega_\varepsilon\} \geq 1 - \varepsilon,$$

ce qui établit le résultat.

1.5 Un critère de continuité en probabilité. Soit X une f.a. vectorielle sur un espace topologique $(T, \mathcal{T})$; alors les propriétés de convergence en probabilité pour X définissent différentes (pseudo)-métriques sur T, sur E_1' ou sur $T \times E_1'$; en particulier, pour tous les couples (s, t) d'éléments de T et (y, z) d'élements de E_1', nous posons:

1.5.1. $\hat{d}(s, t) = \sup_{z \in E_1'} E\{f(\langle X(s) - X(t), z\rangle)\}$,

1.5.2. $D(s, t) = E\{f(\|X(s) - X(t)\|_E)\}$,

1.5.3. $d^t(y, z) = E\{f(\langle X(t), y - z\rangle)\}$,

où $\forall x \in \mathbf{R}$, $f(x) = \frac{|x|}{1+|x|}$.

On notera que la compacité de (E_1', w) montre que pour tout $t \in T$, l'injection $(E_1', w) \to (E_1', d^t)$ est uniformément continue. Suivant la terminologie habituelle, X est *continue en probabilité* si l'injection $(T, \mathcal{T}) \to (T, D)$ est continue. Nous dirons que X *est tendu* si

1.5.4. $\forall \varepsilon > 0$, $\exists K$ compact dans E: $\forall t \in T$, $P\{X(t) \notin K\} \leq \varepsilon$.

Cette dernière notion simplifie l'étude de la continuité en probabilité:

Théorème 1.5.5. *Soit X une f.a. vectorielle sur un espace topologique $(T, \mathcal{T})$; on suppose X tendu. Alors les deux propriétés suivantes sont équivalentes:*

(a) *X est continu en probabilité,*

(b) *Pour tout $z \in E_1'$, $\langle X, z\rangle$ est continu en probabilité.*

Démonstration: Il suffit de prouver que (b) implique (a). Or si (b) est réalisée, pour tout filtre $\mathcal{F}$ convergeant sur $(T, \mathcal{T})$ vers t_0, le filtre des lois des accroissements $\{\mu_{X(s)-X(t_0)}\}$, $s \in \mathcal{F}$, est relativement compact pour la topologie étroite puisque X est tendu et ce filtre n'a qu'une probabilité adhérente, concentrée à l'origine. Il converge donc vers cette probabilité; ceci montre que $\{D(s, t_0)\}$, $s \in \mathcal{F}$, converge vers zéro, c'est la propriété (a).

Corollaire 1.5.6. *Soit X une f.a. vectorielle sur un espace métrique compact T; pour que X soit continu en probabilité, il faut et il suffit qu'il soit tendu et vérifie la propriété 1.5.5 (b).*

Le théorème 1.5.5 et son corollaire sont pour la continuité en probabilité ce qu'est le théorème 1.4.7 pour la continuité des trajectoires. L'analogie est moins étroite entre le théorème 1.2 et l'énoncé suivant:

Théorème 1.5.7. *Soit X une f.a. vectorielle sur un espace topologique $(T, \mathcal{T})$, alors les deux propriétés suivantes sont équivalentes:*

(a) *$\bar{X}$ est continu en probabilité sur $(T \times E_1'; \mathcal{T}, w)$.*

(b) *L'ensemble $\{\langle X, z\rangle, z \in E_1'\}$ est équicontinu en probabilité (i.e. l'injection $(T, \mathcal{T}) \to (T, \hat{d})$ est continue).*

La démonstration en serait basée sur la w-compacité de E_1'. On remarquera que si X est tendu, les conditions 1.5.5 (b) et 1.5.7 (b) sont équivalentes. Si au contraire X n'est pas tendu, même la condition 1.5.7 (b) n'implique pas nécessairement la continuité en probabilité de X.

1.6 Fonctions aléatoires vectorielles gaussiennes. Nous supposons dans ce paragraphe que la *f.a. vectorielle X est gaussienne et centrée.* Dans ces conditions, l'intégrabilité de X permet de lui associer des métriques plus simples et nous posons pour tous les couples (s, t) d'éléments de T et (y, z) d'éléments de E_1'

(1.6.1)
$$d_y(s, t) = E|\langle X(s) - X(t), y\rangle|, \hat{d}(s, t) = \sup\{d_y(s, t), y \in E_1'\}$$
$$D(s, t) = E\|X(s) - X(t)\|_E,$$
$$\bar{d}(s, y; t, z) = E|\langle X(s), y\rangle - \langle X(t), z\rangle|$$
$$d^t(y, z) = E|\langle X(t), y - z\rangle|.$$

Le théorème 1.5.5 et son corollaire fournissent ici:

Théorème 1.6.2. *L'injection $(T, D) \to (T, \hat{d})$ est uniformément continue; si X est tendu au voisinage de t, alors l'injection inverse est continue en t. Enfin supposons (T, D) (quasi)-compact, alors D et $\hat{d}$ sont équivalentes si et seulement si X est tendu sur T.*

Ce théorème exprime donc que si la métrique D est naturellement associée à la continuité en probabilité de X, on peut souvent lui substituer

pour cet usage la métrique plus petite $\hat{d}$. Pourtant $\hat{d}$ est associée plus naturellement à une autre structure liée à X, son espace autoreproduisant.

La f.a. réelle $\bar{X}$ associée à X (cf. 1.1) est une f.a. gaussienne centrée usuelle sur $T \times E_1'$. Suivant la terminologie générale, son espace
autoreproduisant $\bar{H}$ est l'ensemble $\{\bar{h} \in \mathbf{R}^{T \times E_1'} : \exists k \in L^2(\Omega, P),\ \bar{h} = \int k\bar{X}dP\}$; les propriétés d'intégrabilité de k et de X montrent alors que
pour tout couple (t, z) de $T \times E_1'$, on a:

$$\bar{h}(t, z) = \langle h(t), z \rangle, h(t) = \int kXdP \in E;$$

$h(t)$ est aussi égal à l'intégrale $\int (p_K k)XdP$ où p_K est la projection
orthogonale de $L^2(\Omega, P)$ sur le sous–espace K engendré par $\{X(t, z), t \in T, z \in E_1'\}$. On note alors H l'ensemble de ces intégrales:

$$H = \{h \in E^T : h = \int kXdP, k \in K; \|h\|_H = \|k\|_{L^2(\Omega, P)}\}$$

c'est l'espace autoreproduisant de X et l'inégalité de Cauchy-Schwarz
implique:

$$(1.6.3) \qquad \forall (s, t) \in T \times T, \forall h \in H, \|h(s) - h(t)\|_E \leq \|h\|_H \hat{d}(s, t).$$

Le théorème suivant fournit un critère de séparabilité de l'espace H:

Théorème 1.6.4. *$\{H, \|\cdot\|_H\}$ est un espace de Hilbert; les propriétés
suivantes sont équivalentes:*
*(a) H est séparable, (b) $(T \times E_1')$ est $\bar{d}$-séparable, (c) $\forall z \in E_1', (T, d_z)$
est séparable.*

Corollaire 1.6.5. *Si $(T, \hat{d})$ est séparable, alors H est un espace de
Hilbert séparable.*

Éléments de démonstration: Le seul point non immédiat est l'implication (c) $\Rightarrow$ (b). Supposant seulement que pour tout $z \in Z, (T, d_Z)$

est séparable et notant S_z une suite dense, on constate que $\{(s,z), s \in S_z, z \in Z\}$ est dénombrable et $\bar{d}$-dense dans $T \times E_1'$.

Pour tout élément t de T, on peut aussi construire l'espace autoreproduisant $H(t)$ de $X(t)$; c'est un sous-ensemble de E et on constate que l'application $h \to h(t)$ est une contraction de H sur $H(t)$. Que H soit ou non séparable, soient $(h_i, i \in I)$ une de ses bases orthonormales et $(k_i, i \in I)$ la base associée de K; alors (k_i) est une famille normale indépendante et pour tout $t \in T$, l'espace $H(t)$ étant séparable comme (E_1', w), l'ensemble $\{i \in I: h_i(t) = \int k_i X(t) dP \neq 0\}$ est au plus dénombrable; l'indépendance des termes et les propriétés de $H(t)$ montrent que:

$$(1.6.6) \qquad \forall t \in T, \ P\{X(t) = \sum k_i, h_i(t)\} = 1,$$

de sorte que comme dans le cas réel, l'espace autoreproduisant H est bien adapté à l'étude de la f.a. gaussienne vectorielle X. Le chapitre suivant consacré à l'étude des oscillations des f.a. vectorielles l'illustrera.

2. Oscillations

Les notions d'oscillations que nous allons développer dans ce chapitre prolongent et précisent les notions présentées dans le cas réel par Ito et Nisio et par Jaïn et Kallianpur ([9], [10], [6]). Nous définissons parallèlement des oscillations numériques et des oscillations vectorielles; certaines précautions sont nécessaires pour obtenir des notions maniables même dans le cas des fonctions n'ayant pas de modification séparable. Une partie de cet exposé étant présentée, dans un cadre plus restreint, dans [6], nous n'en répéterons pas certaines preuves.

2.1 Définitions, propriétés générales, exemples. Soient (T, δ) un espace métrique, S une partie dense de T, E un espace de Banach séparable et f une application de S dans E; les oscillations que nous associons à f et que nous allons analyser sont les applications de T dans

$\bar{\mathbf{R}}$ (oscillations numériques) ou dans les parties fermées de E (oscillations vectorielles) définies par:

(2.1.1)
$$W_S(f,t) = \limsup_{\varepsilon \downarrow 0}\{\|f(s) - f(s')\|_E, s \in B(t,\varepsilon) \cap S, s' \in B(t,\varepsilon) \cap S\},$$
$$\bar{W}_S(f,t) = \bigcap_{\varepsilon > 0} \overline{\{f(s) - f(s'), s \in B(t,\varepsilon) \cap S, s' \in B(t,\varepsilon) \cap S\}}.$$

Comme dans les situations réelles, l'étude de W et $\bar{W}$ utilise les applications de $T \times \mathbf{R}^+$ dans $\bar{\mathbf{R}}$ ou dans les parties fermées de E définies par:

(2.1.2)
$$V_S(f,t,u) = \limsup_{\varepsilon \downarrow 0}\{\|f(s) - f(s')\|_E, (s,s') \in (B(t,u) \cap S)^2, \delta(s,s') \leq \varepsilon\},$$
$$\bar{V}_S(f,t,u) = \bigcap_{\varepsilon > 0} \overline{\{f(s) - f(s'), (s,s') \in (B(t,u) \cap S)^2, \delta(s,s') \leq \varepsilon\}}.$$

Enfin si f est aussi définie en $t \in T$, on peut lui associer l'élément de $\bar{\mathbf{R}}$ ou la partie fermée de E définis par:

(2.1.3)
$$U_S(f,t) = \limsup_{\varepsilon \downarrow 0}\{\|f(s) - f(t)\|_E, s \in B(t,\varepsilon) \cap S\},$$
$$\bar{U}_S(f,t) = \bigcap_{\varepsilon > 0} \overline{\{f(s) - f(t), s \in B(t,\varepsilon) \cap S\}}.$$

Exemple 2.1.4: L'exemple suivant illustre les difficultés et les anomalies du maniement de ces notions: on pose $T = \bar{\mathbf{N}}$, $S = \bar{\mathbf{N}}$, $E = \ell^2$; (e_n) est la base canonique de E et (a_n) une suite positive ayant une limite L. On définit l'application f de S dans E en posant pour tout $n \in \mathbf{N}$, $f(n) = a_n e_n$; on constate alors qu'on a au point à l'infini de T:

$$W_S(f,\infty) = 2L, \quad \bar{W}_S(f,\infty) = \{0\},$$
$$\sup_{y \in E'_1} W_S(\langle f, y \rangle, \infty) = +\infty \text{ si } L = +\infty, 0 \text{ si } L \neq +\infty.$$

Effectivement si dans tous les cas, les oscillations numériques ou vectorielles vérifient les inégalités et les inclusions suivantes

(2.1.5)
$$\|\bar{W}_S(f,t)\|_E \leq \sup_{y \in E'_1} W_S(\langle f, y \rangle, t) \leq W_S(f,t),$$
$$\bar{U}_S(f,t) + \bar{U}_S(-f,t) \subset \bar{W}_S(f,t),$$

en général les égalités ne sont pas réalisées. Elles le sont pourtant toutes si f est localement (sur S) à trajectoires relativement compactes (dans E).

Les deux propositions ci-dessous énoncent, comme dans le cas réel, les propriétés fondamentales liant les oscillations de f et sa régularité.

Proposition 2.1.6. *Si f est localement (sur S) une application uniformément continue de (S, δ) dans E alors pour tout couple (t, u) de $T \times \mathbf{R}^+$ et toute application g de S dans E, on a:*

$$V_S(g, t, u) = V_S(f + g, t, u), \quad \bar{V}_S(g, t, u) = \bar{V}_S(f + g, t, u).$$

Proposition 2.1.7. *Pour que f soit la restriction à S d'une fonction g continue sur T, il faut et il suffit que pour tout $t \in T$, $W_S(f, t)$ soit nul. Si de plus f est localement à trajectoires relativement compactes, il faut et il suffit que pour tout $t \in T$, $\bar{W}_S(f, t) = \{0\}$.*

On en déduit un critère de continuité pour les f.a. vectorielles:

Corollaire 2.1.8. *Soit X une f.a. vectorielle sur un espace métrique séparable (T, δ); soit de plus S dénombrable et dense dans T; pour que X ait une modification vectorielle à trajectoires continues, il faut et il suffit que les deux propriétés suivantes soient satisfaites:*

(a) $\forall z \in E_1'$, $\langle X, z \rangle$ *est continu en probabilité,*

(b) *il existe une partie négligeable N de Ω telle que*

$$\forall \omega \notin N, \forall t \in T, W_S(X(\omega), t) = 0.$$

2.2 Oscillations de séries de f.a. vectorielles indépendantes.

L'utilisation, pour la définition des oscillations, des seules valeurs de la fonction sur une partie dénombrable S est particulièrement efficace si la fonction est aléatoire. Si X est une f.a. gaussienne, réelle ou vectorielle, il existe (cf. 1.6.6) une partie négligeable de Ω en dehors de laquelle X est *sur S* la somme de l'un de ses développements de Karhunen-Loeve.

Dans ce paragraphe, nous étudierons les oscillations, aux sens précédents, d'une f.a. vectorielle X définie par un développement en série à termes indépendants: il existe un sous-ensemble dénombrable dense S de T, une suite (x_n) de f.a. vectorielles sur S indépendantes et une partie négligeable N de Ω tels que:

$$(2.2.1) \qquad \forall \omega \notin N, \forall t \in S, \ X(\omega, t) = \sum x_n(\omega, t);$$

pour pouvoir utiliser la proposition 2.1.6, *nous supposerons que pour tout entier n et tout $\omega \notin N$, $x_n(\omega)$ est localement uniformément continue sur (S, δ); nous rappelons que $(\Omega, \mathcal{A}, P)$ est supposé P-complet.* Les oscillations vectorielles de X posséderont alors des propriétés très fortes de mesurabilité.

Lemme 2.2.2. *Pour tout $t \in T$, tout $u > 0$ et tout fermé F de E, l'ensemble $\Omega(X, F) = \{\omega \in \Omega : F \subset \bar{V}_S(X(\omega), t, u)\}$ est mesurable pour la tribu engendrée par X; sa probabilité vaut zéro ou un.*

Lemme 2.2.3. *Pour tout $t \in T$, tut $u > 0$ et tout ouvert U de E, l'ensemble $\Omega'(X, U) = \{w \in \Omega : U \cap \bar{V}_S(X(w), t, u) \neq \emptyset\}$ est mesurable, sa probalité vaut zéro ou un et on a:*

$$P\{\Omega'(X, U)\} > 0 \Rightarrow \exists x \in U : P\{\Omega(X, \{x\})\} = 1.$$

La preuve du lemme 2.2.2 se base sur la définition de $\bar{V}$ et l'uniforme continuité des x_n. Le lemme 2.2.3 est plus difficile; sa preuve suit sans modification ([6]) l'argumentation utilisée dans l'étude de la loi du logarithme itéré par G. Pisier ([11]) dont elle est une extension naturelle. Si en effet (Y_n) est une suite de v.a. vectorielles indépendantes et de même loi, si $(a_n)^{-1}$ est la suite de normalisation usuelle, on peut leur associer la f.a. vectorielle X sur $T = \bar{\mathbf{N}}$ définie par

$$\forall t \in S = \mathbf{N}, \ X(t) = \sum a_t Y_n 1_{\{n \leq t\}},$$
$$X(+\infty) = 0;$$

dans ces conditions, X a la structure 2.2.1; l'hypothèse d'uniforme continuité est vérifiée puisque (a_n) converge vers zéro. L'étude de la propriété du logarithme itéré pour la suite (Y_n) est associée ici à l'étude de l'oscillation $\bar{U}_S(X, +\infty)$.

Les deux lemmes ci-dessus permettent d'utiliser le schéma d'analyse ([9]) des oscillations des f.a. gaussiennes réelles pour établir:

Théorème 2.2.4. *Dans les conditions précédentes, soit X une f.a. vectorielle vérifiant les conditions 2.2.1, alors les oscillations de X sont non aléatoires aux sens suivants: il existe une partie négligeable N_S de Ω et pour tout $t \in T$ une partie négligeable $N_S(t)$ de Ω, il existe aussi des applications non aléatoires w_S et u_S de T dans $\bar{R}$ et $\bar{w}_S$, $\bar{u}_S$ de T dans les parties fermées de E telles que:*

$$\forall \omega \notin N_S, \forall t \in T, W_S(X(\omega), t) = w_S(t), \bar{W}_S(X(\omega), t) = \bar{w}_S(t),$$
$$\forall t \in T, \forall \omega \notin N_S(t), U_S(X(\omega), t) = u_S(t), \bar{U}_S(X(\omega), t) = \bar{u}_S(t).$$

A partir du théorème 2.2.4 et du corollaire 2.1.8, on peut alors démontrer:

Corollaire 2.2.5. *Soit X une f.a. vectorielle vérifiant les conditions 2.2.1, on suppose de plus que X est p.s. continu sur (T, δ), alors X a une modification vectorielle à trajectoires continues.*

2.3 Oscillations des f.a. gaussiennes vectorielles. Les hypothèses du théorème 2.2.4 sont en particulier vérifiées si X est une f.a. gaussienne vectorielle pourvu que les éléments de son espace autoreproduisant possèdent les propriétés suffisantes de continuité. Les propriétés 1.6.3 et 1.6.6 permettent d'énoncer:

Théorème 2.3.1. *Soit X une f.a. gaussienne vectorielle sur l'espace métrique séparable (T, δ); on suppose l'injection $(T, \delta) \rightarrow (T, \hat{d})$ localement uniformément continue. Alors les oscillations de X sont non*

aléatoires aux sens 2.2.4. Si de plus X est continue en probabilité sur
(T, δ), alors les applications $w_S, u_S, \bar{w}_S, \bar{u}_S$ ne dépendent pas de S.

2.3.2. L'exemple de la loi du logarithme itéré pourrait faire croire à des
propriétés générales de convexité pour les oscillations vectorielles; il n'en
est rien et le résultat particulier à la L.L.I. est lié à la forme des séries
de f.a. vectorielles qui la définissent. L'exemple suivant le montrera: on
pose $T = \bar{\mathbf{N}}$, $S = \mathbf{N}$, $E = \mathbf{R}^2$; on note (λ_n) une suite gaussienne normale
et on définit la f.a. gaussienne X sur T à partir de la base canonique
(e_1, e_2) de E en posant:

$$X(\infty) = 0, \ X(n) = (\lambda_n/\sqrt{2\log(n+2)})e_1 \text{ si } n \text{ est impair;}$$
$$X(n) = (\lambda_n/\sqrt{2\log(n+2)})e_2 \text{ si } n \text{ est pair;}$$

on vérifie que X est uniformément continu en probabilité sur $\bar{\mathbf{N}}$ de sorte
que le théorème 2.3.1 s'applique. Un calcul direct fournit à l'infini:

$$\bar{u} = [-1, +1]e_1 \cup [-1, +1]e_2;$$

les trajectoires de X sont p.s. relativement compactes et on construit
facilement $\bar{w}$ à partir de $\bar{u}$ en utilisant 2.1.5; on constate que ni $\bar{u}$, ni $\bar{w}$
ne sont ici convexes. En dimension 1, on n'a pas de telles singularités;
on a en effet:

Théorème 2.3.3. *Soit X une f.a. gaussienne réelle localement unifor-*
mément continue en probabilité sur un espace métrique séparable (T, δ),
alors on a pour tout $t \in T$:

$$\bar{w}(t) = [-w(t), +w(t)] \cap \mathbf{R},$$

de sorte que pour tout $t \in T$, $\bar{w}(t)$ est convexe.

La démonstration s'appuiera sur le lemme:

Lemme 2.3.4. *Soit (x_n) une suite gaussienne réelle convergeant en loi vers zéro; on pose $u = \lim_{n \to \infty} E(\sup_{k \geq n} |x_k|)$. Alors pour tout nombre $x \in]0, u[$, on peut extraire une suite partielle (x_{j_k}) telle que $x = \lim_{n \to \infty} E(\sup_{k \geq n} |x_{j_k}|)$.*

Démonstration du lemme: Pour tout $n \in \mathbf{N}$, on note K_n l'ensemble $\{k : E|x_k|^2 \in [4^{-(n+1)}, 4^{-n}[\}$; on sait alors ([7]) qu'on a pour la suite globale et pour toute suite extraite:

$$u = \lim_{n \to \infty} E(\sup_{k \in K_n} |x_k|), \quad \lim_{n \to \infty} E(\sup_{k \geq n} |x_{j_k}|) = \lim_{n \to \infty} E(\sup_{j_k \in K_n} |x_{j_k}|).$$

Puisque $x \in]0, u[$, à partir d'un certain rang $E(\sup_{k \in K_n} |x_k|)$ est supérieur à x; comme K_n est un ensemble fini, il est facile d'en extraire alors un ensemble K_n' tel que $E(\sup_{k \in K_n'} |x_k|)$ appartienne à $[x, x + 2^{-n}]$; la réunion des K_n' forme alors l'extraction cherchée.

Démonstration du théorème 2.3.3: Les propriétés d'intégrabilité de X montrent qu'il existe une suite (s_n) tendant vers t dans (T, δ), telle que $w(t) = 2 \lim_{n \to \infty} E(\lim_{k \geq n} |X_{s_k} - X(t)|)$ et qu'il suffit pour établir le théorème d'en extraire pour tout $x \in]0, w(t)[$ une suite partielle (s_{j_k}) telle que $x = 2 \lim_{n \to \infty} E(\sup_{k \geq n} |X(s_{j_k}) - X(t)|)$; on utilise pour cela le lemme 2.3.4. L'énoncé précédent s'étend partiellement au cas général:

Théorème 2.3.5. *Soit X une f.a. gaussienne vectorielle localement uniformément continue en probabilité sur un espace métrique séparable (T, δ), alors son oscillation vectorielle est équilibrée, c'est à dire que:*

$$\forall t \in T, \forall \lambda \in [-1, +1], \lambda \bar{w}(t) \subset \bar{w}(t).$$

La démonstration est basée sur le lemme suivant:

Lemme 2.3.6. *Soient X et Y deux f.a. gaussiennes vectorielles sur un espace métrique séparable, localement uniformément continues en probabilité et indépendantes; pour tout $t \in T$, on a alors:*

$$\bar{w}(X, t) \subset \bar{w}(X + Y, t).$$

Démonstration: Les ensembles ci-dessus sont déterminés par les lois de X et de Y; on peut donc supposer que ces f.a. sont réalisées sur deux facteurs indépendants Ω_1, Ω_2 de l'espace produit Ω. Soient S une partie dénombrable dense dans T, $t \in T$ et $x \in \bar{w}(X, t)$; soit de plus N la partie négligeable de Ω_1 associée à X par le théorème 2.2.4. Pour tout $\omega_1 \notin N$, il existe une suite double $(s_n(\omega_1), s'_n(\omega_1))$ extraite de $S \times S$ convergeant vers (t, t) suivant laquelle les accroissements de $X(\omega_1)$ convergent vers x; on peut en extraire une suite partielle $(s_{n_k}(\omega_1), s'_{n_k}(\omega_1))$ telle que la série $\sum \{D_Y(s_{n_k}(\omega_1), t) + D_Y(s'_{n_k}(\omega_1), t)\}$ soit convergente; sur cette suite partielle, l'accroissement de Y converge p.s. (sur Ω_2) vers zéro. Le théorème de Fubini met donc en évidence un ensemble négligeable N' dans l'espace produit Ω tel que pour tout $\omega \notin N'$, $\bar{W}(X(\omega_1) + Y(\omega_2); t)$ contienne x; c'est le résultat du lemme.

Démonstration du théorème 2.3.5: Soient $t \in T$, $\lambda \in [-1, +1]$ et $x \in \bar{w}(X, t)$; soit de plus X_1 et X_2 deux copies de X indépendantes; le lemme ci-dessus montre que λx qui appartient à $\bar{w}(\lambda X_1, t)$ appartient aussi à $\bar{w}(\lambda X_1 + \sqrt{1 - \lambda^2} X_2, t) = \bar{w}(X, t)$; c'est la conclusion du théorème. Le théorème 2.3.5 permet de préciser l'un des termes de la célèbre alternative de Belayev ([1]).

Corollaire 2.3.7.   *Soit X une f.a. gaussienne réelle séparable sur $\mathbf{R}$ et stationnaire; on suppose que X est continue en probabilité et n'a pas p.s. ses trajectoires continues; on a alors p.s.: Pour tout ouvert U de $\mathbf{R}$, l'image $\{X(s), s \in U\}$ est partout dense dans $\mathbf{R}$.*

3. Fonctions aléatoires gaussiennes à valeurs vectorielles et à accroissements stationnaires

Dans ce chapitre, nous étudions la régularité des f.a. gaussiennes X sur $\mathbf{R}$ à valeurs vectorielles et à accroissements stationnaires; nous supposons donc que pour tout couple (a, b) d'éléments de $\mathbf{R}$ la f.a.: $t \to X(a + t) - X(b + t)$ est stationnaire. Dans un article précédent [5], j'ai énoncé un critère simple de régularité (théorème 3.2) si X est en fait stationnaire; j'y avais indiqué explicitement que les majorations employées étaient inefficaces pour l'étude des f.a. à seuls accroissements stationnaires. On se propose pourtant d'étendre le même critère à ce cas; nous utiliserons pour cela la représentation spectrale des f.a.g. à valeurs réelles et à accroissements stationnaires qui nous permettra d'obtenir une bonne évaluation de la continuité en probabilité dans le cas vectoriel.

3.2 Lois de f.a.g. à valeurs vectorielles et à accroissements stationnaires; continuité en probabilité. Dans ce paragraphe, nous utiliserons le résultat classique ([4]):

Théorème 3.2.0. *Soit X une f.a.g. à valeurs réelles et à accroissements stationnaires sur $\mathbf{R}$, on suppose X continue en probabilité. Il existe alors deux mesures aléatoires gaussiennes m et $\bar{m}$ à valeurs réelles indépendantes sur les parties disjointes de $\mathbf{R}^+$, de même lois et mutuellement indépendantes telles que pour tout $t \in \mathbf{R}$:*

$$(3.2.0) \quad X(t) - X(0) = \int_{\lambda \neq 0} \{ \frac{\cos \lambda t - 1}{\lambda} \sqrt{1 + \lambda^2} dm(\lambda) +$$
$$\frac{\sin \lambda t}{\lambda} \sqrt{1 + \lambda^2} d\bar{m}(\lambda) \} + t\bar{m}(0) \ p.s.$$

Le théorème 3.2.0 ne s'étend pas directement à la situation générale. Il permet pourtant d'énoncer:

Théorème 3.2.1. *Soit X une f.a.g. à valeurs vectorielles et à accroissements stationnaires sur $\mathbf{R}$, on suppose que pour tout $y \in E'$, la f.a.r. $\langle X, y \rangle$ est continue en probabilité. Dans ces conditions, les propriétés suivantes sont équivalentes:*

(a) *L'application $t \rightarrow E\|X(t)\|_E$ est localement intégrable pour la mesure de Lebesgue sur $\mathbf{R}$.*

(b) *Il existe un vecteur aléatoire gaussien x à valeurs dans E tel que:*

$$\forall t \in [0,1], \forall y \in E', \ E|\langle X(t), y \rangle^2| \leq E|\langle x, y \rangle^2|.$$

(c) *X est localement tendu.*

(d) *X est continu en probabilité.*

Démonstration: L'implication (b) $\Rightarrow$ (c) résulte des inégalités de Slépian; (c) $\Rightarrow$ (d) se déduit du théorème 1.5.5. Sous l'hypothèse (d), l'application $t \rightarrow E\|X(t)\|_E$ est mesurable (th. 1.3) et localement bornée, donc localement intégrable de sorte que (d) $\Rightarrow$ (a). Il reste à établir l'implication (a) $\Rightarrow$ (b); nous supposerons que $X(0) = 0$ et nous utiliserons les mesures aléatoires m_y et $\bar{m}_y$ asociées à $\langle X, y \rangle$, $y \in E'$ par le théorème 3.2.0; leur moment du second ordre est une mesure positive bornée μ_y sur $\mathbf{R}^+$ et la relation 3.2.0 fournit:

$\forall t \in \mathbf{R}, \forall y \in E'$,
$$E|\langle X(t), y \rangle|^2 = t^2 \mu_y(0) + \int_{\lambda \neq 0} \frac{2 - 2 \cos \lambda t}{\lambda^2}(1 + \lambda^2) d\mu_y(\lambda),$$

et donc en particulier:

$$\forall y \in E', E|\langle X(1), y \rangle|^2 \geq \mu_y(0) + 2(1 - \cos 1)\mu_y(]0, 1]);$$

par ailleurs, x possède (théorème 1.3) une modification mesurable Y et sous l'hypothèse (a), Y est intégrable sur $[0, 1]$ pour la mesure de Lebesgue p.s.; il existe donc un vecteur aléatoire u gaussien dans E tel que:
$$\forall y \in E', E|\langle u, y \rangle|^2 = E|\int_0^1 \langle X(t), y \rangle dt|^2,$$

la relation 3.2.0 fournit encore:

$$E|\int_0^1 \langle X(t), y \rangle dt|^2 \geq \int_{\lambda \neq 0} (\frac{\sin \lambda}{\lambda} - 1)^2 (\frac{1 + \lambda^2}{\lambda^2}) d\mu_y(\lambda)$$

et donc:
$$\forall y \in E', E|\langle u, y \rangle|^2 \geq (1 - \sin 1)^2 \mu_y(]1, +\infty).$$

En regroupant, on obtient pour tout $t \in [0, 1]$ et tout $y \in E'$:

$$E|\langle X(t), y \rangle|^2 \leq 8\mu_y(\mathbf{R}) \leq 9E|\langle X(1), y \rangle|^2 + 320E|\langle u, y \rangle|^2,$$

c'est la propriété (b); le théorème est donc démontré.

3.3 Régularité des trajectoires des f.a.g. à accroissements stationnaires. Dans ce paragraphe, X est une f.a.g. à valeurs vectorielles et à accroissements stationnaires sur $\mathbf{R}$; pour tout $y \in E_1'$ et tout $u > 0$, d_y est la distance définie en 1.6.1, $B_y(0, u)$ est l'ensemble $\{t \in \mathbf{R}: d_y(0, t) \leq u\}$, $A(u)$ désigne $[-u, +u]$, $d_y(A(u))$ est son d_y-diamètre; λ est la mesure de Lebesgue normalisée sur $\mathbf{R}$. Toute régularité des trajectoires de X nécessite que l'ensemble $\{\langle X, z \rangle, z \in E_1'\}$ possède la même régularité uniformément. Les propriétés des f.a.g. réelles impliquent donc:

Proposition 3.3.1. *Pour que X ait une modification à trajectoires localement bornées, il faut que les deux conditions équivalentes suivantes soient réalisées:*

(a) *Il existe une modification Y de X et un nombre M tels que*

$$\forall z \in E_1', \ E \sup\{\langle Y(t), z \rangle, t \in A(1)\} \leq M.$$

(b) *Il existe un nombre M tel que pour tout $z \in E_1'$:*

$$d_z(A(u)) + \int \sqrt{\log \frac{2}{\lambda\{A(1) \cap B_z(0, u)\}}} \, du \leq M.$$

Proposition 3.3.2. *Pour que X ait une modification à trajectoires continues, il faut que les deux conditions équivalentes suivantes soient satisfaites:*

(a) *Il existe une modification Y de X telle que:*

$$\lim_{\varepsilon \downarrow 0} \sup_{z \in E_1'} E \sup\{\langle Y(t), z \rangle, t \in A(\varepsilon)\} = 0.$$

(b)

$$\lim_{\varepsilon \downarrow 0} \sup_{z \in E_1'} \left\{ d_z(A(\varepsilon)) + \int \sqrt{\log \frac{2}{\lambda\{A(\varepsilon) \cap B_z(0, u)\}}} \, du \right\} = 0.$$

Comme dans le cas stationnaire, il est remarquable que les conditions
nécessaires de régularité énoncées aux propositions 3.3.1 et 3.3.2 soient
aussi suffisantes. Cette propriété est un corollaire simple du théorème
de M. Talagrand ([12]) caractérisant la régularité des f.a.g. à partir des
mesures majorantes, joint au théorème 3.2.1 ci-dessus:

Théorème 3.3.3.   *Soit X une f.a.g. à valeurs vectorielles et à accroissements stationnaires sur* **R***; dans ces conditions:*

(a) *Pour que X ait une modification à trajectoires localement bornées,
il faut et il suffit que les conditions 3.3.1 (a) ou (b) soient vérifiées.*

(b) *Pour que X ait une modification à trajectoires continues, il faut
et il suffit que les conditions 3.3.2 (a) ou (b) soient vérifiées.*

Nous ne donnons pas la démonstration du théorème qui repète, mu-
tatis mutandis, celle du théorème 3.2 de [5]. Indiquons simplement que,
dans les situations indiquées, pour tout $z \in E_1'$, $\langle X, z \rangle$ est continu en
probabilité de sorte que si $E(\|X(t)\|_E)$ est localement borné, X est con-
tinu en probabilité et vérifie la condition (b) du théorème 3.2.1; soit x le
vecteur gaussien à valeurs dans E défini par cette condition, le théorème
de Talagrand lui associe une mesure majorante μ sur E_1'; on montre alors
que la probabilité $\pi = \frac{1}{2}\lambda \otimes \mu$ sur $A(1) \times E_1'$ est une mesure majorante
pour $\widetilde{X}$. Au cours de cette preuve, on montre qu'il existe une constante
absolue C et une modification X' de X telles que:

$$E\{ \sup_{t \in A(1)} \|X'(t)\|_E \} \le C[E\|X'(0)\|_E + E\|X'(1)\|_E$$

(3.3.4)
$$+ E\|\int_0^1 X'(t)dt\|_E$$
$$+ \sup_{y \in E_1'} E \sup_{t \in A(1)} |\langle X'(t), y \rangle|].$$

**3.4 Représentation spectrale des f.a.g. à valeurs vectorielles
et à accroissements stationnaires sur R.** La démonstration du théorème 3.2.1 et donc celle du théorème 3.3.3 sont liées aux représenta-
tions spectrales des $\langle X, y \rangle$, $y \in E'$; les évaluations effectuées permettent

en fait d'obtenir une représentation spectrale vectorielle pour la f.a.g. à valeurs vectorielles X elle-même; nous la présentons dans ce paragraphe; nous y utilisons un lemme fondamental pour la construction de vecteurs gaussiens:

Lemme 3.4.0. *Soit m une f.a.g. sur le dual E' d'un espace de Fréchet séparable E, à valeurs réelles; on suppose que pour tout $(y, z; \lambda, \mu) \in E' \times E' \times \mathbf{R} \times \mathbf{R}$, on a:*

$$m(\lambda y + \mu z) = \lambda m(y) + \mu m(z) \ p.s.$$

on suppose aussi qu'il existe un vecteur gaussien v à valeurs dans E tel que:

$$(3.4.0.2) \qquad \forall y \in E', \ E|m(y)|^2 \leq E|\langle v, y \rangle^2|;$$

dans ces conditions, il existe un vecteur gaussien M à valeurs dans E tel que:

$$\forall y \in E', \ P\{m(y) = \langle M, y \rangle\} = 1.$$

Éléments de preuve: on utilise l'espace autoreproduisant H de la f.a.g. réelle m sur E'; on montre qu'il est contenu dans E et on construit M à partir du développement en série de m associé à une base orthonormale de H qui converge p.s. dans E.

3.4.1. Nous utiliserons aussi la notion de *mesure aléatoire gaussienne à valeurs vectorielles dans E indépendantes sur les parties disjointes d'un espace mesurable $(\Lambda, T(\Lambda))$.* Suivant la terminologie usuelle, ceci désigne une fonction aléatoire gaussienne centrée M sur l'ensemble $T = T(\Lambda)$ des parties mesurables de Λ, à valeurs dans l'espace de Banach séparable E et possédant la propriété suivante:

3.4.4.1 *Pour toute suite (t_n) d'éléments de T disjoints, la suite $(M(t_n))$ a ses termes mutuellement indépendants et vérifie:*

$$M(\bigcup_{n \in \mathbf{N}} t_n) = \sum_{n \in \mathbf{N}} M(t_n) \ p.s..$$

Dans tout ce paragraphe, $X = \{X(u), u \in \mathbf{R}\}$ est une f.a.g. à valeurs vectorielles et à accroissements stationnaires sur $\mathbf{R}$; on suppose que X est continue en probabilité. On note $\Lambda = \{\lambda > 0\}$ l'ensemble des nombres réels strictement positifs; T est l'ensemble des parties mesurables de Λ. Pour tout $z \in E'$, le théorème 3.2.0 associe à $\langle X, z \rangle$ une représentation spectrale:

3.4.1.2.

$$\langle X(u) - X(0), z \rangle = \int_{\lambda \in \Lambda} \{ \frac{\cos u\lambda - 1}{\lambda} \sqrt{1 + \lambda^2} dm_z(\lambda)$$
$$+ \frac{\sin u\lambda}{\lambda} \sqrt{1 + \lambda^2} d\bar{m}_z(\lambda) \} + ua_z \text{ p.s.}$$

à partir de deux mesures aléatoires m_z et $\bar{m}_z$ et d'une v.a. gaussienne a_z.

Proposition 3.4.2.   *On peut choisir* $(m_z, \bar{m}_z, a_z)$, $z \in E'$, *de telle façon que pour tout* $t \in T$, *les f.a.g.   réelles sur* E' *définies par* $(m_z(t), \bar{m}_z(t), a_z)$, $z \in E'$ *vérifient la propriété 3.4.0.1.*

Démonstration: Puisque X est continue en probabilité, le théorème 1.3 lui associe une version mesurable Y; pour tout $z \in E'$, $\langle Y, z \rangle$ est alors aussi mesurable et définit une distribution aléatoire tempérée dont la dérivée p_z est stationnaire; on peut alors choisir de déterminer $m_z, \bar{m}_z$ et a_z à partir de la transformée de Fourier $q_z = \hat{p}_z$; pour tout $w \in \Omega$, les applications $z \rightarrow m_z(w)$, $\bar{m}_z(w)$, $a_z(w)$ sont alors des applications linéaires de E' dans l'espace S' des distributions tempérées ou dans $\mathbf{R}$; pour tout $t \in T$, $m_z(t)$ et $\bar{m}_z(t)$ sont alors obtenues par des passages à la limite en probabilité qui fournissent le résultat énoncé.

Proposition 3.4.3.   *Pour tout* $t \in T$, *les f.a.g. réelles définies sur* E' *par* $\{m_z(t), \bar{m}_z(t), a_z\}$, $z \in E'$, *vérifient la propriété 3.4.0.2.*

Démonstration: Elle résulte des évaluations du théorème 3.2.1; elles montrent en effet que pour tout $t \in T$ et tout $z \in E'$, on a:

$$E|a_z|^2 + E|m_z(t)|^2 + E|\bar{m}_z(t)|^2 \leq 2E|m_z(\Lambda)|^2 \leq 2\mu_z(\mathbf{R}^+),$$

$$\mu_z(\mathbf{R}^+) \leq 2E|\langle X(1) - X(0), z\rangle|^2 + 40E|\langle \int_0^1 (X(u) - X(0))du, z\rangle|^2.$$

Les propositions 3.4.2 et 3.4.3 permettent d'utiliser le lemme 3.4.0. On peut donc énoncer:

Proposition 3.4.4. *Il existe une variable aléatoire gaussienne A et deux f.a. gaussiennes M et $\bar{M}$ sur T, toutes à valeurs dans E, telles que:*

$$\forall(t,z) \in T \times E',$$
$$P\{a_z = \langle A, z\rangle, m_z(t) = \langle M(t), z\rangle, \bar{m}_z(t) = \langle \bar{M}(t), z\rangle\} = 1.$$

Proposition 3.4.5. *Pour toute suite (t_n) d'éléments de T disjoints, on a:*

$$M(\bigcup_{n \in \mathbf{N}} t_n) = \sum_{n \in \mathbf{N}} M(t_n) \text{ p.s.},$$
$$\bar{M}(\bigcup_{n \in \mathbf{N}} t_n) = \sum_{n \in \mathbf{N}} \bar{M}(t_n) \text{ p.s..}$$

Démonstration: Nous démontrons la première égalité; nous définissons pour cela une fonction aléatoire S sur $\bar{\mathbf{N}}$ en posant $S_n = \sum_{k=1}^n M(t_k)$, $n \in \mathbf{N}$ et $S_\infty = M(\cup_{n \in \mathbf{N}} t_n)$. Le fait que, pour tout $z \in E'$, $\langle M, z\rangle$ est une mesure aléatoire à valeurs indépendantes suffit à montrer que S est tendue sur $\bar{\mathbf{N}}$ et vérifie la propriété 1.5.5 (b); le théorème 1.5.5 montre donc que S est continue en probabilité: ceci signifie que la série du second membre converge en probabilité vers une somme presque sûrement égale au premier membre, c'est le résultat.

3.4.6. Dans la première partie du paragraphe, nous avons utilisé le fait que pour tout $z \in E'$, $\langle X, z\rangle$ est à accroissements stationnaires. En fait la même propriété est vérifiée pour chaque couple $\{\langle X, y\rangle, \langle X, z\rangle\}$, $(y, z) \in E' \times E'$; pour un tel couple, utilisant les notations de la preuve de la proposition 3.4.2, le produit tensoriel $q_y \otimes \bar{q}_z$ des transformées de Fourier des dérivées des distributions aléatoires $\langle Y, y\rangle$ et $\overline{\langle Y, z\rangle}$ a un moment $E\{q_y \otimes \overline{q_z}\}$ dont le support est contenu dans la diagonale de $\mathbf{R} \times \mathbf{R}$; l'inégalité de Cauchy-Schwartz montre que cette distribution complexe est majorée en module par la mesure positive $E\{q_y \otimes \bar{q}_y\} + E\{q_z \otimes \bar{q}_z\}$; c'est donc une mesure complexe portée par la diagonale de $\mathbf{R} \times \mathbf{R}$; les expressions de A, M, $\bar{M}$ en fonction des q_y, $y \in E'$, impliquent alors:

Proposition 3.4.6. *M et $\bar{M}$ ont mêmes lois; A est indépendant de M et $\bar{M}$. Si s et t sont deux parties disjointes de Λ, alors $M(s)$ est indépendant de $M(t)$ et $\bar{M}(t)$. Les f.a. M et $\bar{M}$ sont l'une et l'autre des mesures aléatoires gaussiennes à valeurs vectorielles dans E indépendantes sur les parties disjointes de Λ.*

Remarque: On montrera ultérieurement que M et $\bar{M}$ ne sont pas nécessairement indépendantes (exemple et théorème 3.4.9).

3.4.7. Nous étudions maintenant la possibilité de donner une signification vectorielle globale aux différentes intégrations stochastiques réelles associées aux représentations spectrales des $\langle X, z\rangle$, $z \in E'$.

Soit $u \in \mathbf{R}$ nous lui associons les applications mesurables F et G de Λ dans $\mathbf{R}$ définies par:

$$F(\lambda) = \frac{\cos u\lambda - 1}{\lambda}\sqrt{1 + \lambda^2}, \quad G(\lambda) = \frac{\sin u\lambda}{\lambda}\sqrt{1 + \lambda^2};$$

et notons S la variable aléatoire vectorielle $X(u) - X(0) - u(A)$; le théorème 3.2.0 indique donc que pour tout $z \in E'$, on a:

$$\langle S, z\rangle = \int F d(\langle M, z\rangle) + \int G d(\langle \bar{M}, z\rangle).$$

Par ailleurs pour tout $\lambda \in \Lambda$, on a:

$$|F(\lambda)|^2 + |G(\lambda)|^2 \leq u^2 + 4,$$

il existe donc des suites *bornées* (F_n), (G_n) de fonctions étagées mesurables sur Λ aẁaleurs réelles convergeant simplement sur Λ vers F et G.

Proposition 3.4.7. *Soit* (F_n), (G_n) *un couple de suites bornées de fonctions étagées mesurables sur* Λ *à valeurs réelles convergeant simplement sur* Λ *vers* F *et* G; *alors la suite* $(\int F_n dM + \int G_n d\bar{M})$ *est une suite de vecteurs aléatoires convergeant en probabilité dans* E; *sa limite est p.s. égale à* S.

Démonstration: Nous définissons une f.a. R sur $\bar{\mathbf{N}}$ à valeurs dans E en posant $R_n = \int F_n dM + \int G_n d\bar{M}$, $n \in \mathbf{N}$ et $R_\infty = S$; comme dans la démonstration de la proposition 3.4.5, on constate que R est continu en probabilité sur $\bar{\mathbf{N}}$ et c'est le résultat.

L'ensemble des propositions ci-dessus démontre le théorème:

Théorème 3.4.8. *Soit* X *une f.a.g. à valeurs vectorielles et à accroissements stationnaires sur* $\mathbf{R}$, *continue en probabilité; il existe alors deux mesures aléatoires gaussiennes* M *et* $\bar{M}$ *à valeurs vectorielles dans* E *indépendantes sur les parties disjointes de* Λ, *de mêmes lois, ainsi qu'un vecteur gaussien* A *à valeurs dans* E *indépendant de* M *et* $\bar{M}$ *tels que pour tout* $u \in \mathbf{R}$, *on ait:*

$$X(u) - X(0) =$$

$$\int_{\lambda \in \Lambda} \left\{ \frac{\cos u\lambda - 1}{\lambda} \sqrt{1+\lambda^2} dM(\lambda) + \frac{\sin u\lambda}{\lambda} \sqrt{1+\lambda^2} d\bar{M}(\lambda) \right\} + uA \quad p.s.$$

l'intégration stochastique s'entendant au sens fort de l'intégration stochastique vectorielle des fonctions mesurables et bornées.

3.4.9. *Sur les liaisons entre* M *et* $\bar{M}$. Si l'espace E a la dimension 1, le théorème 3.2.0 indique que M et $\bar{M}$ sont indépendantes. Ce n'est plus nécessairement le cas si $E = \mathbf{R}^2$ comme le montre l'exemple suivant:

Exemple: Soient g_1 et g_2 deux v.a. gaussiennes centrées réduites et indépendantes; on définit une f.a.g. X sur $\mathbf{R}$ à valeurs dans $\mathbf{R}^2$ en posant:

$$\forall u \in \mathbf{R}, \quad X(u) = \sqrt{2} \begin{pmatrix} g_1 \cos u & +g_2 \sin u \\ g_2 \cos u & -g_1 \sin u \end{pmatrix};$$

on constate que X est à accroissements stationnaires. Sa représentation spectrale est définie par

$$A = \begin{pmatrix} 0 \\ 0 \end{pmatrix}, \quad M = \begin{pmatrix} g_1 \\ g_2 \end{pmatrix} \delta_1, \quad \bar{M} = \begin{pmatrix} g_2 \\ -g_1 \end{pmatrix} \delta_1$$

et les mesures aléatoires M et $\bar{M}$ ne sont pas indépendantes. Plus généralement, on peut énoncer:

Théorème 3.4.9. *Pour que M et $\bar{M}$ soient indépendantes, il faut et il suffit que pour tout $(s, t; s', t') \in \mathbf{R}^2 \times \mathbf{R}^2$ et tout $(y, z) \in E' \times E'$, on ait:*

$$E\{\langle X(s) - X(t), y \rangle \langle X(s') - X(t'), z \rangle\} =$$
$$E\{\langle X(s) - X(t), z \rangle \langle X(s') - X(t'), y \rangle\}$$

on notera que c'est par exemple le cas si X est un processus de Wiener.

Références

[1] Yu. K. Belayev, *Continuity and Hölder's conditions for sample functions of stationary Gaussian processes*, Proc. 4th Berkeley Sympos. Math. Statist. and Prob. 2, 23–33. Univ. Calif. Press, 1961.

[2] R. Carmona, *Tensor product of Gaussian measures*, Springer Lecture Notes in Math. **644** (1978), 96–124.

[3] S. Chevet, *Un résultat sur les mesures gaussiennes*, C.R. Acad. Sci. Paris, A **284** (1977), 441–444.

[4] J.L. Doob, *Stochastic processes*, Wiley. New York, 1953.

[5] X. Fernique, *Fonctions aléatoires gaussiennes à valeurs vectorielles.*

[6] —————, *Oscillations de fonctions aléatoires gaussiennes à valeurs vectorielles*, Math. Scand. **60** (1987), 96–108.

[7] —————, *Une majoration des fonctions aléatoires gaussiennes à valeurs vectorielles*, C.R. Acad. Sci. Paris **300**, Série I, 10 (1985), 315–318.

[8] J. Hoffmann-Jørgensen, *Stochastic processes in Polish spaces.* à paraître.

[9] K. Ito et M. Nisio, *On the oscillation functions of Gaussian processes*, Math. Scand. **22** (1968), 209–223.

[10] N.C. Jain et G. Kallianpur, *Oscillation function of a multiparameter Gaussian process*, Nagoya Math. J. **47** (1972), 15–28.

[11] G. Pisier, *Le théorème central-limite et la loi du logarithme itéré dans les espaces de Banach.* Séminaire Maurey-Schwartz, 1975–76, exposés III et IV.

[12] M. Talagrand, *Régularité des processus Gaussiens*, C.R. Acad. Sci. Paris **301**, Série I (1985), 751–753.

Institut de Recherche Mathématique Avancée, Unité Associée no. 01,
Université Louis Pasteur, 7, rue René Descartes, 67084 Strasbourg Cedex (France)

On Functional Limit Theorems
for a Class of Stochastic Processes
Indexed by Pseudo-Metric Parameter Spaces
(with applications to empirical processes)

PETER GAENSSLER AND WILHELM SCHNEEMEIER

1. Introduction and Main Results

Let $T = (T, d)$ be a pseudo-metric space assumed to be totally bounded for the pseudo-metric d. Let $\ell^\infty(T)$ be the space of all bounded real valued functions on T equipped with the supremum norm $\|\cdot\|_T$ (defined by $\|x\|_T := \sup\{|x(t)| : t \in T\}$, $x \in \ell^\infty(T)$) and let $S_0 := U^b(T, d)$ be the subspace of $\ell^\infty(T)$ consisting of all uniformly d-continuous functions on T; note that S_0 is separable and closed in $(\ell^\infty(T), \|\cdot\|_T)$; cf. Corollary 2 in Section 2 below.

Now, given any S such that $S_0 \subset S \subset \ell^\infty(T)$, let $\mathcal{B}(S)$ be the σ-algebra of all Borel sets in $(S, \|\cdot\|_T)$ and let $\mathcal{B}_b(S)$ be the sub-σ-algebra of $\mathcal{B}(S)$ generated by the open $\|\cdot\|_T$-balls in S, assuming here (for simplicity) that

(1.1) $\mathcal{B}_b(S) \subset \mathcal{B} := \sigma\{\pi_t : t \in T\} \subset \mathcal{B}(S)$ where $\sigma(\{\pi_t : t \in T\})$ denotes the σ-algebra in S generated by the coordinate projections $\pi_t : S \to \mathbf{R}$, defined by $\pi_t(x) := x(t)$ for $x \in S$.

Note that the second inclusion in (1.1) is automatically fulfilled due to the $\|\cdot\|_T$-continuity of each π_t; also $\mathcal{B}_b(S)$ is, in general (due to non-separability of $(S, \|\cdot\|_T)$), strictly smaller than $\mathcal{B}(S)$.

Then, given any sequence $(\beta_n)_{n\in\mathbf{N}}$ of random elements β_n in $(S, \mathcal{B})$ (i.e. $\mathcal{A}, \mathcal{B}$-measurable random functions $\beta_n : \Omega \to S$, defined on some basic p-space $(\Omega, \mathcal{A}, \mathbf{P})$), and a random element β_0 in $(S, \mathcal{B}_b(S))$, defined on the same p-space $(\Omega, \mathcal{A}, \mathbf{P})$, $(\beta_n)_{n\in\mathbf{N}}$ is said to converge in law to β_0 (denoted by $\beta_n \xrightarrow{\mathcal{L}_b} \beta_0$) iff (cf. [8], (34) p. 65 and (28) p. 47/48).

(1.2) (i) $\mathbf{P}(\beta_0 \in S_0) = 1$ and (ii) $\lim_{n\to\infty} \int f \circ \beta_n \, d\mathbf{P} = \int f \circ \beta_0 \, d\mathbf{P}$ for all $f \in U_b^b(S) := \{g\colon S \to \mathbf{R}\colon g$ bounded, uniformly $\|\cdot\|_T$-cts. and $\mathcal{B}_b(S)$-mb.$\}$.

Furthermore, $(\beta_n)_{n\in\mathbf{N}}$ is said to be relatively $\mathcal{L}_b$-sequentially compact iff for any subsequence $(\beta_{n'})$ of (β_n) there exists a further subsequence $(\beta_{n''})$ of $(\beta_{n'})$ and a limiting random element β_0 in $(S, \mathcal{B}_b(S))$ such that $\beta_{n''} \xrightarrow{\mathcal{L}_b} \beta_0$ as $n'' \to \infty$.

Denoting with $w_x(\cdot)$ the oscillation modulus defined for any $x \in \ell^\infty(T)$ and $\delta > 0$ by

$$w_x(\delta) := \sup\{|x(t_1) - x(t_2)|\colon t_1, t_2 \in T \quad \text{s.t.} \quad d(t_1, t_2) \leq \delta\}$$

the following criterion generalizes Theorem 1.17 in [9] (being concerned there with the classical situation $T = [0.1]$, $d(t_1, t_2) = |t_1 - t_2|$, $S_0 = C[0,1] \subset S = D[0,1] \subset \ell^\infty([0,1])$).

Theorem 1. *Given a sequence $(\beta_n)_{n\in\mathbf{N}}$ of random elements in $(S, \mathcal{B})$, the following four statements are equivalent:*

(i) *$(\beta_n)_{n\in\mathbf{N}}$ fulfills the following two conditions (a) and (b):*

   (a) *For any sequence $(f_m)_{m\in\mathbf{N}}$ in $U_b^b(S)$ with $f_m \downarrow 0$ as $m \to \infty$ one has*

$$\limsup_{n\to\infty} \int f_m \circ \beta_n \, d\mathbf{P} \to 0 \quad \text{as } m \to \infty;$$

   (b) $\liminf_{n\to\infty} \int f \circ \beta_n \, d\mathbf{P} \geq 1$ *for all $f \in U_b^b(S)$ with $f \geq 1_{S_0}$.*

(ii) *$(\beta_n)_{n\in\mathbf{N}}$ is relatively $\mathcal{L}_b$-sequentially compact.*

(iii) *$(\beta_n)_{n\in\mathbf{N}}$ fulfills*

   (A) $\lim_{\delta\downarrow 0} \limsup_{n\to\infty} \mathbf{P}^*(w_{\beta_n}(\delta) \geq \varepsilon) = 0$ *for each $\varepsilon > 0$ (where $\mathbf{P}^*$ denotes the outer measure pertaining to $\mathbf{P}$), and*

   (B) $\lim_{K\to\infty} \limsup_{n\to\infty} \mathbf{P}(|\beta_n(t)| \geq K) = 0$ *for all $t \in T$.*

(iv) *$(\beta_n)_{n\in\mathbf{N}}$ fulfills*

   (A) *as in (iii), and*

   (B') $\lim_{K\to\infty} \limsup_{n\to\infty} \mathbf{P}(\|\beta_n\|_T \geq K) = 0$.

This theorem implies especially theorem (1.2) in [4] and Theorem B in [8], respectively; cf. [8], p. 118, according to which it follows that under (A) the other condition (B') is automatically fulfilled in case of empirical C-processes $\beta_n = (\beta_n(C))_{C \in C}$ indexed by classes C of sets being totally bounded for $d = d_\mu$ defined by $d_\mu(C_1, C_2) := \mu(C_1 \triangle C_2)$, μ being the law according to which the underlying observations constituting the empirical C-processes are distributed.

The crucial step in the proof of Theorem 1 is the part verifying that (iv) implies (i): As we will see in Section 3 below this can be done (avoiding the concept of δ-tightness still used in [8], Prop. B_2, p. 117) along the same lines as in [9] based on straightforward generalizations of Lemma 1.15 and Lemma 1.16 in [9] together with a fundamental approximation lemma to be presented in the next section.

One should note that the present approach does not only yield an appropriate frame for Donsker's Functional CLT for empirical C-processes in the classical situation, choosing $(T, d) = (C, d_\mu)$ with $C := \{(-\infty, t]: t \in \mathbf{R}\}$, but at the same time for its higher-dimensional analogue, i.e. with $C = \{(-\infty, \underline{t}]: \underline{t} \in \mathbf{R}^k\}$ being the class of all lower left orthants in $\mathbf{R}^k$, $k \geq 1$, cf. [3], [15] and [17] for a different approach based on a generalization of Skorokhod's $D[0, 1]$-space (and its topology) to higher dimensions.

One should also note that the present approach is appropriate as well when studying empirical processes indexed by classes $\mathcal{F}$ of functions; cf. [2], [5], [6], [7], [11], [14], and [16]. (Here again it follows for $\mathcal{F} \subset \mathcal{L}_2(\mu)$ (as in [8], p. 118) that under (A) the other condition (B') is automatically fulfilled in case of empirical $\mathcal{F}$-processes $\beta_n = (\beta_n(f))_{f \in \mathcal{F}}$ indexed by classes $\mathcal{F}$ of functions being totally bounded for $d_{\mu, \mathcal{F}}$ defined by

$$d^2_{\mu, \mathcal{F}}(f_1, f_2) := \int (f_1 - f_2)^2 d\mu - (\int (f_1 - f_2) d\mu)^2, \quad f_1, f_2 \in \mathcal{F};$$

cf. Theorem 4.1.1 in [7] and Theorem 2.12 in [11].

Another independent and more general approach to investigate functional CLT's for stochastic processes is contained in [1], based on results from the book "Stochastic Processes on Polish Spaces" by Hoffmann–Jørgensen [13], from which we learned on occasion of the 6th International Conference on Probability in Banach Spaces (held at Sandbjerg, Denmark, June 16–21, 1986) where our present approach was contributed to.

Of course, to apply Theorem 1 in connection with e.g. empirical C- (or F-) processes mentioned before, the main problem for obtaining a functional limit theorem, i.e. the statement that $\beta_n \xrightarrow{\mathcal{L}_b} \beta_0$, is to verify (iii)(A), since usually convergence of the finite dimensional distributions (fidis) pertaining to β_n follows from classical results, thus yielding that the law of any accumulation point β_0 of (β_n) (w.r.t. $\mathcal{L}_b$-convergence) is uniquely determined by its fidis and hence also on $\mathcal{B}_b(S)$ according to (1.1).

To verify (iii)(A) in specific situations the following result together with its corollary presents some new sufficient conditions (which turn out to be especially useful to derive in a straightforward way functional limit theorems for empirical C-processes indexed by Vapnik–Chervonenkis classes C of sets (and its bootstrapped versions; cf. [10])).

For this, let for any $\varepsilon > 0$ $N(\varepsilon, T, d) := \inf\{n \in \mathbf{N}$: there exist $t_1, \ldots, t_n \in T$ s.t. for each $t \in T$ $d(t, t_i) \leq 2\varepsilon$ for some t_i, $1 \leq i \leq n\}$. $\log N(\varepsilon, T, d)$ is called the metric entropy of (T, d); note that the assumed total boundedness of (T, d) is equivalent with the fact that $N(\varepsilon, T, d) < \infty$ for each $\varepsilon > 0$.

Now, the sufficient conditions in question comprise the behaviour of the increments of β_n, $n \in \mathbf{N}$, and the behaviour of $N(\varepsilon, T, d)$ as ε tends to zero in the following way (where the underlying method of proof is based on the well known chaining argument; see section 3 below):

Theorem 2. *Given a sequence $(\beta_n)_{n \in \mathbf{N}}$ of random elements in $(S, \mathcal{B})$, suppose that the following two conditions (1) and (2) are fulfilled:*

   (1) $\mathbf{P}(|\beta_n(t_1) - \beta_n(t_2)| \geq \varepsilon) \leq f(d(t_1, t_2), n, \varepsilon)$ *for each $t_1, t_2 \in T$, $\varepsilon > 0$ and $n \in \mathbf{N}$, where $f : \mathbf{R}_+^3 \to \mathbf{R}_+$ is assumed to be mono-*

*tone increasing in its first and monotone decreasing in its second
component.*

(2) *There exist functions* $r: (0, \infty) \to (0, \infty)$ *and* $g: (0, \infty) \to (0, \infty)$
such that

(i) r *is strictly decreasing and one-to-one with* $\lim_{t \downarrow 0} r(t) = \infty$

(ii) g *is monotone increasing s.t.* $\int_0^2 g(u) u^{-1} du < \infty$

(iii) $\int_0^1 u^{-1} N\left(\frac{u}{2}, T, d\right)^2 f(6u, r(4u), \varepsilon g(u)) du < \infty$ *for all* $\varepsilon > 0$.

Then, for each $1 \geq \delta > 0$, $\varepsilon > 0$, *and* $n \geq r(\delta)$ *one has*

$$\mathbf{P}^*(w_{\beta_n}(\delta) \geq 5\varepsilon) \leq 2 \int_{\frac{1}{4} r^*(n)}^{\delta} u^{-1} N\left(\frac{u}{2}, T, d\right)^2 f(6u, r(4u), \frac{\varepsilon}{2K_1} g(u)) du$$
$$+ N(\delta, T, d)^2 f(5\delta, n, \varepsilon) + \mathbf{P}^*(w_{\beta_n}(r^*(n)) \geq \varepsilon),$$

where $\infty > K_1 \geq \int_0^{2\delta} g(u) u^{-1} du$ *and* $r^*(n)$ *is determined by* $r(r^*(n)) = n$.

From this one obtains easily

Corollary 1. *Given a sequence* $(\beta_n)_{n \in \mathbb{N}}$ *of random elements in* (S, B)
*satisfying (1) and (2), suppose that, in addition, the following conditions
(3) and (4) are fulfilled:*

(3) $\limsup\limits_{n \to \infty} \mathbf{P}^*(w_{\beta_n}(r^*(n)) \geq \varepsilon) = 0$ *for each* $\varepsilon > 0$, *and*

(4) $\limsup\limits_{n \to \infty} \mathbf{P}(|\beta_n(t_0)| \geq K) \downarrow 0$ *as* $K \to \infty$ *for some* $t_0 \in T$. *and*
$\limsup\limits_{n \to \infty} f(u, n, K) \to 0$ *as* $K \to \infty$ *for any* $u \in (0, \infty)$.

Then $(\beta_n)_{n \in \mathbb{N}}$ *is relatively* $\mathcal{L}_b$-*sequentially compact (thus yielding a
functional limit theorem, i.e. the statement that* $\beta_n \overset{\mathcal{L}_b}{\longrightarrow} \beta_0$ *for some
limiting random element* β_0 *in* $(S, B_b(S))$, *provided that convergence of
the fidis of* β_n, $n \in \mathbb{N}$, *to those of* β_0 *also takes place).*

Remarks: Condition (2)(iii) is readily verified in situations where
$N(\cdot, T, d)$ is (essentially) of polynomial behaviour as it is in the case
$T = C$, $d = d_\mu$, C being a Vapnik-Chervonenkis class of sets (crf. [4],
(7.13)), and where, in addition, an appropriate Bernstein-type inequality is available (cf. [8], Lemma 4(i)). Therefore in these cases it only

remains to verify condition (3), i.e. to get the oscillation of β_n on small sets (of order $r^*(n)$) under control; note that the other condition (4) is trivially fulfilled for empirical C-processes taking $t_0 = \emptyset$.

In some sense Theorem 2 and Corollary 1, respectively, mimic Corollary 1.19 in [9] (being concerned there with the classical situation $T = [0,1]$, $d(t_1, t_2) = |t_1 - t_2|$, $S_0 = C[0,1] \subset S = D[0,1] \subset \ell^\infty([0,1])$ as it is the case with Theorem 1 and Theorem 1.17 in [9].

2. Auxiliary Lemmata

The results of section 1 are based on the following lemmata, the first one being of fundamental importance in the present situation, whereas the other two are straightforward generalizations of Lemma 1.15 and Lemma 1.16 in [9].

Lemma 1. *Let $T = (T, d)$ be a pseudo-metric space assumed to be totally bounded for the pseudo-metric d and let $S_0 := U^b(T, d)$ be the space of all (real valued and bounded) uniformly d-continuous functions on T. Then there exists a countable subset S_1 of S_0 such that for any $x \in \ell^\infty(T)$ with $w_x(\delta) \le \varepsilon$ for some $\delta > 0$ and $\varepsilon > 0$ there exists an $y \in S_1$ with $\|x - y\|_T \le 5\varepsilon$.*

Remark: In the special situation $T = [0,1]$, $d(t_1, t_2) = |t_1 - t_2|$, $S_0 = C[0,1] \subset S = D[0,1]$, instead of Lemma 1 it was sufficient in [9] to refer to the simple fact (cf. (**) on p. 67 in [9]) that whenever $x \in D[0,1]$ satisfies $w_x(\delta) \le \varepsilon$ for some $\delta > 0$ and $\varepsilon > 0$, then $x \in C[0,1]^{2\varepsilon}$.

From Lemma 1 it follows easily

Corollary 2. *$(U^b(T, d), \|\cdot\|_T)$ is a separable space if and only if (T, d) is totally bounded for d.*

Lemma 2. (cf. Lemma 1.15 in [9]). *Let $(K_n)_{n \in \mathbb{N}}$ be a decreasing sequence of subsets of S having the following property (K):*

(K) *Each sequence $(x_n)_{n \in \mathbb{N}}$ with $x_n \in K_n$ for every $n \in \mathbb{N}$ has an accumulation point in $(S, \|\cdot\|_T)$.*

Then, for any sequence $(f_n)_{n \in \mathbb{N}} \subset C(S) := \{f \in \mathbb{R}^S : f \ \|\cdot\|_T\text{-conti-}$
nuous$\}$ with $f_n \downarrow 0$ one has $\sup\{f_n(x): x \in K_n\} \downarrow 0$ as $n \to \infty$.

Lemma 3.   (cf. Lemma 1.16 in [9]). *Let $(\varepsilon_k)_{k \in \mathbb{N}}$ and $(\delta_k)_{k \in \mathbb{N}}$ be*
sequences of nonnegative real numbers such that $\varepsilon_k \downarrow 0$ as $k \to \infty$. Let
$K > 0$ be some constant and

$$K_n := \bigcap_{k=1}^{n} \{x \in S: w_x(\delta_k) < \varepsilon_k\} \cap \{x \in S: \|x\|_T < K\}, \quad n \in \mathbb{N}.$$

Then any sequence $(x_n)_{n \in \mathbb{N}}$ with $x_n \in K_n$ for every $n \in \mathbb{N}$ has an
accumulation point in $(S_0, \|\cdot\|_T)$, whence $(K_n)_{n \in \mathbb{N}}$ has the property (K)
stated in Lemma 2.

3. Proof of the Results.

Proof of Theorem 1: (i) $\Rightarrow$ (ii): The proof of this part goes by
algebraic induction starting with the following set of functions in $U_b^b(S)$

$$\{\min(\|\cdot -x_n\|_T, 1): n \in \mathbb{N}\},$$

where $\{x_n : n \in \mathbb{N}\}$ is dense in S_0, using a result of [12] on certain generalizations of the Weierstraß approximation theorem, and the Daniell-
Stone representation theorem to arrive at

$$\lim_{k \to \infty} \int f \circ \beta_{n_k} d\mathbf{P} = \int f \circ \beta_0 d\mathbf{P} \quad \text{for all} \quad f \in U_b^b(S)$$

for a certain subsequence $(\beta_{n_k})_{k \in \mathbb{N}}$ of $(\beta_n)_{n \in \mathbb{N}}$ (obtained by the diagonal
method) and a certain limiting random element β_0 in $(S, \mathcal{B}_b(S))$ which
proves the assertion after having shown that $\mathbf{P}(\beta_0 \in S_0) = 1$ according
to condition (b). For details we refer to the proof of Theorem 11a) in
[8].

(ii) $\Rightarrow$ (iii): We will show that (A) is fulfilled (the proof that also (B)
holds true runs similarly):

For an indirect proof assume to the contrary that there exists an $\varepsilon > 0$ such that for each $m \in \mathbf{N}$

$$\limsup_{n \to \infty} \mathbf{P}^*(w_{\beta_n}(\tfrac{1}{m}) \geq \varepsilon) \geq \sigma > 0. \quad \text{for some} \quad \sigma > 0.$$

Then, for each $m \in \mathbf{N}$ there exists an $n_m \in \mathbf{N}$ (with $n_m > n_{m-1}$, $n_0 := 1$) such that

$$(+) \qquad \mathbf{P}^*(w_{\beta_{n_m}}(\tfrac{1}{m}) \geq \varepsilon) > \sigma \quad \text{for every} \quad m \in \mathbf{N}.$$

On the other hand, it follows from (ii) that there exists a subsequence $(n_{m_k})_{k \in \mathbf{N}}$ of $(n_m)_{m \in \mathbf{N}}$ and a limiting random element β_0 in $(S, \mathcal{B}_b(S))$ such that $\mathbf{P}(\beta_0 \in S_0) = 1$ and $\beta_{n_{m_k}} \xrightarrow{\mathcal{L}_b} \beta_0$ as $k \to \infty$.

But from this it follows (cf. the Portmanteau-Theorem in [8] and the fact that for each $\varepsilon, \delta > 0$ $\{x \in S : w_x(\delta) \geq \varepsilon\}$ is $\| \cdot \|_T$-closed in S) that for each $\delta > 0$

$$\limsup_{k \to \infty} \mathbf{P}^*(w_{\beta_{n_{m_k}}}(\tfrac{1}{m_k}) \geq \varepsilon) \leq \limsup_{k \to \infty} \mathbf{P}^*(w_{\beta_{n_{m_k}}}(\delta) \geq \varepsilon)$$

$$\leq \mathbf{P}(\{w_{\beta_0}(\delta) \geq \varepsilon\} \cap \{\beta_0 \in S_0\}) \to 0 \quad \text{as } \delta \to 0,$$

which contradicts (+).

(iii) $\Rightarrow$ (iv): Since (T, d) is totally bounded, there exists for any $\delta > 0$ a finite subset E_δ of T having the following property: For each $t \in T$ there exists an $e = e_t \in E_\delta$ such that $d(t, e) \leq \delta$. Now, given any $\varepsilon, \delta, K > 0$, we obtain

$$\{x \in S : w_x(\delta) < \varepsilon\} \cap \bigcap_{e \in E_\delta} \{x \in S : |x(e)| < K\}$$

$$\subset \{x \in S : \|x\|_T < K + \varepsilon\},$$

and therefore it follows from (A) and (B) that for each $\delta > 0$

$$(\overset{=}{B}) \quad \lim_{K \to \infty} \limsup_{n \to \infty} \mathbf{P}(\|\beta_n\|_T \geq K)$$

$$\leq \lim_{K \to \infty} \limsup_{n \to \infty} [\mathbf{P}^*(w_{\beta_n}(\delta) \geq 1) + \sum_{e \in E_\delta} \mathbf{P}(|\beta_n(e)| \geq K - 1)]$$

$$\limsup_{n \to \infty} \mathbf{P}^*(w_{\beta_n}(\delta) \geq 1) \to 0 \quad \text{as } \delta \downarrow 0 \text{ according to (A)).}$$

This proves (B') (which in turn obviously implies (B)).

(iv) $\Rightarrow$ (i): At first we will show

Part 1: For any $\eta > 0$ there exist sequences $(\varepsilon_k)_{k\in\mathbb{N}}$ and $(\delta_k)_{k\in\mathbb{N}}$ of nonnegative real numbers with $\varepsilon_k \downarrow 0$ as $k \to \infty$ and a $K > 0$ (depending on η) such that for $n \geq n_1(\eta)$

$$\mathbf{P}^* \left(\bigcup_{k=1}^{n} \{ w_{\beta_n}(\delta_k) \geq \varepsilon_k \} \cup \{ \|\beta_n\|_T \geq K \} \right) \leq \eta/2.$$

For this, let $\eta > 0$ be arbitrary but fixed and let $(\varepsilon_k')_{k\in\mathbb{N}}$ be a sequence of nonnegative real numbers with $\varepsilon_k' \downarrow 0$ as $k \to \infty$. Then, it follows according to (A) that for any $k \in \mathbb{N}$ there exists a $\delta_k' > 0$ such that

$$\limsup_{n\to\infty} \mathbf{P}^* (w_{\beta_n}(\delta_k') \geq \varepsilon_k') \leq (\eta/8) \cdot 2^{-k}.$$

Hence, for any $N \in \mathbb{N}$, we have

$$\limsup_{n\to\infty} \mathbf{P}^* \left(\bigcup_{k=1}^{N} \{ w_{\beta_n}(\delta_k') \geq \varepsilon_k' \} \right) \leq \eta/8.$$

Therefore, for any $N \in \mathbb{N}$ there exists an $n_N \in \mathbb{N}$ with $n_N > n_{N-1}$, $n_0 := 0$, such that

$$(+) \qquad \mathbf{P}^* \left(\bigcup_{k=1}^{N} \{ w_{\beta_n}(\delta_k') \geq \varepsilon_k' \} \right) \leq \eta/4 \quad \text{for all} \quad n \geq n_N.$$

Now, put $\varepsilon_k := \varepsilon_1'$, $\delta_k := \delta_1'$ for $k \in \{1, \ldots, n_1 - 1\}$ and $\varepsilon_k := \varepsilon_N'$, $\delta_k := \delta_N'$ for $k \in \{n_N, \ldots, n_{N+1} - 1\}$. Then, for any $n \geq n_1$ there exists a unique $N \in \mathbb{N}$ with $n \in \{n_N, \ldots, n_{N+1} - 1\}$, whence

$$\mathbf{P}^* \left(\bigcup_{k=1}^{n} \{ w_{\beta_n}(\delta_k) \geq \varepsilon_k \} \right) = \mathbf{P}^* \left(\bigcup_{k=n_1}^{n_{N+1}-1} \{ w_{\beta_n}(\delta_k) \geq \varepsilon_k \} \right)$$

$$= \mathbf{P}^* \left(\bigcup_{i=1}^{N} \{ w_{\beta_n}(\delta_i') \geq \varepsilon_i' \} \right) \leq \eta/4 \quad \text{by} \quad (+).$$

Together with (B') we thus obtain the statement of part 1.

Part 2: Here we shall make use of the auxiliary lemmata of section 2. We will show at first that (a) holds true:

For this, let $(f_m)_{m \in \mathbb{N}}$ be a sequence in $U_b^b(S)$ with $f_m \downarrow 0$ as $m \to \infty$ and assume w.l.o.g. $f_1 \leq 1$. According to Part 1 and Lemma 3 there exists for any $\eta > 0$ a decreasing sequence $(K_n)_{n \in \mathbb{N}}$ of subsets of S (namely $K_n = \bigcap_{k=1}^n \{x \in S \colon w_x(\delta_k) < \varepsilon_k\} \cap \{x \in S \colon \|x\|_T < K\}$) having the property (K) (stated in Lemma 2) and being such that

$$\mathbf{P}^*(\beta_n \in \complement K_n) \leq \eta/2 \quad \text{for all} \quad n \geq n_1(\eta).$$

Therefore, by Lemma 2, there exists an $m_1 \in \mathbb{N}$, $m_1 \geq n_1(\eta)$, such that for all $m \geq m_1$

$$\sup\{f_m(x) \colon x \in K_m\} \leq \eta/2.$$

Thus it follows for all n and m with $n \geq m \geq m_1$

$$\int f_m \circ \beta_n \, d\mathbf{P} = \int^* f_m \circ \beta_n \, d\mathbf{P}$$

$$\leq \int^* f_m \circ \beta_n \cdot 1_{\{\beta_n \in K_n\}} d\mathbf{P} + \int^* f_m \circ \beta_n \cdot 1_{\{\beta_n \in \complement K_n\}} d\mathbf{P}$$

$$\leq \eta/2 + \mathbf{P}^*(\beta_n \in \complement K_n) \leq \eta,$$

whence $\limsup_{n \to \infty} \int f_m \circ \beta_n \, d\mathbf{P} \leq \eta$ for m sufficiently large.

Since $\eta > 0$ was chosen arbitrarily, this proves (a). The proof will be concluded by verifying (b), i.e. that $\liminf_{n \to \infty} \int f \circ \beta_n \, d\mathbf{P} \geq 1$ for all $f \in U_b^b(S)$ with $f \geq 1_{S_0}$. For this, consider any $f \in U_b^b(S)$ with $f \geq 1_{S_0}$. Let $\eta > 0$ be arbitrary but fixed; then, since f is uniformly continuous, there exists a $\delta = \delta(\eta) > 0$ such that

$$f \geq (1 - \eta/2) 1_{S_0^\delta},$$

where $S_0^\delta := \{x \in S \colon \|x - y\|_T < \delta \text{ for some } y \in S_0\}$; note that $S_0^\delta \in \mathcal{B}_b(S)$ (cf. [8], Lemma 11).

Furthermore, as in Part 1, there exist sequences $(\varepsilon_k)_{k\in\mathbb{N}}$ and $(\delta_k)_{k\in\mathbb{N}}$ of nonnegative real numbers with $\varepsilon_k \downarrow 0$ as $k \to \infty$ such that

(*) $\mathbf{P}^*(w_{\beta_n}(\delta_n) \geq \varepsilon_n) \leq \eta/2$ for all $n \geq n_1 = n_1(\eta)$.

On the other hand it follows from Lemma 1:

(**) Whenever $x \in S$ satisfies $w_x(\delta) < \varepsilon$ for some $\delta > 0$ and $\varepsilon > 0$, then $x \in S_0^{5\varepsilon}$.

Now, (*) together with (**) imply

$$\mathbf{P}(\beta_n \in \complement S_0^{5\varepsilon_n}) \leq \eta/2 \quad \text{for all} \quad n \geq n_1.$$

Finally, choose $n_2 \in \mathbb{N}$, $n_2 \geq n_1$, such that $5\varepsilon_{n_2} \leq \delta = \delta(\eta)$; then for all $n \geq n_2$ we obtain

$$\int f \circ \beta_n \, d\mathbf{P} = \int f \circ \beta_n \cdot 1_{\{\beta_n \in S_0^{5\varepsilon_n}\}} \, d\mathbf{P} + \int f \circ \beta_n \cdot 1_{\{\beta_n \in \complement S_0^{5\varepsilon_n}\}} \, d\mathbf{P}$$

$$\underset{(f\geq 0)}{\geq} (1 - \eta/2)\mathbf{P}(\beta_n \in S_0^{5\varepsilon_n}) \geq (1 - \eta/2)(1 - \eta/2) \geq 1 - \eta,$$

whence

$$\liminf_{n\to\infty} \int f \circ \beta_n \, d\mathbf{P} \geq 1 - \eta.$$

Since $\eta > 0$ was chosen arbitrarily, this proves (b) concluding the proof of Theorem 1.

Proof of Theorem 2: Let $1 \geq \delta > 0$, $\varepsilon > 0$ and $n \geq r(\delta)$ be arbitrary but fixed. Put $\delta_k := \delta \cdot 2^{1-k}$, $k \in \mathbb{N}$.

Since (T, d) is totally bounded, there exists for each $k \in \mathbb{N}$ a finite subset T_k of T having the following properties (α) and (β):

(α) For each $t \in T$ there exists an $s(t, k) \in T_k$ such that $d(t, s(t, k)) \leq 2\delta_k$.

(β) $|T_k|$ is minimal among all subsets of T having the property (α), i.e. $|T_k| = N(\delta_k, T, d)$.

Then we have

$$
\begin{aligned}
w_{\beta_n}(\delta) &= \sup\{|\beta_n(t_1) - \beta_n(t_2)| : d(t_1, t_2) \le \delta\} \\
&\le \sup\{|\beta_n(t_1) - \beta_n(s(t_1, 1))| + |\beta_n(s(t_1, 1)) - \beta_n(s(t_2, 1))| \\
&\quad + |\beta_n(s(t_2, 1)) - \beta_n(t_2)| : d(t_1, t_2) \le \delta\} \\
&\le 2\sup\{|\beta_n(t) - \beta_n(s(t, 1))| : t \in T\} \\
&\quad + \sup\{|\beta_n(s_1) - \beta_n(s_2)| : s_1, s_2 \in T_1 \text{ and } d(s_1, s_2) \le 5\delta\} \\
&= 2S_n' + S_n'', \text{ say.}
\end{aligned}
$$

To obtain an upper estimate for S_n' we are using the so-called chaining argument (cf. e.g. [16] VII.2.). For this, let

$$
m(n) := \sup\{k \in \mathbf{N} : r(2\delta_k) \le n\}
$$

(noticing that $\{\cdots\} \ne \emptyset$ since $n \ge r(\delta) = r(\delta_1) \ge r(2\delta_1)$); then

$$
\begin{aligned}
S_n' &\le \sup \left\{ \sum_{k=1}^{m(n)} |\beta_n(s(t, k)) - \beta_n(s(t, k+1))| \right. \\
&\quad \left. + |\beta_n(s(t, m(n) + 1)) - \beta_n(t)| : t \in T\right\} \\
&\le \sum_{k=1}^{m(n)} \sup\{|\beta_n(s(t, k)) - \beta_n(s(t, k+1))| : t \in T\} + w_{\beta_n}(2\delta_{m(n)+1}) \\
&= S_n''' + w_{\beta_n}(2\delta_{m(n)+1}), \text{ say.}
\end{aligned}
$$

Thus, since $r(2\delta_{m(n)+1}) > n$ and therefore (by (2)(i)) $2\delta_{m(n)+1} < r^*(n)$, we obtain

$$
(+) \qquad\qquad w_{\beta_n}(\delta) \le 2S_n''' + S_n'' + 2w_{\beta_n}(r^*(n)).
$$

Now, for every $k \in \mathbf{N}$, let $\varepsilon_k' := g(\delta_k)$; then, according to (2)(ii) (with $\delta_0 := 2\delta$) we have

$$
0 < \sum_{k \in \mathbf{N}} \varepsilon_k' = 2 \sum_{k \in \mathbf{N}} \int_{\delta_k}^{\delta_{k-1}} \frac{g(\delta_k)}{\delta_{k-1}} du \le 2 \int_0^{2\delta} \frac{g(u)}{u} du \le 2K_1 < \infty.
$$

Therefore, putting

$$\varepsilon_k := \frac{\varepsilon'_k}{2K_1}\varepsilon, k \in \mathbf{N}, \quad \text{and} \quad N(u) := N(u, T, d),$$

we get

$$\mathbf{P}(S'''_n \geq \varepsilon)$$

$$\leq \mathbf{P}\Big(S'''_n \geq \sum_{k=1}^{m(n)} \varepsilon_k\Big)$$

$$= \mathbf{P}\Big(\sum_{k=1}^{m(n)} \sup\{|\beta_n(s(t, k)) - \beta_n(s(t, k+1))| : t \in T\} \geq \sum_{k=1}^{m(n)} \varepsilon_k\Big)$$

$$\leq \sum_{k=1}^{m(n)} \mathbf{P}(\sup\{|\beta_n(s') - \beta_n(s'')| : s', s'' \in T_k \cup T_{k+1}, d(s', s'') \leq 3\delta_k\} \geq \varepsilon_k)$$

$$\leq \sum_{k=1}^{m(n)} \sum_{\substack{s', s'' \in T_k \cup T_k \\ d(s', s'') \leq 3\delta_k}} \mathbf{P}(|\beta_n(s') - \beta_n(s'')| \geq \varepsilon_k)$$

$$= 2 \sum_{k=1}^{m(n)} \int_{k+1}^{\delta_k} \delta_k^{-1} \sum_{\substack{s', s'' \in T_k \cup T_k \\ d(s', s'') \leq 3\delta_k}} \mathbf{P}(|\beta_n(s') - \beta_n(s'')| \geq \frac{\varepsilon}{2K_1}g(\delta_k))du$$

$$\leq 2 \sum_{k=1}^{m(n)} \int_{\delta_k}^{\delta_k} u^{-1} \sum_{\substack{s', s'' \in T_k \cup T_k \\ d(s', s'') \leq 3\delta_k}} \mathbf{P}(|\beta_n(s') - \beta_n(s'')| \geq \frac{\varepsilon}{2K_1}g(u))du$$

$$\underset{(1)}{\leq} 2 \sum_{k=1}^{m(n)} \int_{\delta_k}^{\delta_k} u^{-1} N(\delta_{k+1})^2 f(3\delta_k, n, \frac{\varepsilon}{2K_1}g(u))du$$

$$\leq \sum_{k=1}^{m(n)} \int_{\delta_k}^{\delta_k} u^{-1} N\big(\frac{u}{2}\big)^2 f(6u, n, \frac{\varepsilon}{2K_1}\overline{g(u)})du$$

$$\leq \sum_{k=1}^{m(n)} \int_{\delta_k}^{\delta_k} u^{-1} N\big(\frac{u}{2}\big)^2 f(6u, r(2\delta_k), \frac{\varepsilon}{2K_1}g(u))du$$

$$\leq \sum_{k=1}^{m(n)} \int_{\delta_k}^{\delta_k} u^{-1} N\big(\frac{u}{2}\big)^2 f(6u, r(4u), \frac{\varepsilon}{2K_1}g(u))du$$

$$\leq 2 \int_{\frac{1}{4}r^*(n)}^{\delta} u^{-1} N(\frac{u}{2})^2 f(6u, r(4u), \frac{\varepsilon}{2K_1} g(u)) du.$$

Finally, since

$$P(S_n'' \geq \varepsilon) \leq N(\delta)^2 f(5\delta, n, \varepsilon),$$

in view of (+) the assertion of Theorem 2 is proved.

Proof of Corollary 1: Since $g(t) \to 0$ as $t \downarrow 0$, concerning the proof of Theorem 2 we even have that for any $\varepsilon > 0$ and for $\delta > 0$ small enough

$$P(S_n'' \geq \varepsilon) \leq P(S_n'' \geq g(\delta)) \leq N(\delta)^2 f(5\delta, n, g(\delta)),$$

whence

$$\limsup_{n \to \infty} P(S_n'' \geq \varepsilon) \leq N(\delta)^2 f(5\delta, r(\delta), g(\delta)) \leq \delta G(\delta)$$

with

$$G(\delta) := \delta^{-1} N(\frac{\delta}{2})^2 f(6\delta, r(4\delta), g(\delta)).$$

Since G is integrable by (2)(iii), there exists a sequence $(\delta_\ell)_{\ell \in \mathbb{N}}$ of non-negative real numbers tending to zero such that $\delta_\ell G(\delta_\ell) \to 0$ as $\ell \to \infty$, and therefore it follows by Theorem 2 that for ℓ sufficiently large

$$\limsup_{n \to \infty} P^*(w_{\beta_n}(\delta_\ell) \geq 5\varepsilon) \leq 2 \int_0^{\delta_\ell} u^{-1} N(\frac{u}{2})^2 f(6u, r(4u), \frac{\varepsilon}{2K_1} g(u)) du$$
$$+ \delta_\ell G(\delta_\ell) + \limsup_{n \to \infty} P^*(w_{\beta_n}(r^*(n)) \geq \varepsilon)$$
$$\underset{(3)}{=} 2 \int_0^{\delta_\ell} u^{-1} N(\frac{u}{2})^2 f(6u, r(4u), \frac{\varepsilon}{2K_1} g(u)) du + \delta_\ell G(\delta_\ell)$$
$$\to 0 \quad \text{as} \quad \ell \to \infty.$$

This shows that the condition (iii)(A) of Theorem 1 is fulfilled.

It remains to show that also the other condition (B) is fulfilled.

For this, given any $t \in T$, using (4) we get

$$\limsup_{n \to \infty} \mathbf{P}(|\beta_n(t)| \geq 2K)$$

$$\leq \limsup_{n \to \infty} [\mathbf{P}(|\beta_n(t) - \beta_n(t_0)| \geq K) + \mathbf{P}(|\beta_n(t_0)| \geq K)]$$

$$\underset{(1)}{\leq} \limsup_{n \to \infty} [f(d(t, t_0), n, K) + \mathbf{P}(|\beta_n(t_0)| \geq K)]$$

$$\leq \limsup_{n \to \infty} f(d(t, t_0), n, K) + \limsup_{n \to \infty} \mathbf{P}(|\beta_n(t_0)| \geq K)$$

$$\to 0 \quad \text{as} \quad K \to \infty.$$

4. Proof of the Lemmata.

Proof of Lemma 1: Since (T, d) is totally bounded, there exists for each $n \in \mathbf{N}$ a finite subset T_n of T having the following property:

For each $t \in T$ there exists an $s \in T_n$ such that $d(t, s) < n^{-1}$. Now, denoting by $\mathbf{Q}$ and $\mathbf{Q}_+$, respectively, the set of rational and nonnegative rational numbers, respectively, we are going to show that

$$S_1 := \bigcup_{n \in \mathbf{N}} \bigcup_{q \in \mathbf{Q}_+} \{t \to \sup\{x(s) - q \cdot d(t, s) : s \in T_n\} : x \in \mathbf{Q}^{T_n}\}$$

has the properties stated in Lemma 1.

First, it follows from the total boundedness of (T, d) that for each $s \in T$, $d(\cdot, s) \in S_0$ whence $S_1 \subset S_0$.

Now, consider any $x \in \ell^\infty(T)$ with $w_x(\delta) \leq \varepsilon$ for some $\delta > 0$ and $\varepsilon > 0$. Choose $\delta' \in (0, \delta] \cap \mathbf{Q}$, $\varepsilon' \in [\varepsilon, \frac{5}{4}\varepsilon] \cap \mathbf{Q}$ and $M \in [2\|x\|_T, \infty) \cap \mathbf{Q}$, and consider the straight line given by

$$\ell(u) := q \cdot u, \quad u \geq 0, \quad \text{with } q := \frac{\max(\varepsilon', M)}{\delta'} \in \mathbf{Q}_+.$$

Then one easily shows

$$\text{(a)} \qquad w_x(u) \leq \ell(u) \quad \text{for all} \quad u \geq \tilde{\delta} := \frac{\varepsilon' \cdot \delta'}{\max(\varepsilon', M)}.$$

Next, choose $k \in \mathbf{N}$ such that $k^{-1} \leq \tilde{\delta}$ and consider the function y defined by

$$y(t) := \sup\{\tilde{x}(s) - \ell(d(t,s)) : s \in T_k\}, \quad t \in T,$$

with an $\tilde{x} \in \mathbf{Q}^{T_k}$ such that $\sup\{|\tilde{x}(s) - x(s)| : x \in T_k\} \leq \varepsilon'/2$.

Then $y \in S_1$ and

(b) $$\sup\{|y(s) - x(s)| : s \in T_k\} \leq 2\varepsilon'.$$

As to (b), let $s \in T_k$ be arbitrary but fixed; then

$$0 \leq \tilde{x}(s) - x(s) - \ell(d(s,s)) + \varepsilon'/2 \leq y(s) - x(s) + \varepsilon'/2$$
$$= \sup\{\tilde{x}(s') - x(s) - \ell(d(s,s')) : s' \in T_k\} + \varepsilon'/2$$
$$\leq \sup\{w_x(d(s,s')) + \tilde{x}(s') - x(s') - \ell(d(s,s')) : s' \in T_k\} + \varepsilon'/2$$
$$\leq \sup\{w_x(d(s,s')) - \ell(d(s,s')) : s \in T_k\} + \varepsilon' \leq 2\varepsilon',$$

where the last inequality results from the fact that for $d(s,s') \leq \tilde{\delta}$ one has $w_x(d(s,s')) \leq w_x(\tilde{\delta}) \leq \varepsilon'$, whereas for $d(s,s') > \tilde{\delta}$ $w_x(d(s,s')) \leq \ell(d(s,s'))$ by (a).

This shows that $\sup\{|y(s) - x(s)| : s \in T_k\} \leq 2\varepsilon'$ proving (b).

Next, one easily verifies

(c) $$w_y(\tilde{\delta}) \leq \varepsilon'.$$

Finally, for any $t \in T$ there exists an $s = s(t) \in T_k$ such that $d(t,s) < k^{-1} \leq \tilde{\delta}$ and therefore we obtain

$$|y(t) - x(t)| \leq |y(t) - y(s)| + |y(s) - x(s)| + |x(s) - x(t)| \underset{(b)}{\leq}$$
$$w_y(\tilde{\delta}) + 2\varepsilon' + w_x(\tilde{\delta}) \underset{(c)}{\leq} 4\varepsilon' \leq 5\varepsilon.$$

Since $t \in T$ was chosen arbitrarily, this shows that $\|x - y\|_T \leq 5\varepsilon$.

Proof of Lemma 2: Let $\varepsilon > 0$ be arbitrary but fixed and suppose w.l.o.g. that $f_n \leq 1$ (otherwise consider $f'_n := \min(1, f_n) \in C(S)$); then, for every $n \in \mathbb{N}$ there exists an $x_n \in K_n$ such that

$$\sup\{f_n(x) \colon x \in K_n\} \leq f_n(x_n) + \varepsilon.$$

According to (K) there exists an $x \in S$ and a subsequence $(x_{n_k})_{k \in \mathbb{N}}$ of $(x_n)_{n \in \mathbb{N}}$ such that

$$\|x_{n_k} - x\|_T \to 0 \quad \text{as} \quad k \to \infty.$$

Now, since $f_n \downarrow 0$ and since the f_n's are $\| \cdot \|_T$-continuous, it follows that for each $m \in \mathbb{N}$

$$\lim_{n \to \infty} \sup\{f_n(x) \colon x \in K_n\} \leq \limsup_{k \to \infty} f_{n_k}(x_{n_k}) + \varepsilon \leq$$
$$\limsup_{k \to \infty} f_m(x_{n_k}) + \varepsilon = f_m(x) + \varepsilon \downarrow \varepsilon \quad \text{as} \quad m \to \infty.$$

Since $\varepsilon > 0$ was chosen arbitrarily, this proves Lemma 2.

Proof of Lemma 3: Since (T, d) is totally bounded, there exists a countable subset A of T whose closure A^c in (T, d) coincides with T. Furthermore, since $[-K, K]$ is compact there exists (by the diagonal method) a subsequence $(x_{n_k})_{k \in \mathbb{N}}$ of $(x_n)_{n \in \mathbb{N}}$ (with $n_1 < n_2 < \ldots$) such that

$$\lim_{k \to \infty} x_{n_k}(t) \quad \text{exists (in } \mathbb{R}) \text{ for every} \quad t \in A.$$

Thus, for every $t \in A$, $\tilde{x}(t) := \lim_{k \to \infty} x_{n_k}(t)$ is well defined.

We are going to show that $\tilde{x}$ is uniformly continuous on A:

For this, given any $\varepsilon > 0$ there exists a $k_0 = k_0(\varepsilon) \in \mathbb{N}$ such that $\varepsilon_{n_{k_0}} \leq \varepsilon$. then, for any $t, t' \in A$ with $d(t, t') \leq \delta_{n_{k_0}}$ we get for all $k \geq k_0$,

$$|\tilde{x}(t) - \tilde{x}(t')| \leq |\tilde{x}(t) - x_{n_k}(t)| + |x_{n_k}(t) - x_{n_k}(t')| + |x_{n_k}(t') - \tilde{x}(t')|$$
$$\leq |\tilde{x}(t) - x_{n_k}(t)| + w_{x_{n_k}}(\delta_{n_{k_0}}) + |x_{n_k}(t') - \tilde{x}(t')|$$
$$\leq |\tilde{x}(t) - x_{n_k}(t)| + \varepsilon_{n_{k_0}} + |x_{n_k}(t') - \tilde{x}(t')|,$$

whence for $k \to \infty$

$$(*) \qquad\qquad |\tilde{x}(t) - \tilde{x}(t')| \leq \varepsilon_{n_{k_0}} \leq \varepsilon,$$

This proves that $\tilde{x}$ is uniformly continuous on A.

Therefore (since $A^c = T$) there exists a uniformly continuous extension $x : T \to \mathbf{R}$ of $\tilde{x}$. The proof will be concluded if we finally show that

$$(+) \qquad\qquad \|x_{n_k} - x\|_T \to 0 \text{ as } k \to \infty.$$

For this, given an arbitrary $\varepsilon > 0$, choose $k_0 = k_0(\varepsilon) \in \mathbf{N}$ such that $\varepsilon_{n_{k_0}} \leq \varepsilon/2$ (whence by (*) $w_x(\delta_{n_{k_0}}) \leq \varepsilon/2$). Since (T, d) is totally bounded, there exists a finite subset E of A having the following property:

For each $t \in T$ there exists an $e_t \in E$ such that $d(t, e_t) \leq \delta_{n_{k_0}}$.

Therefore, for any $t \in T$ and all $k \geq k_0$ we get

$$|x_{n_k}(t) - x(t)| \leq |x_{n_k}(t) - x_{n_k}(e_t)| + |x_{n_k}(e_t) - x(e_t)| + |x(e_t) - x(t)|$$
$$\leq \varepsilon + \sup\{|x_{n_k}(e) - \tilde{x}(e)| : e \in E\} \to \varepsilon \text{ as } k \to \infty.$$

Since $\varepsilon > 0$ was chosen arbitrarily, this proves $(+)$.

References

[1] N.T. Andersen and V. Dobrić, *The Central Limit Theorem for Stochastic Processes*, Ann. Probability 15 (1987), 164–177.

[2] K.S. Alexander, *The Central Limit Theorem for Empirical Processes on Vapnik-Červonenkis classes*, Ann. Probability 15 (1987), 178–203.

[3] R.F. Bass and R. Pyke, *The space $D(A)$ and weak convergence for set-indexed processes*, Ann. Probability 13 (1985), 860–884.

[4] R.M. Dudley, *Central limit theorems for empirical measures*, Ann. Probability 6 (1978), 899–929. Correction, ibid. 7, 1979, pp. 909–911

[5] —————, *Vapnik-Červonenkis Donsker classes of functions*, In: Aspects Statistiques et aspects physiques des processus gaussiens (Proc. Colloque C.N.R.S. St. Flour, 1980), C.N.R.S., Paris (1981), 251–269.

[6] —————, *Donsker classes of functions*, In: Statistics and Related Topics (Proc. Symp. Ottawa, 1980), North Holland, N.Y. (1981), 341–352.

[7] —————, *A course on empirical processes*, pp. 1–142 in "École d'Été de Probabilités de Saint-Flour XII-1982", Lecture Notes in Math. **1097**. Springer Verlag 1984

[8] P. Gaenssler, *Empirical Processes*, IMS Lecture Notes - Monograph Series **3** (1983). 179 pp.

[9] P. Gaenssler, E. Haeusler and W. Schneemeier, *Selected Topics on Empirical Processes*, In: Proceedings of the Third Prague Symposium on Asymptotic Statistics. Ed. by P. Mandl and M. Hušková - Amsterdam-New York-Oxford, Elsevier Science Publishers B.V. (1984), 57–91.

[10] P. Gaenssler, *Bootstrapping empirical measures indexed by Vapnik-Chervonenkis classes of sets*, pp. 467–481 in "Probability Theory and Mathematical Statistics, Vol. 1" (Vilnius 1985), VNU Sci. Press, Utrecht 1987.

[11] E. Giné and J. Zinn, *Some limit theorems for empirical processes*, Ann. Probability **12** (1984), 929–989.

[12] E. Hewitt, *Certain generalizations of the Weierstrass Approximation Theorem*, Duke Math. J. **14** (1947), 419–427.

[13] J. Hoffmann-Jørgensen, *Stochastic processes on Polish spaces.* (to appear)

[14] V.I. Kolčinskii, *Functional Limit Theorems and Empirical Entropy I*, in Russian, Theor. Probability Math. Statist. (Kiev) **33** (1985), 31–42.

[15] G. Neuhaus, *On weak convergence of stochastic processes with multidimensional time parameter*, Ann. Math. Statist. **42** (1971), 1285–1295.

[16] D. Pollard, *Convergence of Stochastic Processes*, Springer Verlag, New York (1984). 215 pp.

[17] M.L. Straf, *Weak convergence of stochastic processes with several parameters*, In: Proceedings 6th Berkeley Symposium on Math. Statistics and Probability, Vol. 2, Berkeley - Los Angeles, Univ. California Press (1971), 187–221.

Mathematical Institute, University of Munich, Theresienstraße 39, D–8000 Munich 2, W.–Germany

Random Martingale Transform Inequalities

D.J.H. GARLING

1. Notation

Certain inequalities play a fundamental role in the theory of martingales. In order to describe these, let us begin by describing the notation that we use. Because we wish to transform martingales in a random way, the setting is a little more complicated than usual.

We suppose that (Ω, Σ, P) is a probability space, that $(F_n)_{n \geq 0}$ is a filtration of Ω (with $F = \sigma(\cup_n F_n)$) and that G is a sub-σ-field of Σ independent of F. We set $G_n = \sigma(F_n, G)$. Suppose that $m = (m_n)_{n \geq 0}$ is a martingale, adapted to the filtration (F_n), taking values in a Banach space E. If $1 \leq p \leq \infty$ we set $\|m\|_p = \sup_n \|m_n\|_p$. We denote the corresponding martingale difference sequence by (d_n) (so that $d_0 = m_0$, $d_n = m_n - m_{n-1}$ for $n > 0$). As usual, we set $m_n^* = \sup_{j \leq n} \|m_j\|$ and $m^* = \sup_j \|m_j\|$; m^* is the maximal function of m. Suppose now that (ν_n) is a real-valued process, predictable with respect to the filtration (G_n), in the sense that ν_0 is constant, and ν_n is G_{n-1}-measurable for $n > 0$.

If $\|\nu_n\|_\infty < \infty$ for each n then the sequence $(\nu_n d_n)$ is a martingale difference sequence with respect to (G_n): the corresponding martingale is denoted by $\nu.m$ and called the transform of m by ν. If ν is a predictable process, we set $\|\nu\|_\infty = \sup_n \|\nu_n\|_\infty$.

We shall denote by $\varepsilon = (\varepsilon_n)$ a Bernoulli sequence of random variables — that is, an independent sequence of random variables each taking the values 1 and -1 with probability $1/2$ — which is G-measurable. If m is a martingale, we can construct the transform $\varepsilon.m$. We set

$$B^{(p)}(m) = \sup_n (E(\|(\varepsilon.m)_n\|^p | F))^{1/p}.$$

2. Preliminaries

The first inequalities that we consider (Doob's maximal inequalities) hold for all martingales (cf. [7] VII Theorems 3.2 and 3.4)

$$(1.1w) \qquad cP(m^* \geq c) \leq \int_{(m^* \geq c)} \|m\| dP \leq \|m\|_1, \quad \text{for } c > 0$$

and

$$(1.p) \qquad \|m^*\|_p \leq p' \|m\|_p, \quad \text{for } 1 < p < \infty$$

(where $1/p + 1/p' = 1$).

The other inequalities that we consider hold for real-valued martingales.

If $1 < p < \infty$ there exists a finite positive constant K_p such that if ν is a predictable process then

$$(2.p) \qquad \|\nu.m\|_p \leq K_p \|\nu\|_\infty \|m\|_p$$

(Burkholder's transform inequalities). (In fact $\max(p, p') - 1$ is the best possible value for K_p [6].)

If $1 \leq p \leq \infty$ there exist positive constants c_p and C_p such that

$$(3.p) \qquad c_p \|(\Sigma d_j^2)^{1/2}\|_p \leq \|m^*\|_p \leq C_p \|(\Sigma d_j^2)^{1/2}\|_p$$

(the Burkholder-Davis-Gundy inequalities). (Suitable values for c_p and C_p are given in [8].)

To what extent do inequalities (2.p) and (3.p) extend to vector valued martingales? A Banach space E is said to be an MT-space if for some $1 < p < \infty$ there exists a K_p for which (2.p) holds for all transforms of E-valued martingales. In fact if (2.p) holds for one value of p then it holds for all (with different constants) ([3] Theorem 1.1); it holds if and only if it holds for all non-random transforms ([3] Theorem 2.2) and indeed if and only if it holds when $(\nu_n) = (-1)^n$ ([5] Lemma 2.1).

What about inequality (3.p)? One way forward is to consider $(\Sigma\|d_j\|^p)^{1/p}$ in place of $(\Sigma d_j^2)^{1/2}$; corresponding inequalities then hold (on the left and on the right) if and only if E can be renormed to be uniformly p-convex or uniformly p-smooth [12]. It is however possible to proceed differently. Suppose that E is a Banach lattice and that $x_1, \ldots, x_n \in E$. It is then possible to define $(\sum_{j=1}^n |x_j|^2)^{1/2}$ (cf. [10] Theorem 1.d.1). Further if E is q-concave for some $q < \infty$ (and this is a condition which we shall see arises naturally) then for $1 \le p < \infty$ there exists positive constants c_p and C_p such that if $\varepsilon_1, \ldots, \varepsilon_n$ are Bernoulli random variables then

(4.p)
$$c_p\|(\sum_{j=1}^n |x_j|^2)^{1/2}\| \le \|\sum_{j=1}^n \varepsilon_j x_j\|_p$$
$$\le C_p\|(\sum_{j=1}^n |x_j|^2)^{1/2}\|$$

(this follows from [10] Theorem 1.d.6 and Kahane's inequality [9] pp 18–24).

Bearing this in mind, let us say that a Banach space E satisfies lower estimates for random martingale transforms (LERMT) if for some $1 < p < \infty$ there exists K_p such that

(6.p)
$$\|m\|_p \le K_p\|\varepsilon.m\|_p$$

for all transforms of E-valued martingales by (independent) Bernoulli sequences and satisfies upper estimates for random martingale transforms (UERMT) if for some $1 < p < \infty$ there exists K_p such that

(7.p)
$$\|\varepsilon.m\|_p \le K_p\|m\|_p$$

for all transforms of E-valued martingales by independent Bernoulli sequences. Then for $1 < p < \infty$ the inequalities (3.p) can be considered as the real-valued version of (6.p) and (7.p).

Burkholder ([4] Lemma 1) has observed that E is an MT-space if and only if it satisfies both (6.p) and (7.p). In this paper we shall consider spaces which satisfy one of these estimates: these spaces appear to behave rather differently.

3. Fundamental properties

First we observe that the properties of satisfying LERMT and UERMT are superproperties. Next we establish a duality result:

Theorem 1. *If a Banach space E satisfies UERMT then its dual satisfies LERMT. If E' satisfies UERMT then E satisfies LERMT.*

Proof: We shall prove the second statement: the proof of the first is similar. Suppose that E' satisfies UERMT, and that (7p) holds. By approximation, it is sufficient to consider the case where the σ-fields F_n are finite and $m = (m_1, \ldots, m_n)$ is an E-valued martingale of finite length. There exists an F_n-measurable E'-valued function ϕ such that

$$\|\phi\|_p = 1 \quad \text{and} \quad E(\phi(m_n)) = \|m_n\|_{p'}.$$

Let

$$\phi_j = E(\phi|F_j) \quad \text{and let } \delta_j = \phi_j - \phi_{j-1}.$$

Then

$$
\begin{aligned}
\|m_n\|_{p'} &= E(\phi(m_n)) \\
&= \sum_{j=1}^{n} E(\delta_j(d_j)) \\
&= \sum_{j=1}^{n} E(\varepsilon_j \delta_j)(\varepsilon_j d_j) \\
&= E((\varepsilon.\phi)(\varepsilon.m)) \\
&\leq \|\varepsilon.\phi\|_p \|\varepsilon.m\|_{p'} \leq K_p \|\varepsilon.m\|_{p'}.
\end{aligned}
$$

Thus E satisfies LERMT.

We now give a rather easy example to show that c_0 does not satisfy LERMT. Let N be a positive integer, and let $D = \prod_{n=1}^{N}\{-1,1\}$, considered as an abelian group with Haar measure.

Let $\Omega = D \times D'$, where D' is a copy of D.

If $\omega = (d, d') \in \Omega$, let $\pi_n(\omega) = d_n$, $\varepsilon_n(\omega) = d'$.

Let $F_0 = \{\Omega, \phi\}$, Let $F = \sigma(\pi_1, \ldots, \pi_n)$ for $1 \leq n \leq N$, and let $G = \sigma(\varepsilon_1, \ldots, \varepsilon_N)$. Notice that $(\varepsilon_1, \ldots, \varepsilon_N)$ is a finite Bernoulli sequence.

Now if $e \in D$ and $1 \leq n \leq N$ let

$$\phi_{(e,n)}(d, d') = \begin{cases} 1 & \text{if } e_j = d_j \text{ for } j \leq n, \\ -1 & \text{if } e_j = d_j \text{ for } j < n \text{ and } e_n \neq d_n \ . \\ 0 & \text{otherwise} \end{cases}$$

Note that $\phi_{(e,n)}$ is F_n-measurable and that

$$E(\phi_{(e,n)}|F_{n-1}) = 0.$$

Let $\ell_\infty = \ell_\infty(D)$, with the supremum norm.

If $1 \leq n \leq N$ and $\omega = (d, d') \in \Omega$, let

$$\delta_n(\omega) = \big(\phi_{(e,n)}(d)\big)_{e \in D} \ .$$

Then δ_n is an F_n-measurable $\ell_\infty(D)$-valued random variable, and $E(\delta_n|F_{n-1}) = 0$; thus $(\delta_n)_{1 \leq n \leq N}$ is an $\ell_\infty(D)$-valued martingale difference sequence. Let $(m_n)_{1 \leq n \leq N}$ be the corresponding martingale. If $\omega = (d, d')$ then

$$|(m_N(\omega))_e| \leq N$$
$$\text{and } (m_N(\omega))_d = N,$$

so that $\|m_N\|_p = N$ for $1 \leq p \leq \infty$.

On the other hand, if $\omega = (d, d')$,

$$((\varepsilon.m)_n(\omega))_e = \sum_{j=1}^{n} \varepsilon_j(d')\phi_{(e,j)}(d),$$

so that

$$((\varepsilon.m)_n(\omega))_e = 0 \text{ or } \sum_{j=1}^{n} \varepsilon_j(d') \text{ or } \sum_{j=1}^{k} \varepsilon_j(d') - \varepsilon_k(d')$$

for some $k \leq n$. Consequently

$$\|(\varepsilon.m)_n(\omega)\|_\infty \leq 2s_n^*(\omega)$$

where $s_n = \sum_{j=1}^{n} \varepsilon_j$. Thus

$$\|(\varepsilon.m)_N\|_p \leq 2\|s_N^*\|_p \leq K_p N^{1/2}.$$

Consequently c_0 does not satisfy LERMT.

This has important consequences.

Theorem 2.   *If E satisfies LERMT, then E is of cotype q, for some $q < \infty$. If E satisfies UERMT then E is super-reflexive.*

Proof: If E is not of cotype q, for some $q < \infty$, then c_0 is finitely represented in E [11], and so E does not satisfy LERMT.

It follows from Theorem 1 that ℓ_1 does not satisfy UERMT. Thus if E satisfies UERMT, ℓ_1 is not finitely represented in E, and so once again E must be of cotype q, for some $q < \infty$. We now use an argument due to Aldous [1]. Suppose that $\delta > 0$ and that $m = (m_1, \ldots, m_n)$ is an E-valued martingale taking values in the unit ball of E, and with the property that $\|d_j(\omega)\| \geq \delta$ for all $1 \leq j \leq n$ and all ω. Then

$$n^{1/q}\delta \leq (\sum_{j=1}^{n} \|d_j\|^q)^{1/q} \leq c_{p,q}(E(\|\sum_{j=1}^{n} \varepsilon_j d_j\|^p | F_n))^{1/p}.$$

where $c_{p,q}$ is a suitable cotype constant, so that

$$n^{1/q}\delta \leq c_{p,q} K_p \|m_n\|_p$$

and $n \leq (c_{p,q} K_p/\delta)^q$. This means that E is super-reflexive.

4. The equivalence of certain inequalities

In this section, we show that certain inequalities are equivalent to (6p) and to (7p). Many of the arguments are variations of standard ones, and we only sketch the details.

Theorem 1. *Suppose that E is a Banach space. The following statements (to hold for all E-valued martingales and independent Bernoulli sequences ε) are equivalent, for all $1 \leq p < \infty$.*

$(8.p^*)$. *There exists a constant c_p such that*

$$\lambda^p P((\|(\varepsilon.m)^*\| > \lambda) \leq c_p \|m^*\|_p^p.$$

$(9.p^*)$. *There exists a constant c_p such that*

$$\lambda^p P(\|(\varepsilon.m)_n\| > \lambda) \leq c_p \|m^*\|_p^p$$

for all n.

$(10.p^*)$. *There exists a constant c_p such that*

$$\lambda^p P(B^{(p)}(m) > \lambda) \leq c_p \|m^*\|_p^p.$$

$(11.p^*)$. *There exists a constant c_p such that*

$$\|(\varepsilon.m)^*\|_p \leq c_p \|m^*\|_p.$$

$(12.p^*)$. *There exists a constant c_p such that*

$$\|B^{(p)}(m)\|_p = \|\varepsilon.m\|_p \leq c_p \|m^*\|_p.$$

(13^*). *There exists a constant K such that if $B^{(1)}(m) > 1$ almost everywhere then $\|m^*\|_1 > K$.*

(14^*). *There exists a constant K such that if $(\varepsilon.m)^* > 1$ almost everywhere then $\|m^*\|_1 > K$.*

Statements (8.p), (9.p), (10.p), (13) and (14), where in each case m^* is replaced by m.

(15). *If $\|m\|_1 < \infty$, then $\sum_{n=1}^{\infty} \varepsilon_n d_n$ converges almost everywhere.*

(16). *E satisfies UERMT.*

Proof: By (1.p), (16) is equivalent to (11.p*) for some $1 < p < \infty$, and for $1 < p < \infty$ the starred and unstarred statements are equivalent. The following implications are obvious:

$(8.1) \Rightarrow (9.1) \Rightarrow (10.1) \Rightarrow (13) \Rightarrow (13^*)$

$(11.p^*) \Rightarrow (8.p^*) \Rightarrow (9.p^*) \Rightarrow (10.p^*)$

$(11.p^*) \Rightarrow (12.p^*)$

$(12.1^*) \Rightarrow (13^*).$

It is therefore sufficient to establish the implications

$(10.p^*) \Rightarrow (11.q^*)$ for any $1 \leq p, q < \infty$

$(13) \Rightarrow (14) \Rightarrow (8.1)$ and

$(13^*) \Rightarrow (14^*) \Rightarrow (8.1^*).$

$(12.1^*) \Rightarrow (15) \Rightarrow (14).$

The proof that (10.p*) implies (11.q*) is standard: use the argument of [3] pp. 1000–1001, together with Remark 1.2, replacing f by m, g by $\varepsilon.m$ and g^* by $B^{(p)}(m)$.

Suppose that (x_i) is a sequence in E such that $\sup_n \|\sum_{j=1}^{n} \varepsilon_j x_j\| > 1$ almost everywhere. Let

$$\tau = \inf\{n : \|\sum_{j=1}^{n} \varepsilon_j x_j\| > 1\}.$$

Then

$$E(\|\sum_{j=1}^{n} \varepsilon_j x_j\|) \geq E(\|\sum_{j=1}^{n \wedge \tau} \varepsilon_j x_j\|) \geq P(\tau \geq n)$$

for all n, so that $\sup_n E\|\sum_{j=1}^{n} \varepsilon_j x_j\| \geq 1$. Conditioning on F it follows that if $(\varepsilon.m)^* \geq 1$ almost everywhere then $B^{(1)}(m) \geq 1$ almost everywhere; thus (13) implies (14) and (13*) implies (14*).

The argument of Bollobas described in [3] pp. 999–1000 shows that (14) implies (8.1) and, with an obvious modification, (14*) implies (8.1*). If (12.1*) is satisfied, E satisfies UERMT, and is therefore super-reflexive. We can use the arguments of [3] p. 1001 to deduce that (12.1*) implies (15). Finally the proof that (15) implies (14) is the same as that given on page 999 of [3].

In the same way, we can establish the following theorem. This time, we omit the details.

Theorem 2. *Suppose that E is a Banach space. Then the following statements (to hold for all E-valued martingales m and independent Bernoulli sequences ε) are equivalent, for all $1 \leq p < \infty$.*
(15.p) There exists a constant c_p such that

$$\lambda^p P(m^* > \lambda) \leq c_p \|\varepsilon.m\|_p^p.$$

(16) There exists a constant K such that if $m^* > 1$ almost everywhere then $\|\varepsilon.m\|_1 \geq K$.
(17.p) There exists a constant c_p such that

$$\|m^*\|_p \leq c_p \|(\varepsilon.m)^*\|_p.$$

(18) E satisfies LERMT.

5. Spaces finitely represented in ℓ_1

We can now show that the class of spaces satisfying LERMT is much larger than the class of MT spaces.

Theorem 3. *If E is finitely represented in ℓ_1, E satisfies LERMT.*

Proof: Since LERMT is a super-property, it is sufficient to consider martingales taking values in ℓ_1. Suppose that $m = (m_n)$ is such a martingale. Then

$$m_n^* = \sup_{1 \leq j \leq n} \|m_j\| = \sup_{1 \leq j \leq n} \sum_t |m_j(t)|$$
$$\leq \sum_t \sup_{1 \leq j \leq n} |m_j(t)| = \sum_t (m_n(t))^*.$$

Thus

$$\|m_n^*\|_1 \leq \sum_t E(m_n(t)^*)$$

$$\leq \sqrt{10} \sum_t E(\sum_{j=1}^n d_j(t)^2)^{1/2}$$

$$\leq \sqrt{20} \sum_t E|\sum_{j=1}^n \varepsilon_j d_j(t)|$$

$$= \sqrt{20}\|\varepsilon.m\|_1,$$

using Burgess Davis' inequality (cf. [8] II.2.8) and Khintchin's inequality. The result now follows from Theorem 2.

Note that this example shows that it does not follow from the fact that E satisfies LERMT that E' satisfies UERMT.

6. An example

The fact that a space which is finitely represented in ℓ_1 satisfies LERMT shows that this class of spaces is more comprehensive than the class of UMD spaces. Bourgain [2] has given an example of a super-reflexive lattice which is not a UMD space: in fact, since Bourgain works with the square function in his calculations, it follows from the remarks in section 2 that Bourgain's example fails to satisfy UERMT. In this section, we shall give an example, based on that of Bourgain (but considerably more complicated) to show that not every super-reflexive lattice satisfies LERMT.

Theorem 4. *Suppose that $q > 4$. Then there exists a lattice X which satisfies an upper 2 and lower-q estimate, and which does not satisfy LERMT.*

Before we begin the proof, let us make a few remarks. Given $\varepsilon > 0$, we construct a finite-dimensional lattice X with upper-2 and lower-q constants equal to 1 on which there exists a martingale f for which

$$\|(\Sigma|d_j|^2)^{1/2}\|_{L_2(X)} \leq \varepsilon\|f\|_{L_2(X)};$$

the result then follows by taking an ℓ_2-direct sum.

The example is constructed in the same way as Bourgain's example. We consider a finite group G with Haar measure P, and a suitable subset T of G. We define a norm $\|\ \|_X$ on the real-valued functions on G by setting

$$\|\phi\|_X = \sup \left(\sum_i \left(\int_{A_i} \phi^2 dP \right)^{q/2} \right)^{1/q}$$

where the supremum is taken over all partitions $G = A_1 \cup \cdots \cup A_t$, where each A_s is contained in a translate of T. Then X has upper-2 and lower-q constants equal to 1, $\|f\|_X = \|f_g\|_X$ for all g in G and

$$\|\phi\|_X \leq \|x\|_2^{2/q} \left(\int_T \phi_g^2 dP \right)^{1/2 - 1/q}$$

(here ϕ_g is the g-translate of ϕ).

The group G will be a product $G = G_0 \times \cdots \times G_N$, equipped with its natural filtration $\Sigma_0, \Sigma_1, \ldots, \Sigma_N$, where Σ_N is the σ-field generated by sets of the form $\{g_0\} \times \cdots \times \{g_j\} \times \prod_{j < i \leq N} G_i$.

Suppose that $\phi \in X$. Let $\phi_j = E(\phi | \Sigma_j)$ and let $\delta_j = \phi_j - \phi_{j-1}$. If $g \in G$, let $\Phi(g) = \phi_g$. Then $\Phi \in L_2(G; X) = L_2(X)$; since $\|\Phi(g)\|_X = \|\phi\|_X$ for all $g \in G$, $\|\Phi\|_{L_2(X)} = \|\phi\|_X$.

Similarly, if $\Phi_j = E(\Phi | \Sigma_j)$ and $\Delta_j = \Phi_j - \Phi_{j-1}$, $\Delta_j(g) = (\delta_j)_g$, so that

$$\|(\Sigma \Delta_j^2)^{1/2}\|_{L_2(X)} = \|(\Sigma \delta_j^2)^{1/2}\|_X.$$

From all this it follows that it is sufficient to show that there are absolute constants A_1, A_2 and A_3 such that given a large integer R, we can construct a group $G = G_0 \times \cdots \times G_N$, a subset T of G and a function f on G such that

(6.1) $$\|f\|_2 \leq A_1 R$$

(6.2) $$\int_T f^2 dP \geq A_2 R$$

and

(6.3) $$\int_T (Sf)_g^2 dP \le A_3 \quad \text{for all } g \text{ in } G.$$

For then $\|f\|_X \ge (A_2 R)^{1/2}$, while $\|Sf\|_X \le A_4 R^{2/q}$, (where $A_4^q = A_1^2 A_3^{q-2}$).

We now turn to the construction. In what follows, $A_1, \ldots, A_{18}$ denote absolute constants. Suppose that R is a large positive integer. We define an increasing sequence $1 = k_1 < \cdots < k_R$ of integers such that if

(6.4) $$\beta_r = \frac{1}{3(k_r + 1)r} \quad \text{and} \quad \alpha_r = 1 - \beta_r$$

then

(6.5) $$\alpha_r^{k_r+1} \log\left(\frac{1}{\alpha_r}\right) < \frac{1}{R^{1/2}(k_r + 1)}$$

for $r = 2, \ldots, R$. Let $N = k_R!$

Let us set $\ell(0) = 0$, $\ell(j) = 1 + \frac{1}{2} + \cdots + \frac{1}{j}$. Let C be an integer such that

(6.6) $$C > R k_{R+1} \ell(N).$$

We set $G_0 = H_1 \times \cdots \times H_R$, where each H_r is a cyclic group of order $CR^2 + 1$, and for $1 \le j \le N$ we take G_j to be a cyclic group of order $(N!)^2$. Let $G = G_0 \times \cdots \times G_N$, and let P be Haar measure on G.

For each $1 \le r \le R$, let

$$K_r = \{g \in G : g_0 \in H_r, g_0 \ne e\}.$$

Thus the K_r are disjoint, and

(6.7) $$P(K_r) = CR^2/(CR^2 + 1)^R = \pi, \text{ say.}$$

Notice that if $1 \le r < s < R$ and if $g \in G$ then

$$(6.8) \qquad\qquad P(K_r \cap K_s g) = 0 \text{ or } \pi/CR^2.$$

For each $1 \le j \le N$, let B_j be a subgroup of G_j of order $|G_j|/j(\ell(N) - \ell(j-1))$, and let $A_j = G_j \backslash B_j$, so that

$$|A_j| = |G_j| \frac{\ell(N) - \ell(j)}{\ell(N) - \ell(j-1)}.$$

For $1 \le j \le N$ and $1 \le r \le R$, let

$$C_{j,r} = K_r \cap \{g \in G : g_i \in A_i \text{ for } 1 \le i \le j\}$$

and let

$$D_{j,r} = K_r \cap \{g \in G : g_i \in A_i \text{ for } 1 \le i < j, g_j \in B_j\}.$$

Note that

$$(6.9) \qquad P(C_{j,r}) = \frac{(\ell(N) - \ell(j))\pi}{\ell(N)}, \quad P(D_{j,r}) = \frac{\pi}{j\ell(N)}.$$

Next we define some normalising factors. We define w_r by

$$w_r^2 = r\ell(N) \left[\pi \sum_{j=1}^{N} \{j^{2k_r}(\ell(N) - \ell(j)) + j^{2k_r+1}(\ell(N) - \ell(j))^2\} \right]^{-1}.$$

We shall see the purpose of these factors later. For the moment, note that

$$\sum_{j=1}^{N} j^{2k_r}(\ell(N) - \ell(j)) \sim N^{2k_r+1} \int_0^1 x^{2k_r} \log(\frac{1}{x}) dx$$

$$\sim \frac{N^{2k_r+1}}{(2k_r+1)^2}$$

and

$$\sum_{j=1}^{N} j^{2k_r+1}(\ell(N)-\ell(j))^2 \sim N^{2k_r+2}\int_0^1 x^{2k_r+1}(\log(\frac{1}{x}))^2 dx$$

$$\sim \frac{2N^{2k_r+2}}{(2k_r+2)^3}.$$

Thus

(6.10) $$\frac{A_5 N^{2k_r+2}\pi}{(k_r+1)^3\ell(N)r} \le \frac{1}{w_r^2} \le \frac{A_6 N^{2k_r+2}\pi}{(k_r+1)^3\ell(N)r}.$$

We now define a martingale difference sequence $d_0, d_1, \ldots, d_N$ by setting $d_0 = 0$, and setting

$$d_j = \sum_{r=1}^{R} w_r \left(j^{k_r}\chi(C_{j,r}) - j^{k_r+1}(\ell(N)-\ell(j))\chi(D_{j,r}) \right).$$

Let $f_j = \sum_{i\le j} d_i$, so that $(f_i)_{i=0}^{N}$ is the corresponding martingale.
First notice that

$$E(f_N^2) = \sum_{j=1}^{N} E(d_j^2)$$

$$= \sum_{r=1}^{R} w_r^2 \sum_{j=1}^{N}\{j^{2k_r}P(C_{j,r}) + j^{2k_r+2}(\ell(N)-\ell(j))^2 P(D_{j,r})\}$$

$$= \sum_{r=1}^{R} \frac{w_r^2\pi}{\ell(N)} \sum_{j=1}^{N}\{j^{2k_r}(\ell(N)-\ell(j)) + j^{2k_r+1}(\ell(N)-\ell(j))^2\}$$

$$= \sum_{r=1}^{R} r = \frac{R(R+1)}{2}$$

so that (6.1) is established.

Now let $T_r = C_{\alpha_r N, r}$ (α_r is defined in (6.4)) and let $T = \cup_{r=1}^{R} T_r$.
Note that

(6.11) $$P(T_r) = \frac{\ell(N)-\ell(\alpha_r N)}{\ell(N)}\pi \le A_7\frac{\beta_r\pi}{\ell(N)}$$

and that on T_r

$$f_N \geq w_r \left[\sum_{j=1}^{\alpha_r N} j^{k_r} - N^{k_r+1}(\ell(N) - \ell(\alpha_r N)) \right]$$
$$\geq \frac{w_r N^{k_r+1}}{k_r + 1} \left[(1 - \frac{1}{3(k_r + 1)r})^{k_r+1} - 2\beta_r \right]$$
$$\geq \frac{A_7 w_r N^{k_r+1}}{(k_r + 1)}.$$

Thus

$$\int_T f_N^2 dP \geq \sum_{r=1}^{R} \frac{A_7^2 w_r^2 N^{2k_r+2} \pi}{3(k_r + 1)^3 r \ell(N)}$$
$$\geq A_2 R, \quad \text{by (6.10)}$$

and we have established (6.2).

It remains to establish (6.3), and to do this we need to consider several cases. First note that if $1 \leq j \leq N$, and k is a positive integer then

(6.12) $$j^k(\ell(N) - \ell(j)) \leq A_9 \frac{N^k}{k}.$$

Thus on $D_{j,r}$

$$(Sf)^2 = w_r^2 \left(\sum_{i=1}^{j-1} i^{2k_r} + j^{2k_r+2}(\ell(N) - \ell(j))^2 \right)$$
$$\leq A_{10} \frac{w_r^2 N^{2k_r+2}}{\pi(k_r + 1)^2}$$
$$\leq A_{11}(k_r + 1)\ell(N)r/\pi, \quad \text{by (6.10)},$$

and on $C_{N,r}$,

$$(Sf)^2 = w_r^2 \sum_{i=1}^{N} i^{2k_r} \leq A_{12} \frac{(k_r + 1)^2 \ell(N)r}{\pi N}.$$

Consequently

(6.13) $\|(Sf)^2\chi(K_r)\|_\infty^2 \le A_{13}(k_r + 1)\ell(N)r/\pi.$

We consider $g_0 = (h_1, \ldots, h_R)$.

Case 1. There are two or more indices r for which h_r is not the identity. For each r and s,

$$P(T_r \cap (K_s)_g) \le \pi/CR^2, \quad \text{by (6.8)}$$

so that

$$\int_T (Sf)_g^2 dP = \sum_{r=1}^R \sum_{s=1}^R \int_{T_r} ((Sf)^2\chi(K_s))_g dP$$
$$\le \frac{\pi\|S(f)\|_\infty^2}{C} \le A_3,$$

by (6.6) and (6.13).

Case 2. There is exactly one index t such that h_t is not the identity. In this case,

$$T_r \cap (K_s)_g = \phi,$$

unless $r = s = t$. Thus

$$\int_T (Sf)_g^2 dP = \int_{T_t} ((Sf)^2\chi(K_t))_g dP$$
$$\le \|Sf\chi(X_t)\|_\infty^2 P(T_t)$$
$$\le A_3$$

by (6.4), (6.11) and (6.13).

Case 3. g_0 is the identity in G_0. This is the most interesting and difficult case, since $(K_r)_g = K_r$, for each $1 \le r \le R$.

Now if $j \le \alpha_r N$,

$$P(T_r \cap (D_{j,r})_g) = \begin{cases} 0 & \text{if } g_j = e \\ \lambda_j P(T_r) & \text{if } g_j \ne e \end{cases},$$

where

$$\lambda_j = \left[\prod_{\substack{i<j \\ g_i \neq e}} \frac{\ell(N) - 2\ell(i)}{\ell(N) - \ell(i)} \right] \frac{\ell(N) - \ell(j)}{j}.$$

The actual value of λ_j is not important; what is important is that λ_j does not depend upon r. We set

$$X_r = \cup_{j \leq \alpha_{r-1} N} \ D_{j,r},$$

$$Y_r = \cup_{\alpha_{r-1} N < j \leq \alpha_r N} \ D_{j,r}$$

and

$$Z_r = (\cup_{\alpha_r N < j \leq N} D_{j,r}) \cup C_{N,r}.$$

Now

$$\int_T (Sf)_g^2 = \sum_{r=1}^R \int_{T_r} ((Sf)^2 \chi(K_r))_g dP$$

$$= \Sigma_1 + \Sigma_2 + \Sigma_3,$$

where

$$\Sigma_1 = \sum_{r=1}^R \int_{T_r \cap (X_r)_g} (Sf)_g^2 dP$$

$$\Sigma_2 = \sum_{r=1}^R \int_{T_r \cap (Y_r)_g} (Sf)_g^2 dP$$

and

$$\Sigma_3 = \sum_{r=1}^R \int_{T_r \cap (Z_r)_g} (Sf)_g^2 dP.$$

Now on X_r, Sf attains its maximum on $D_{\alpha_{r-1}N,r}$, and

$$(Sf)^2 \leq w_r^2 \left(N^{2k_r+1} \alpha_{r-1}^{2k_r+2} + N^{2k_r+2} \alpha_{r-1}^{2k_r+2} (\log \frac{1}{a_r - 1})^2 \right)$$

$$\leq \frac{A_{14} r \ell(N)(k_r + 1)}{R\pi}$$

by (6.5) and (6.19). Thus

$$\int_{T_r \cap (X_r)_g} (Sf)_g^2 dP \leq A_{15} \beta_r r(k_r + 1)/R$$

$$\leq A_{15}/R$$

by (6.11) and (6.4), and so $\Sigma_1 \leq A_{15}$. Next,

$$\int_{T_r \cap (Y_r)_g} (Sf)_g^2 dP \leq \|(Sf)^2 \chi(K_r)\|_\infty P(T_r \cap (Y_r)_g)$$

$$= (\sum_{\alpha_{r-1}N < j \leq \alpha_r N} \lambda_j) P(T_r) \|(Sf)^2 \chi(K_r)\|_\infty$$

$$\leq A_{16} (\sum_{\alpha_{r-1}N < j \leq \alpha_r N} \lambda_j)$$

by (6.11), (6.13) and (6.4), so that

$$\Sigma_2 \leq (\sum_{j=1}^{N} \lambda_j) A_{16} \leq A_{16}.$$

Finally, on Z_r

$$(Sf)^2 \leq w_r^2 (\sum_{i=1}^{N} i^{2k_r} + N^{2k_r+2}(\ell(N) - \ell(\alpha_r N))^2)$$

$$\leq A_{17} w_r^2 N^{2k_r+2} \beta_r^2$$

so that

$$\int_{T_r \cap (Z_r)_g} (Sf)_g^2 dP \leq A_{17} w_r^2 N^{2k_r+2} \beta_r^2 P(T_r)$$

$$\leq A_{18}/r^2 \quad \text{by (6.10), (6.4) and (6.11),}$$

so that $\Sigma_3 \leq 2A_{18}$.

This completes the proof.

References

[1] D.J. Aldous, *Unconditional bases and martingales in* $L_p(F)$, Math. Proc. Cambridge Phil. Soc. **85** (1979), 117–123.

[2] J. Bourgain, *Some remarks on Banach spaces in which martingale difference sequences are unconditional*, Arkiv Math. **21** (1983), 163–168.

[3] D.L. Burkholder, *A geometric characterization of Banach spaces in which martingale difference sequences are unconditional*, Annals of Probability **9** (1981), 997–1011.

[4] —————————, *A geometric condition that implies the existence of certain singular integrals of Banach-space valued functions*, Proceedings of a conference on Harmonic Analysis, Volume 1. (Belmont, Wadsworth International, 1983).

[5] —————————, *Boundary value problems and sharp inequalities for martingale transforms*, Annals of Probability **12** (1984), 647–702.

[6] —————————, *An elementary proof of an inequality of R.E.A.C. Paley*, Bull. London Math. Soc. **17** (1985), 474–478.

[7] J.L. Doob, *Stochastic Processes.* Wiley, New York, 1953.

[8] A. Garsia, *Martingale inequalities: seminar notes on recent progress.* (Reading, Benjamin, 1973).

[9] J.-P. Kahane, *Some random series of functions.* (2nd Edn., Cambridge, C.U.P., 1985).

[10] J.- Lindenstrauss and L. Tzafriri, *Classical Banach space II.* Springer-Verlag, Berlin, 1979.

[11] B. Maurey and G. Pisier, *Series de variables aleatoires vectorielles independantes et proprietes geometriques des espaces de Banach*, Studia Math. **58** (1976), 45–90.

[12] G. Pisier, *Martingales with values in uniformly convex spaces*, Israel J. Math. **20** (1975), 326–350.

On Random Multipliers
in the Central Limit Theorem
with p-stable Limit, $0 < p < 2$*

EVARIST GINÉ, MICHAEL B. MARCUS AND JOEL ZINN

1. Introduction.

Let B be a separable Banach space with dual space B^*. Let X be a B-valued random variable, $\{X_j\}_{j=1}^\infty$ independent identically distributed (i.i.d.) copies of X, $\{\varepsilon_j\}_{j=1}^\infty$ a Rademacher sequence independent of $\{X_j\}_{j=1}^\infty$, and $\{g_j\}$ an orthogaussian sequence independent of $\{X_j\}_{j=1}^\infty$. It is well known ([5]) that X satisfies the central limit theorem in B if and only if εX satisfies the central limit theorem in B, i.e. if and only if the sequence

$$(1.1) \qquad \{n^{-1/2} \sum_{j=1}^n \varepsilon_j X_j\}_{n=1}^\infty$$

converges in distribution, and also if and only if ([6]) gX satisfies the central limit theorem in B, i.e.

$$(1.2) \qquad \{n^{-1/2} \sum_{j=1}^n g_j X_j\}_{n=1}^\infty$$

converges in distribution. Actually the sequence $\{g_j\}$ in (1.2) can be replaced by any i.i.d. sequence $\{\xi_j\}_{j=1}^\infty$ independent of $\{X_j\}$ as long as ξ is symmetric, non-degenerate and $\xi \in L_{2,1}(\Omega, \mathcal{F}, P)$, where $L_{2,1}$ is the Lorentz space of real random variables ξ verifying

$$(1.3) \qquad \int_0^\infty [P\{|\xi| > t\}]^{1/2} dt < \infty$$

*This research has been partially supported by NSF grants.

(this is an observation of Pisier published in [6]). Moreover, Ledoux and Talagrand [11] proved that there are B-valued random variables X in some Banach spaces B that satisfy the CLT but such that the sequence $\{n^{-1/2} \sum_{j=1}^{n} \xi_j X_j\}$ does not converge in distribution unless $\xi \in L_{2,1}(\Omega, \mathcal{F}, P)$, thus answering the question of what "independent multipliers" ξ preserve the CLT property. In recent years Gaussian and Rademacher randomization have proved extremely useful in the study of the CLT and the LIL in Banach spaces (e.g. [8], [6], [12]) as well as in the study of processes that are not Gaussian (e.g. [17]). (For a negative result in the case of non-independent multipliers in the CLT see [1].)

The object of this note is to study multiplier questions for the p-stable limit case, $0 < p < 2$. Some of the results below were obtained several years ago (c. 1982). It seems appropriate to us, at this time, to collect, complete and publish them as a potentially useful complement to the $p = 2$ case.

The most immediate extension of the multiplier problem for Banach space valued random variables that are in the domain of attraction of a p-stable law is the following: If, for $0 < p < 2$,

$$\{n^{-1/2} \sum_{j=1}^{n} \varepsilon_j X_j\}_{n=1}^{\infty}$$

converges in distribution, then for which i.i.d. sequences $\{\xi_j\}_{j=1}^{\infty}$ independent of $\{X_j\}_{j=1}^{\infty}$ does

$$(1.4) \qquad \{n^{-1/p} \sum_{j=1}^{n} \xi_j X_j\}_{j=1}^{n}$$

converge in distribution? The answer to this question, in the case $1 < p < 2$, in complete analogy to the case $p = 2$, is that ξ must be in $L_{p,1}$ (i.e. satisfy (1.3) with 2 replaced by p). This result is given in Section 4. However, we must add that the proof is exactly the same as in the case

$p = 2$. We also show in Section 4 that for $0 < p < 1$ all that is required is that $E|\xi|^p < \infty$.

What is more interesting, and what marks the point of departure of this paper, is that there are variations of the multiplier problem in the p-stable case $0 < p < 2$ that do not exist (or are trivial) in the Gaussian case. The one that first attracted our attention is the case of p-stable multipliers. For $0 < p < 2$ we define the canonical symmetric p-stable random variable θ by

$$(1.5) \qquad E\, e^{i\lambda\theta} = e^{-c_p|\lambda|^p}, \quad -\infty < \lambda < \infty$$

where c_p is consistent with the condition

$$(1.6) \qquad \lim_{t\to\infty} t^p P[|\theta| > t] = 1,$$

and let $\{\theta_j\}_{j=1}^{\infty}$ be i.i.d. copies of θ. We pose the problem: Characterize the Banach space valued random variables X such that

$$(1.7) \qquad \{n^{-1/p} \sum_{j=1}^{n} \theta_j X_j\}_{j=1}^{n}$$

converges in distribution. The limit if it exists and is non-degenerate is necessarily a p-stable law of B. In Section 2 we show that (1.7) converges in distribution to the law of a p-stable random variable Z if and only if X is a p-stable generating variable for Z. That is, if and only if the function $f \to \exp\{-c_p E|f(X)|^p\}$, $f \in B^*$, is the characteristic functional of Z. (Note that since $E\|X\|^p < \infty$, given Z, it is always possible to find a p-stable generating variable for Z with norm 1. The relevance of this observation will become clear later in this Introduction.) It is surprising that this problem has such a complete and simple solution. In the case $p = 2$ this "problem" is a restatement of the general problem of characterizing those B-valued random variables which satisfy the CLT (i.e. for which (1.2) converges in distribution.) In (1.7) θ can be replaced by any symmetric variable ξ in the domain of normal attraction of θ with identical results.

We generalize our study of the convergence in distribution of (1.4) by posing the following question: Suppose that ξ is in the general domain of attraction of a p-stable random variable θ, i.e. that there exist real numbers $\{a_n\}_{n=1}^{\infty}$ such that

$$(1.7) \qquad a_n^{-1} \sum_{j=1}^{n} \xi_j \xrightarrow{d} \theta$$

(we use "$\xrightarrow{d}$" to denote convergence in distribution to the law of the random variable at the right of the arrow) where $\{\xi_j\}_{j=1}^{\infty}$ are i.i.d. copies of ξ. Then for which X does

$$(1.8) \qquad \{a_n^{-1} \sum_{j=1}^{n} \xi_j X_j\}_{j=1}^{\infty}$$

converge in distribution? It is clear that the limit law, if it exists, is a B-valued random variable Z for which X is a p-stable generating variable. However, even when B is the real line (1.8) does not hold for all ξ satisfying (1.7). The counterexample for $B = \mathbf{R}$ fails if we require $E|X|^{p+\varepsilon} < \infty$ for some $\varepsilon > 0$. (In $\mathbf{R}$, X is p-stable generating if and only if $E|X|^p < \infty$.) With this condition on $\|X\|$ we show in Theorem 1.2 that (1.8) holds for ξ satisfying (1.7) if and only if B is of stable type p.

When B is not of stable type p we consider the question of the weak convergence of (1.8) in several ways. In Corollary 3.4 we show that, with additional conditions on the distribution of ξ, (1.8) converges in distribution for all p-stable generating random variables X and, of course, the limit is the p-stable variable generated by X. The second way in which we consider this question is to require that (1.8) converges in distribution for all ξ satisfying (1.7) but only for a more restricted class of processes X. A result of this type is given in Corollary 3.6. We show that if X satisfies the standard CLT, i.e. if (1.1) converges in distribution, then (1.8) converges in distribution for all ξ satisfying (1.7). In fact, let $0 < p < 2$ denote the stable index of the limit in (1.7). Then (1.8) converges to the B valued p-stable random variable generated by X. We also have results of a mixed type which involve additional conditions on both ξ and X and which are related to the ξ-radial processes studied in [16].

We will often use the comparison principle ([9]) — or the weaker contraction principle [10]. What we will precisely refer to is the following: Let $\{X_{nj} : j = 1, \ldots, n, \ n \in \mathbb{N}\}$ be a triangular array of row-wise independent B-valued random variables and let $\{\xi_{nj}\}$, $\{\eta_{nj}\}$ be triangular arrays of row-wise independent symmetric real random variables, independent of $\{X_{nj}\}$, and such that for all $t > 0$

$$(1.9) \qquad P\{|\xi_{nj}| > t\} \leq P\{|\eta_{nj}| > t\}.$$

Then

$$\left\{ \sum_{j=1}^{n} \eta_{nj} X_{nj} \right\}_{n=1}^{\infty} \text{ tight } \Rightarrow \left\{ \sum_{j=1}^{n} \xi_{nj} X_{nj} \right\}_{n=1}^{\infty} \text{ tight.}$$

This result is well-known. It is essentially contained in [9], Lemma 5.10, but truncation and use of Hoffmann–Jørgensen's inequality on Minkowski functionals of the sums relative to compact sets is required to complete the proof. (Since the proof is standard we will omit it.)

On notation: $\{X_j\}_{j=1}^{\infty}$ denotes always an i.i.d. sequence of B-valued random variables with law equal to the law of X. $\{\varepsilon_j\}$, $\{\theta_j\}$, $\{\xi_j\}$, etc. will be i.i.d. sequences of real random variables respectively distributed as ε, θ, ξ, and always *independent* of $\{X_j\}$; we also assume ε, θ, ξ to be *independent* of X. Moreover, $P\{\varepsilon = 1\} = P\{\varepsilon = -1\} = 1/2$ and the Rademacher sequence $\{\varepsilon_j\}$ is always independent of any other sequences of random variables that appear in the same expressions as $\{\varepsilon_j\}$.

2. Multipliers in domains of normal attraction of stable laws.

In this section we consider the following question: For what B-valued random variables X does the sequence

(2.1)
$$\left\{ n^{-1/p} \sum_{j=1}^{n} \xi_j X_j \right\}_{n=1}^{\infty}$$

converge in distribution if ξ is a symmetric real random variable in the domain of normal attraction of a (non-degenerate) p-stable law? (ξ is in the domain of normal attraction of a p-stable law if $a_n = n^{-1/p}$ in (2.1) ([2]).)

2.1 Lemma. *If the sequence (2.1) converges in distribution, with ξ in the domain of normal attraction of a non-degenerate p-stable law, then $E\|X\|^p < \infty$.*

Proof: Convergence of (2.1) implies ([2]) that

$$\lim_{t \to \infty} t^p P\{\|\xi X\| > t\} = c < \infty.$$

But

$$t^p P\{\|\xi X\| > t\} = E[\|X\|^p (t^p \|X\|^{-p} P_\xi\{|\xi| > t\|X\|\})]$$
$$\geq c' \, E\|X\|^p I(\|X\| \leq t)$$

where $c' = \inf_{u \geq 1} u^p P\{|\xi| > u\} \neq 0$ and P_ξ denotes conditional probability given X. Hence $E\|X\|^p < \infty$. $\blacksquare$

2.2 Definition. *A B-valued random variable X is p-**stable** generating if there exists a B-valued symmetric p-stable random variable Z whose characteristic functional is*

(2.2)
$$E \, \exp\{if(Z)\} = \exp\{-c_p E|f(X)|^p\}, \quad f \in B^*,$$

with c_p as in (1.5). We then say that Z is generated by X.

2.3 Remarks: (1) Given a symmetric p-stable B-valued random variable Z, there exists a unique symmetric finite measure σ on the unit sphere S of B such that

$$E \ \exp\{if(Z)\} = \exp\{-c_p \int_S |f(s)|^p d\sigma(s)\} \quad \text{for all } f \in B^*$$

or equivalently,

$$\mathcal{L}(Z) = \text{Pois}(\sigma \times p dr/r^{1+p}))$$

([2]). σ is called the **spectral measure** of Z. If Z is generated by X, then obviously the spectral measure of Z is

$$(2.3) \qquad \sigma(A) = E[I(X/\|X\| \in A) + I(X/\|X\| \in -A)]\|X\|^p/2,$$

$A \subset S$ a Borel set. This shows, in particular, that if X is p-stable generating then $E\|X\|^p < \infty$.

(2) We recall that any finite symmetric measure σ on S is the spectral measure of a p-stable law for $p < 1$ ([2], Theorem 3.7.9). So, for $p < 1$, any B-valued random variable X satisfying $E\|X\|^p < \infty$ is p-stable generating.

The solution to the multiplier problem for multipliers in the domain of attraction of a p-stable law is as follows:

2.4 Theorem. *Let $0 < p < 2$. Let B be a separable Banach space, X a B-valued random variable and ξ any symmetric real valued random variable in the domain of normal attraction of the standard p-stable law. Then the sequence (2.1) converges in distribution to a (necessarily p-stable) random variable Z if and only if X is p-stable generating and Z is generated by X.*

Proof: Suppose the sequence (2.1) converges to a B-valued random variable Z. Then, for all $f \in B^*$, $\xi f(X)$ is in the domain of normal attraction of a real valued p-stable law. By the statement of Lemma 2.1 and its proof $t^p P\{|\xi f(X)| > t\} \to E|f(X)|^p$. Hence

$$\sum_{j=1}^{n} \xi_j f(X_j)/(n(E|f(X)|^p)^{1/p} \xrightarrow[d]{} \theta$$

and therefore $E \exp\{if(X)\} = \exp\{-c_p E|f(X)|^p\}$. This shows that Z is p-stable, and X is p-stable generating for Z.

Conversely, assume that X is p-stable generating for Z. Since for every $f \in B^*$, the sequence $\{n^{-1/p} \sum_{j=1}^n \xi_j f(X_j)\}$ converges weakly by the domains of attraction theorem in $\mathbf{R}$, it is enough to prove tightness of the sequence (2.1). For this purpose we can assume by the comparison principle that

$$P\{|\xi| > t\} = t^{-p}, \quad t \geq 1.$$

Then the measure

(2.4) $$\nu_n := n \, \mathcal{L}(\xi X/n^{1/p})$$

satisfies

$$(2.5) \quad \nu_n\{x: x/\|x\| \in A, \|x\| > t\}$$
$$= 2^{-1} E[I(X/\|X\| \in A) + I(X/\|X\| \in -A)] \left(n \wedge \frac{\|X\|^p}{t^p} \right)$$

for all $t > 0$ and for all Borel sets $A \subset S$, the unit ball of B. Therefore the measures ν_n increase to the measure $\nu := \sigma \times p dr/r^{1+p}$, with σ as in (2.3), on the rectangles $\{x: x/\|x\| \in A, s < \|x\| < t\}$ of $(\mathbf{R}\setminus\{0\}) \times S = B\setminus\{0\}$. Therefore, for all Borel subsets C of $B \setminus \{0\}$ we have $\nu_n(C) \uparrow \nu(C)$. Hence, by elementary facts on Lévy measures ([2], ex. 4, p. 123), it follows that

$$\text{Pois } \nu_n \underset{w}{\longrightarrow} \text{Pois } \nu$$

(here $\underset{w}{\longrightarrow}$ stands for weak convergence of probability measures, as usual). But since (2.1) is the sequence of row sums of a triangular array of row-wise i.i.d. symmetric random variables with accompanying laws ν_n, its tightness is equivalent to tightness of $\{\text{Pois } \nu_n\}$ by Theorem 4.3 in [2], p. 122 (the word "symmetric" is missing in the statement there). Hence, the sequence (2.1) is tight. ∎

An equally simple proof of the above theorem can be given using the Le Lage, Woodroofe and Zinn representation for stable laws ([13], [18]).

Second proof of Theorem 2.4: (sufficiency). Assume that X is p-stable generating for Z. Then so is εX (since $E|f(X)|^p = E|\varepsilon f(X)|^p$, $f \in B^*$) and by e.g. [18], Lemma 1.4, a version of Z is the (convergent) series

$$(2.6) \qquad \sum_{j=1}^{\infty} \varepsilon_j \Gamma_j^{-1/p} X_j$$

where $\Gamma_j = Y_1 + \cdots + Y_j$, for $\{Y_j\}$, i.i.d. with $P[Y_1 < \lambda] = e^{-\lambda}$ and independent of $\{\varepsilon_i\}$ and of $\{X_i\}$. Note that by the strong law of large numbers and the contraction principle the series in (2.6) converges a.s. if and only if

$$(2.6a) \qquad \sum_{j=1}^{\infty} \varepsilon_j X_j / j^{1/p}$$

converges a.s. This observation will be used frequently.

Let $\{(\Gamma_{jk}, \varepsilon_{jk}, X_{jk})_{j=1}^{\infty}\}_{n=1}^{\infty}$ be independent copies of the random vectors $(\Gamma_j, \varepsilon_j, X_j)_{j=1}^{\infty}$. Then,

$$(2.7)$$

$$Z =_d n^{-1/p} \sum_{k=1}^{n} \left(\sum_{j=1}^{\infty} \varepsilon_{jk} \Gamma_{jk}^{-1/p} X_{jk} \right)$$

$$= n^{-1/p} \sum_{k=1}^{n} \varepsilon_{lk} \Gamma_{lk}^{-1/p} X_{lk} + n^{-1/p} \sum_{k=1}^{n} \left(\sum_{j=2}^{\infty} \varepsilon_{jk} \Gamma_{jk}^{-1/p} X_{jk} \right)$$

$$:= S_n' + S_n''.$$

The sums S_n' and S_n'' are conditionally independent and symmetric given $\{\Gamma_{jk}\}$, so that given $\{\Gamma_{jk}\}$ both $S_n' + S_n''$ and $S_n' - S_n''$ have the same conditional distribution. Therefore, for every convex symmetric set $K \subset B$,

$$P\{S_n' \in K^c\} \le 2\, P\{Z \in K^c\}.$$

This shows that the sequence $\{S'_n\}$ is tight (and so is $\{S''_n\}$). Now, since $P(\Gamma_1^{-1/p} > n) \sim n^{-p}$ as $n \to \infty$, $\Gamma_1^{-1/p}$ belongs to the domain of normal attraction of the canonical p-stable law and therefore, by the one dimensional central limit theorem, $f(S'_n) \xrightarrow{d} f(Z)$ for all $f \in B^*$. Hence, $S'_n \xrightarrow{d} Z$ and, by the contraction principle, so does the sequence (2.1) for any symmetric real random variable ξ in the domain of normal attraction of a p-stable law.   ∎

Next we derive some consequences of Theorem 2.4. In [15] necessary and sufficient conditions are obtained for the convergence in distribution of the empirical characteristic function. Analogous results can be obtained for the p-stable weighted empirical characteristic function defined below by employing Theorem 2.4 and the results in [15]. let η be a real valued random variable and let $\{\eta_j\}$ be i.i.d. copies of η. Consider, for $1 < p < 2$,

$$(2.8) \qquad T_n = n^{-1/p} \sum_{j=1}^{n} \theta_j \exp\{i\eta_j t\}, \quad t \in [-1, 1],$$

where $\{\theta_j\}$ are i.i.d. canonical p-stable independent of $\{\eta_j\}$. By Theorem 2.4, T_n converges in distribution in $C[-1, 1]$ if and only if $e^{e\eta t}$ is a p-stable generating $C[-1, 1]$-valued random variable. It follows from [18], Theorem A, that this happens if and only if

$$(2.9) \qquad \int_0^\infty [\ln N([-1, 1], d_p, \varepsilon)]^{1/q} d\varepsilon < \infty$$

where, for all $s, t \in [-1, 1]$,

$$d_p(s, t) = (E|e^{i\eta s} - e^{i\eta t}|^p)^{1/p},$$

$1/p + 1/q = 1$ and $N([-1, 1], d_p, \varepsilon)$ denotes the minimum number of balls of radius ε in the pseudo-metric d_p needed to cover $[-1, 1]$. So we have, with the previous notation:

2.5 Corollary.   *For $1 < p < 2$, the p-stable weighted empirical characteristic function T_n converges in distribution in $C[-1,1]$ if and only if the entropy integral (2.9) is finite.*

Corollary 2.5, for $p = 2$, was proved in [14] on account of the convergence equivalence of the sequences (1.1) and (1.2). Incidentally, it follows from [17] and [18] that, for certain random variables η, T_n can converge in distribution for all $p \leq p_0$, $p_0 < 2$, but not converge for any $p' \in (p_0, 2]$. (Note that in each case $\{\theta_j\}_{j=1}^{\infty}$ is p-stable with the same value p that appears in norming $n^{-1/p}$.) Therefore, in studying the empirical characteristic function it is possible that the sequence $\{n^{-1/2} \sum_{j=1}^{n} \varepsilon_j \exp(i\eta_j t)\}$ does not converge in distribution in $C[-1,1]$, but that T_n does for some $p < 2$. This comment provides some justification for studying p-stable weighted empirical processes.

Now we point out another consequence of Theorem 2.4. Kronecker's lemma (e.g. [3]) gives the following implication:

$$(2.10) \quad \sum_{j=1}^{\infty} \varepsilon_j X_j / j^{1/p} \text{ converges a.s.} \Rightarrow \text{a.s.-} \lim_{n \to \infty} n^{-1/p} \sum_{j=1}^{n} \varepsilon_j X_j = 0.$$

Combining Theorem 2.4 with the representations (2.6) and (2.6a) for p-stable random variables we actually obtain an equivalence:

2.6 Corollary.   *For $0 < p < 2$ and for all B-valued random variables X we have the following equivalence:*

$$(2.11) \quad \sum_{j=1}^{\infty} \varepsilon_j X_j / j^{1/p} \text{ converges a.s.}$$

$$\Leftrightarrow \{n^{-1/p} \sum_{j=1}^{n} \theta_j X_j\}_{n=1}^{\infty} \text{ converges in distribution,}$$

where θ is the canonical p-stable random variable.

Statement (2.10) can be completed in another direction. In doing so, we relate the p-stable generating property with domains of attraction.

2.7 Theorem. *Let $0 < p < 2$. Then:*

$$\{\|n^{-1/p'} \sum_{j=1}^{n} \varepsilon_j X_j\|\}_{n=1}^{\infty} \text{ stochastically bounded for some } p' \in (p, 2]$$

$$\Rightarrow \sum_{j=1}^{\infty} \varepsilon_j X_j / j^{1/p} \text{ converges a.s.}$$

$$\Rightarrow \|n^{-1/p} \sum_{j=1}^{n} \varepsilon_j X_j\| \to 0 \text{ a.s.}$$

Proof: The second implication is given in (2.10). For the first, notice that the stochastic boundedness of the sequence $\{n^{-1/p'} \sum_{j=1}^{n} \varepsilon_j X_j\}$ implies

$$(2.12) \qquad \sup_{t>0} t^{p'} P\{\|X\| > t\} < \infty,$$

and in particular $E\|X\|^{p''} < \infty$ for all $p'' < p'$. Hence if $p < 1$ then, by Remark 2.3(2), X is p-stable generating or equivalently, $\sum_{j=1}^{\infty} \varepsilon_j X_j / j^{1/p}$ converges a.s. Therefore we restrict to $p \geq 1$. In this case (2.12) implies $E \max_{j \leq n} n^{-1/p''} \|X_j\| < \infty$ with $p < p'' < p'$, and we can apply Hoffmann-Jørgensen's inequality ([7]) to obtain

$$(2.13) \qquad \sup_{n} n^{-1/p''} E\| \sum_{j=1}^{n} \varepsilon_j X_j\| < \infty.$$

For m fixed and $n > m$, set $T_m = 0$, $T_k = \sum_{j=m+1}^{k} \varepsilon_j X_j$, $k \leq n$. Then, by summation by parts and the triangle inequality we obtain:

$$E\| \sum_{j=m+1}^{n} \varepsilon_j X_j / j^{1/p}\|$$

$$= E\| \sum_{j=m+1}^{n} (T_j - T_{j-1})/j^{1/p}\|$$

$$= E\| \sum_{j=m+1}^{n} T_j / j^{1/p} - \sum_{j=m}^{n-1} T_j /(j+1)^{1/p}\|$$

$$= E\|T_n/n^{1/p} + \sum_{j=m+1}^{n-1} T_j(j^{-1/p} - (j+1)^{-1/p})\|$$

$$\leq E\| \sum_{j=m+1}^{n} \varepsilon_j X_j\|/n^{1/p}$$

$$+ (\sup_{k>m} E\| \sum_{j=m+1}^{k} \varepsilon_j X_j\|/k^{1/p''})(\sum_{j=m+1}^{n-1} (j^{-1/p} - (j+1)^{-1/p})j^{1/p''})$$

which converges to zero as $n, m \to \infty$ by (2.13). ∎

The following is an immediate consequence of Theorem 2.7.

2.8 Corollary. *If X is a p-stable generating B-valued random variable, then it is also r-stable generating for all $0 < r < p$. If X satisfies the bounded CLT for some $p \in (0, 2]$ in the sense that there exists a slowly varying function L such that the sequence $\{\|L(n)n^{-1/p} \sum_{j=1}^{n} \varepsilon_j X_j\|\}_{n=1}^{\infty}$ is stochastically bounded, then X is p'-stable generating for all $p' \in /0, p)$ and, equivalently, the sequence $\{n^{-1/p} \sum_{j=1}^{n} \xi_j X_j\}_{j=1}^{n}$ converges in distribution for any real valued symmetric random variable ξ in the domain of normal attraction of a p-stable law.*

This follows from Theorem 2.7, the fact that for all $\varepsilon > 0$, $L(n) \leq n^{\varepsilon}$ for n large enough, and the comparison principle. Note that the first part of this corollary can be easily derived in other ways: It is obvious from the representations (2.6) and (2.6a); it is also obvious from domination of Lévy measures ([2], ex. 4, p. 123).

3. Multipliers not necessarily in the domain of normal attraction of a stable law

Suppose that ξ is a symmetric real valued random variable in the domain of attraction of a canonical p-stable random variable θ, i.e. that there exists a sequence $\{a_n\}$, $a_n \uparrow \infty$, such that

$$(3.1) \qquad a_n^{-1} \sum_{j=1}^{n} \xi_j \xrightarrow{d} \theta.$$

the question we address in this section is: For what B-valued random variables X do we also have

$$(3.2) \qquad a_n^{-1} \sum_{j=1}^{n} \xi_j X_j \xrightarrow[d]{} Z?$$

(Note that Z is necessarily p-stable.) We have a complete answer to this question for $E\|X\|^{p+\varepsilon} < \infty$, $\varepsilon > 0$. (This occurs if and only if B is of type p-stable.) We also have various partial answers as discussed in the introduction.

The situation here is more complicated than in the case of ξ in the domain of normal attraction of a p-stable law: for (3.2) to hold, $|\xi|$ and $\|\xi X\|$ must have comparable tails, but if the tail of $|\xi|$ is of smaller order than t^{-p} and $\|X\|$ is "barely" p-integrable then $\|\xi X\|$ may have a larger tail; this interference of X on the tail of $\|\xi X\|$ disappears if $E\|X\|^{p+\varepsilon} < \infty$ for some $\varepsilon > 0$. This occurs already in $\mathbf{R}$, as shown in the following elementary proposition.

Recall that (3.1) holds if and only if the function

$$(3.3) \qquad H(t) := P\{|\xi| > t\}$$

is regularly varying of index $-p$, and that

$$(3.4) \qquad n\, H(a_n) \simeq 1.$$

Recall also that, on $\mathbf{R}$, X is p-stable generating if and only if $E|X|^p < \infty$ (by Definition 2.2).

3.1 Proposition. *(1) If X and ξ are independent real valued random variables, $E|X|^{p+\varepsilon} < \infty$ for some $\varepsilon > 0$ and ξ is symmetric and in the domain of attraction of a p-stable law, then the sequence in (3.2) converges in distribution (with the normalizing constants a_n as in (3.4)).*
(2) If $E|X|^p < \infty$ and $P\{|X| > t\} = L(t)/t^p$ with $L(t)$ slowly varying

(and necessarily $L(t) \to 0$ as $t \to \infty$), then there exists ξ symmetric independent of X and in the domain of attraction of a p-stable law such that the sequence in (3.2) does not converge weakly. Actually, with a_n as given by (3.4),

$$n\, P\{|X| > a_n\} \to \infty.$$

Proof: Convergence of (3.2) is equivalent to convergence of $n\, P\{|\xi X| > t a_n\}$ for all $t > 0$, and we have

$$(3.5) \quad n\, P\{|\xi X| > t a_n\} = n\, E\, H(t a_n/|X|) \simeq E\, H(t a_n/|X|)/H(a_n).$$

By regular variation, $\lim_{n\to\infty} H(t a_n/|X|)/H(a_n) = |X|^p/t^p$, so that part (1) of the proposition will be proved if we show that we can interchange expectation and limit in the term at the right side of (3.5). Consider $L(t) := t^p H(t)$, $t \geq 0$. L is slowly varying and bounded on bounded intervals. Then, by the representation in e.g. [4], p. 282, for every $\varepsilon > 0$ there exists $u_0 < \infty$ such that if $u, v \geq u_0$, then $L(u)/L(v) \leq 2\exp\{\varepsilon|ln\ v - ln\ u|\}$, and $L(u) \geq u^{-\varepsilon}$. So, if $K_0 = \sup_{u \leq u_0} L(u)$, we have for $u \geq u_0$,

$$L(u/|X|)/L(u) \leq \begin{cases} 2(|X| \vee |X|^{-1})\varepsilon & \text{if } u/|X| > u_0 \\ K_0 u_0^\varepsilon |X|^\varepsilon & \text{if } u/|X| \leq u_0. \end{cases}$$

Hence for every $\varepsilon > 0$ there are $K < \infty$ and $n_0 < \infty$ (depending on t) such that

$$H(t a_n/|X|)/H(a_n) \leq K|t^{-1}X|^p(|t^{-1}X|^\varepsilon \vee |t^{-1}X|^{-\varepsilon}).$$

So, we can apply dominated convergence in (3.5) to obtain

$$\lim_{n\to\infty} n\, P\{|\xi X| > t a_n\} = E|X|^p/t^p.$$

Part (1) is proved. To prove part (2), by taking $L^*(t) = \inf_{1 \leq t' < t} L(t')$, $t \geq 1$, we can assume $L(t) \downarrow 0$ monotonically for $t \geq 1$. Take $a_n =$

$n^{1/p}U(n)$ and $U(n) = [L(n^{1/p-\epsilon})]^{2/p}$ with $\epsilon < 1/p$. Then, for n large enough,

$$(3.6) \quad n\, P\{|X| > a_n\} = L(n^{1/p}u(n))/L^2(n^{1/p-\epsilon}) \leq 1/L(n^{1/p-\epsilon}) \to \infty.$$

Given these a_n's it is easy to construct ξ symmetric, in the domain of attraction of a p-stable law and such that $P\{|\xi| > a_n\} = 1/n$. If (3.6) holds, then $\{a_n^{-1}\sum_{j=1}^n \epsilon_j X_j\}$ is not tight and, by the contraction principle, neither is $\{a_n^{-1}\sum_{j=1}^n \xi_j X_j\}$. ∎

There are other ways to ensure that dominated convergence applies in (3.5): for instance one may impose no conditions on X, but assume $H(ua_n) \leq u^{-p}H(a_n)$ for $u \geq 1$ (this is essentially what we do in Corollary 3.4).

A good reference for type p-stable Banach spaces, $0 < p < 2$, is [19]. Recall that in a type p-stable Banach space any finite measure σ on the unit sphere S is the spectral measure for a p-stable law in B ([2]); hence, any B-valued random variable X with $E\|X\|^p < \infty$ is p-stable generating. Also, all Banach spaces are of stable type p for $p < 1$; and for $p \geq 1$ B is type p-stable if and only if ℓ_p is not finitely representable in B ([19]). (In particular for $1 \leq p < 2$ $\ell_{p'}$ is type p-stable if and only if $p' > p$.)

3.2 Theorem. *If B is type p-stable, X is a B-valued random variable such that $E\|X\|^{p+\epsilon} < \infty$ for some $\epsilon > 0$, and ξ is a real symmetric random variable in the domain of attraction of a p-stable law, with $P\{|\xi| > t\} = H(t)$ and $n\, H(a_n) \simeq 1$, then*

$$(3.7) \qquad a_n^{-1}\sum_{j=1}^n \xi_j X_j \xrightarrow[d]{} Z$$

where Z is the p-stable variable generated by X. Conversely: If B is a Banach space such that (3.7) holds whenever (i) ξ is a symmetric real valued random variable in the domain of attraction of a p-stable law and

(ii) X is a p-stable generating variable with $\|X\| = 1$, then B is type p-stable.

Proof: By Theorem 3.7.10 and Remark 1 after 3.7.11 in [2], if B is of type p-stable then (3.7) holds if and only if for each $\delta > 0$ the sequence of measures

$$(3.8) \qquad \{n \, \mathcal{L}(\xi X/a_n)|_{\{x \in B \, : \, \|x\| > \delta\}}\}_{n=1}^{\infty}$$

converges weakly to $\nu|_{\{\|x\| > \delta\}}$, where ν is the Lévy measure of Z. By Remark 2.3 (see also the definition of ν in the proof of Theorem 2.4, after (2.5)), for all $A \subset S$ a Borel set, and $t > 0$,

$$(3.9) \quad \nu(A \times [t, \infty)) = 2^{-1} E[I(X/\|X\| \in A) + I(X/\|X\| \in -A)]\|X\|^p/t^p.$$

Note that if we let ν_n be the measures in (3.8), then
(3.10)
$$\nu_n(A \times [t, \infty)) = 2^{-1} E[I(X/\|X\| \in A) + I(X/\|X\| \in -A)]n \, H(ta_n/\|X\|).$$

Now, since $E\|X\|^{p+\varepsilon} < \infty$, we can pass to the limit under the "E" sign in (3.10), as in the proof of Proposition 3.1. But

$$n \, H(ta_n/\|X\|) = n \, H(a_n)[H(ta_n/\|X\|)/H(a_n)] \to \|X\|^p/t^p \quad \text{as } n \to \infty.$$

Therefore, the limit of (3.10) is (3.9), i.e. for all Borel sets $A \subset S$ and $t > 0$,

$$(3.11) \qquad \nu_n(A \times [t, \infty)) \to \nu(A \times [t, \infty)).$$

Since the measure σ on S, $\sigma(A) = 2^{-1} E[I(X/\|X\| \in A) + I(X/\|X\| \in -A)]\|X\|^p$, is finite, hence tight, it follows that for every $\varepsilon > 0$ there exists a compact subset K of S and a positive number $t > 0$ such that $\nu(K^c \times [t, \infty)) < \varepsilon$, and therefore by (3.11), $\limsup_n \nu_n(K \times [\delta, t])^c < \varepsilon$, showing that $\{\nu_n|_{\{\|x\|>\delta\}}\}$ is a tight sequence. Now, weak convergence of $\{\nu_n|_{\{\|x\|>\delta\}}\}$ to $\nu|_{\{\|x\|>\delta\}}$ follows from (3.11). Hence (3.7) holds.

For the converse, we only need to consider $p \geq 1$. By the finite representability property, it is enough to show that for each $p \in [1,2)$ there is a p-stable generating random variable X on ℓ_p and a real valued symmetric random variable ξ in the domain of attraction of a p-stable law such that (3.7) does not hold. If $\{e_j\}$ is the canonical basis for ℓ_p, take $X = e_j$ with probability $p_j = c/j(ln(e \vee j))^3$, $j \in \mathbf{N}$, with $\sum p_j = 1$, and take ξ symmetric such that $P\{|\xi| > t\} = 1/t^p(ln(e \vee t))^4$, $t \geq 1$. Then X generates the p-stable random vector

$$Z = \sum_{j=1}^{\infty} p_j^{1/p}\theta_j e_j$$

(Z converges since $\|Z\|^p = \sum_{j=1}^{\infty} p_j|\theta_j|^p$ is a convergent series by the three series theorem.) Let $X_{(\alpha)}$ denote the α-th coordinate of X. Define

$$\delta_{n,\alpha} = \inf[t : n\, P\{|\xi|X_{(\alpha)}I(|\xi|X_{(\alpha)} \leq a_n) > t\} \leq 1/8.3^p].$$

A necessary condition for (3.3) to hold is [5], Theorem 6.9

$$(3.12) \qquad \lim_{m \to \infty} \limsup_{n \to \infty} a_n^{-p} \sum_{\alpha=m}^{\infty} \delta_{n,\alpha} = 0.$$

a_n satisfies the equation $n^{-1} = P\{|\xi| > a_n\} = 1/a_n^p(ln\, a_n)^4$, so that $a_n \simeq n^{1/p}/(ln\, n)^{4/p}$. As for $\delta_{n,\alpha}$, note

$$n\, P\{|\xi|X_{(\alpha)}I(|\xi|X_{(\alpha)} \leq a_n) > t\} \simeq np_\alpha\left[\frac{1}{t^p(ln(e \vee t))^4} - \frac{1}{n}\right], \quad t \geq 1,$$

so that we need to solve for t in $np_\alpha/t^p(ln\, t)^4 \simeq 1/8.3^p + p_\alpha$. At least for α large enough independent of n, we obtain

$$\delta_{n,\alpha} > \frac{6(np_\alpha)^{1/p}}{(ln\, 6^p np_\alpha)^{4/p}}I(np_\alpha > 1).$$

So, for large n there are non-negative constants c' and c'' such that

$$a_n^{-p} \sum_{\alpha=m}^{\infty} \delta_{n,\alpha}^p \geq c' \frac{(\ln n)^4}{n} \, \text{Card}\{\alpha: 1 < np_\alpha \leq 2\}$$

$$\geq c'' \ln n \to \infty \quad \text{as } n \to \infty.$$

Hence, (3.12) does not hold for this choice of X and ξ, and consequently neither does (3.3). ∎

The following theorem and its corollaries hold with no restrictions on the banach space B: the restrictions are placed instead on X and ξ. The main ingredients in the proof will be the contraction principle and an argument using order statistics ([**13**], p. 625, and [**18**], p. 256). In what follows, given $H: \mathbf{R}^+ \to [0,1]$, we let $H^{-1}(u) = \inf\{t: H(t) \leq u\}$, $u \in [0,1]$, $\inf(\emptyset) = +\infty$.

3.3 Theorem. *Let $0 < p < 2$. Let ξ be a symmetric real random variable with tail function $P\{|\xi| > t\} := H(t)$ regularly varying of index $-p$. Let $a_n = H^{-1}(1/n)$, $n \in \mathbf{N}$. Let $\lambda(t)$, $t > 0$, be a positive function such that*

$$(3.13) \qquad H^{-1}(st) \leq \lambda(s)H^{-1}(t), \quad st \leq 1, \quad s \geq 1.$$

Let X be a p-stable generating B-valued random variable such that the series

$$(3.14) \qquad \sum_{j=1}^{\infty} \varepsilon_j \lambda(\Gamma_j) X_j$$

converges a.s., where Γ_j is defined in (2.6). Then (3.1) holds, and the limit Z of $\{a_n^{-1} \sum_{j=1}^{n} \xi_j X_j\}_{n=1}^{\infty}$ is the p-stable B-valued random variable generated by X.

Proof: By previous arguments, it is enough to show that the sequence in (3.1) is tight. Since ξ has the same law as $H^{-1}(U)$, U uniform on

[0, 1], following the arguments in page 625 of [13] and 256 of [18], for all n and any $m < n$ we have

$$a_n^{-1} \sum_{j=1}^{n} \xi_j X_j =_d a_n^{-1} \sum_{j=1}^{n} \varepsilon_j H^{-1}(\Gamma_j/\Gamma_{n+1}) X_j$$

$$= a_n^{-1} \sum_{j=1}^{n} \varepsilon_j H^{-1}(\Gamma_j/\Gamma_{n+1}) I_{[\Gamma_j>1]} X_j$$

$$+ a_n^{-1} \sum_{j=1}^{m} \varepsilon_j H^{-1}(\Gamma_j/\Gamma_{n+1}) I_{[\Gamma_j\leq1]} X_j$$

$$+ a_n^{-1} \sum_{j=m+1}^{n} \varepsilon_j H^{-1}(\Gamma_j/\Gamma_{n+1}) I_{[\Gamma_j\leq1]} X_j := I + II + III.$$

Since $P\{III \neq 0\} = P\{\Gamma_m \leq 1\} \to 0$ as $m \to \infty$ independently of n, tightness of the sequence in (3.1) will follow from tightness of the sequences I and II for all m. Note that we can define $\{a_n\}$ by $a_n = H^{-1}(\frac{1}{n})$. In proving the tightness of I we use (3.13) with $s = \Gamma_j > 1$ and $st = \Gamma_j/\Gamma_{n+1}$. We have

$$a_n^{-1} H^{-1}(\Gamma_j/\Gamma_{n+1}) \leq \lambda(\Gamma_j) a_n^{-1} H^{-1}(1/\Gamma_{n+1})$$

$$= \lambda(\Gamma_j) H^{-1}(1/\Gamma_{n+1})/H^{-1}(1/n).$$

By the law of large numbers $P\{1/2 \leq \inf_{n\geq n_0} \Gamma_{n+1}/n \leq \sup_{n\geq n_0} \Gamma_{n+1}/n \leq 3/2\} \to 1$ as $n_0 \to \infty$. Call these sets B_{n_0}. By regular variation and monotonicity there exists $C < \infty$ such that in B_{n_0}, $H^{-1}(1/\Gamma_{n+1})/H^{-1}(1/n) \leq C < \infty$ for all n sufficiently large. (Note that H^{-1} is regularly varying at zero with exponent $-1/p$: see [20], p. 24.) Hence, for every $\varepsilon > 0$ there is $C < \infty$ and $n_0 > 0$ such that

$$a_n^{-1} H^{-1}(\Gamma_j/\Gamma_{n+1}) I(\Gamma_j > 1) \leq C\lambda(\Gamma_j), \quad n > n_0,$$

on B_{n_0}, and $P B_{n_0} \geq 1-\varepsilon$. Thus, the tightness of the sequence I follows from the convergence of the series

$$\sum_{j=1}^{\infty} \varepsilon_j \lambda(\Gamma_j) X_j$$

and the contraction principle applied conditionally on B_{n_0}.

In proving the tightness of II we cannot use (3.13) since $\Gamma_j \leq 1$. (The significance of the condition $s \geq 1$ can be seen in the example following Corollary 3.3.) Instead we write

$$a_n^{-1} H^{-1}(\Gamma_j/\Gamma_{n+1}) = a_n^{-1} H^{-1}(1/n) \frac{H^{-1}(\Gamma_j/\Gamma_{n+1})}{H^{-1}(\Gamma_j/n)} \frac{H^{-1}(\Gamma_j/n)}{H^{-1}(1/n)}$$

and note that $a_n^{-1} H^{-1}(1/n) = 1$, that for every $\delta > 0$, on the set $A_\delta = \{\omega : \Gamma_1(\omega) > \delta\}$, $\frac{H^{-1}(\Gamma_j/n)}{H^{-1}(1/n)} I(\Gamma_j \leq 1) \to \Gamma_j^{-1/p} I(\Gamma_j \leq 1)$ as $n \to \infty$ uniformly in j and that for every $\delta > 0$ and $n_0 \in \mathbf{N}$, on $B_{\delta,n_0} = A_\delta \cap B_{n_0}$, $\frac{H^{-1}(\Gamma_j/\Gamma_{n+1})}{H^{-1}(\Gamma_j/n)} I(\Gamma_j \leq 1) \to 1$ as $n \to \infty$ uniformly in j. In conclusion, for all $\delta > 0$ and $n_0 \in \mathbf{N}$, there are a constant $C > 0$ and a natural number n_1 such that

$$a_n^{-1} H^{-1}(\Gamma_j/\Gamma_{n+1}) I(\Gamma_j \leq 1) \leq C \Gamma_j^{-1/p}$$

for all $n \geq n_1$ and all $\omega \in B_{\delta,n_0}$. By the law of large numbers for every $\varepsilon > 0$ we can find δ and n_0 such that $P B_{\delta,n_0} > 1 - \varepsilon$. Now, tightness of I follows from the fact that the series $\sum_{j=1}^{\infty} \varepsilon_j \Gamma_j^{-1/p} X_j$ converges a.s. (since X is a p-stable generating random variable) and by the contraction principle applied conditionally on B_{δ,n_0}. ∎

In applications of Theorem 3.3 it is more convenient to be able to verify conditions on H as opposed to H^{-1}. Suppose, for example, that

(3.13a) $$H(xy) \geq \alpha(x) H(y) \quad y > 0, x \geq 1$$

for some function $\alpha(x)$. If g is $N(0,1)$ independent of ξ then for $\bar{H}(t) := P\{|g\xi| > t\}$ we still have

$$\bar{H}(xy) \geq \alpha(x) \bar{H}(y)$$

as is easy to verify. But $\bar{H}$ is strictly increasing and continuous. So, taking $\alpha(x) \geq 1/s$ and $y = \bar{H}^{-1}(st)$ we obtain

$$\bar{H}^{-1}(st) \leq \lambda(x) \bar{H}^{-1}(t)$$

with

$$(3.15) \qquad \lambda(s) = \inf\{z : \frac{1}{\alpha(\frac{1}{z})} \le s\}.$$

The conclusion of Theorem 3.3 is then that $\{a_n^{-1} \sum_{j=1}^n \varepsilon_j |g_j| \xi_j X_j\}$ converges. Hence, by contraction, so does $\{a_n^{-1} \sum_{j=1}^n \varepsilon_j \xi_j X_j\}$. Note that $\lambda(s) \le 1/\alpha^{-1}(1/s)$ ($\lambda(s)$ coincides with $1/\alpha^{-1}(1/s)$ at continuity points, but $\lambda(s)$ is right continuous whereas $1/\alpha^{-1}(1/s)$ is left continuous).

In the proof of Theorem 3.2 we constructed examples of ξ in the domain of attraction of a p-stable random variable and p-stable generating X such that (3.1) does not hold for ξX if B is not type p-stable. The tail of ξ had the form $t^{-p} L(t)$ where $L(t) \downarrow 0$ as $t \uparrow \infty$. Corollary 3.4 below shows that the hypothesis that B is type p-stable is not needed if L is non-decreasing. The next two corollaries give other partial results if no assumption is imposed on B.

Let $\lambda(t) = t^{-1/p}$. Then inequality (3.13) for $\bar{H}$ holds if $H(xy) \ge x^{-p} H(y)$ for all $y > 0$ and $x \ge 1$. Let $L(x) = x^p H(x)$, $x > 0$. Then this inequality becomes $L(xy) \ge L(y)$ for all $y > 0$ and $x \ge 1$, i.e. L is non-decreasing. (This shows why s is taken to be greater than or equal to 1 in (3.13).) Then Theorem 3.3 gives:

3.4 Corollary. *Let $0 < p < 2$. If X is a B-valued p-stable generating random variable and if ξ is a symmetric real valued random variable such that the function*

$$L(t) := t^p P\{|\xi| > t\}, \quad t > 0$$

is slowly varying and non-decreasing, and if $\{a_n\}$ satisfies $P\{|\xi| > a_n\} \simeq 1/n$, then

$$a_n^{-1} \sum_{j=1}^n \xi_j X_j \xrightarrow[d]{} Z$$

where Z is the B-valued p-stable random variable generated by X.

3.5 Remark: Actually Corollary 3.4 has an easier proof: By standard arguments it suffices to show the tightness of $\{a_n^{-1} \sum_{j=1}^n \xi_j I(|\xi_j| \le$

$ca_n)X_j\}$ for all $c > 1$. But it turns out that if $P\{|\theta| > t\} = 1/t^p$, the condition on L implies that the tails of a constant multiple of $\theta/n^{1/p}$ dominate those of $\xi I(|\xi| \leq ca_n)/a_n$ and the result follows from the contraction principle and Theorem 2.4.

If we take $\lambda(t) = t^{-1/p'}$ for some $p' \in (p, 2)$, (3.13a) becomes $H(xy) \geq x^{-p'}H(y)$ for all $y > 0$ and $x \geq 1$, or $L(xy) \geq x^{-(p'-p)}L(y)$. Obviously, for every slowly varying L this inequality holds for all $x \geq x_0$, for some $x_0 < \infty$. But a modification of H on a finite interval is not relevant regarding convergence in (3.1). We thus have:

3.6 Corollary. *Let $0 < p < p' < 2$. If X is p'-stable generating then (3.1) holds for any real valued symmetric random variable ξ in the domain of attraction of a p-stable random variable (and with a_n such that $P\{|\xi| > a_n\} \simeq 1/n$). In particular by Corollary 2.8 this conclusion holds if X satisfies the bounded CLT for some $p' \in (p, 2]$.*

Taking

$$(3.16) \qquad\qquad \alpha(x) = H(x)$$

condition (3.13a) becomes

$$(3.17) \qquad\qquad H(xy) \geq H(x)H(y), \quad x, y \geq 1.$$

These are the kind of regularly varying tails for ξ that obstruct convergence for X p-stable generating (see both Proposition 3.1(2) and the second part of the proof of Theorem 3.2). A typical example of a regularly varying function H satisfying (3.17) is

$$(3.18) \qquad H(x) = x^{-p}(\ln(e \vee x))^{-\beta}, \quad \beta \geq 1, \ x \geq 1.$$

In this case $H^{-1}(s) \simeq [s^{1/p}(\log 1/s)^{\beta/p}]^{-1}$, $s \leq 1$, and

$$\lambda(j) = j^{-1/p}(\log j)^{\beta/p}.$$

One can see that the tail distribution in (3.18) does not satisfy the requirement of Corollary 3.4. Another way of comparing Corollaries 3.4 and 3.7 is to note that the sequence $\{a_n\}$ in Corollary 3.4 are greater or equal to $\{n^{1/p}\}$ whereas those in (3.21) below are less than or equal to $\{n^{1/p}\}$.

Corollary 3.7. *Let $0 < p < 2$ and let ξ be a symmetric real valued random variable in the domain of attraction of a p-stable law with tail probability function $H(t) = P\{|\xi| > t\}$ which satisfies*

$$(3.19) \qquad H(xy) \geq H(x)H(y), \quad x, y \geq 1.$$

Let $a_n = H^{-1}(1/n)$, $n \in \mathbf{N}$, and suppose that for X a B-valued random variable

$$(3.20) \qquad \sum_{j=1}^{\infty} \varepsilon_j X_j / H^{-1}(1/\Gamma_j)$$

converges a.s., where $\{X_j\}_{j=1}^{\infty}$ are i.i.d. copies of X. Then

$$(3.21) \qquad a_n^{-1} \sum_{j=1}^{n} \xi_j X_j \xrightarrow{d} Z$$

where Z is the p-stable random variable generated by X.

Proof: This is an immediate consequence of Theorem 3.3 with λ given by (3.15) and (3.16). (Note that $\lambda(j) \geq j^{1/p}$ and therefore that (3.13) implies that X is a p-stable generating random variable.) ∎

Actually Corollary 3.7 can also be reinterpreted as a result on generating random variables but this time on η-radial generating variables. Let η be a symmetric infinitely divisible random variable with
(3.22)

$$E\,e^{it\eta} = \exp\{-\int_0^{\infty} (\cos st - 1)d\tau[s, \infty)\} := e^{-\Phi(|t|)}, \quad -\infty < t < \infty,$$

where τ is the Lévy measure of η. Let $\|X\| = 1$. If

(3.23) $E\, e^{if(Z)} = \exp\{-E\Phi(|f(x)|)\}, \quad f \in B^*,$

defines a B-valued random variable Z then we say that X is a η-radial generating process for Z. (In [16] these processes are called ξ-radial processes. We change the notation here because ξ is used to denote random variables that, except in the stable case, are not the same as η.) The next theorem is a restatement of Corollary 3.7, for $\|X\| = 1$, in the terminology of η-radial processes.

3.8 Corollary. *Let ξ be a real valued symmetric random variable that satisfies $a_n^{-1} \sum_{j=1}^n \xi_j \xrightarrow[d]{} \theta$ for θ p-stable $0 < p < 2$. Let X be a η-radial generating process, i.e. (3.22) and (3.23) are satisfied, and let τ be the associated Lévy measure. Assume that $H(x)$ is continuous and strictly decreasing where*

(3.24) $H(x) = P(|\xi| > x) \geq 1/\tau[x^{-1}, \infty), \quad x \geq x_0$

for some x_0 sufficiently large and that

(3.25) $H(xy) \geq H(x)H(y), \quad x, y \geq 1.$

Then

(3.26) $$a_n^{-1} \sum_{j=1}^n \xi_j X_j \xrightarrow[d]{} Z$$

where Z is the B-valued p-stable random variable generated by X and $\{X_j\}_{j=1}^\infty$ are i.i.d. copies of X.

Proof: As we noted above, (3.25) implies

$$H^{-1}(st) \leq H^{-1}(t)/H^{-1}(1/s) \quad st \leq 1, s \geq 1.$$

But (3.24) implies

$$1/H^{-1}(1/\Gamma_j) \le \tau^{-1}(\Gamma_j)$$

(where here we define $\tau^{-1}(x) = \sup[u : \tau[u, \infty) > x\}]$) and the fact that X is a η-radial generating process implies, by Lemma 3.2 [16], that the series

$$\sum_{j=1}^{\infty} \epsilon_j \tau^{-1}(\Gamma_j) X_j$$

converges a.s. Thus we can use Theorem 3.3 with $\lambda(s) = 1/H^{-1}(1/s)$.
∎

Note that by (3.25) and (3.24), $\tau^{-1}(x) \ge x^{-p}$. Therefore the η-ardial generating processes X described in Corollary 3.8 are also p-stable generating processes. Thus we can use Corollary 3.4 to find other random variables ξ for which the limit in (3.26) is valid.

Finally let us compare Corollary 3.8 with Corollary 3.6. Corollary 3.6 gives us that (3.26) holds for all $\{\xi_i\}$ such that $a_n^{-1} \sum_{j=1}^{n} \xi_j \xrightarrow{d} \theta$ for θ p-stable if X is a p-stable generating variable for $p' > p$, i.e. if $\sum_{j=1}^{\infty} \epsilon_j X_j / j^{1/p'}$ converges. However the η-radial generating variables considered in Corollary 3.8 only require (besides (3.25)) that $\sum_{j=1}^{\infty} \epsilon_j X_j / j^{1/p} L(j)$ converge, where $L(j)$ is slowly varying at infinity. ((3.25) implies that $L(j) \le 1$.) So, the class of η-radial generating variables in Corollary 3.8 is larger than the class of p'-stable generating variables. On the other hand the class of multipliers ξ is larger in Corollary 3.6 than in Corollaries 3.7 and 3.8.

4. Multipliers that preserve domains of normal attraction.

Let us write $X \in CLT_p$ if X is in the domain of normal attraction of a p-stable law, $0 < p < 2$, that is, if the sequence

$$\{n^{-1/p} \sum_{j=1}^{n} (X_j - E X_j I(\|X_j\| \le n^{1/p}))\}_{n=1}^{\infty}$$

converges in distribution. By Lemma 2.6 in [5], $X \in CLT_p$ if and only if the sequence

$$\{n^{-1/p} \sum_{j=1}^{n} \varepsilon_j X_j\}_{n=1}^{\infty}$$

converges in distribution. We are interested in determining for what ξ do we have $\xi X \in CLT_p$ whenever $X \in CLT_p$. The answer is different for $p < 1$ and for $1 < p \leq 2$. The case $p > 1$ is very similar to the case $p = 2$ already described in the introduction, both in results and proofs.

We begin with the case $p < 1$.

4.1 Theorem. *Let $0 < p < 1$. then, $\xi X \in CLT_p$ whenever $X \in CLT_p$ if and only if $E|\xi|^p < \infty$. The same is true for $1 \leq p < 2$ if B type p-stable.*

Proof: The condition $E|\xi|^p < \infty$ is necessary for $\xi X \in CLT_p$, as in the proof of Lemma 2.1. For $p < 1$, the condition

$$(4.1) \qquad n\, P\{X/\|X\| \in A, \quad |\xi|\|X\| > n^{1/p}t\} \longrightarrow \sigma(A)/t^p$$

for some measure σ on the unit sphere S of B, and for all Borel subsets A of S and $t > 0$, is (necessary and) sufficient for ξX to belong to CLT_p (e.g. [2], Theorem 3.7.11, since all Banach spaces are of type p-stable, $p < 1$). But if $E|\xi|^p < \infty$ and if for all $A \subset S$ Borel and $t > 0$,

$$(4.2) \qquad n\, P\{X/\|X\| \in A, \|X\| > n^{1/p}t\} \longrightarrow \tau(A)/t^p,$$

then the limit of the left hand side of (4.1) is $\tau(A)\, E|\xi|^p/t^p$ by Fubini's theorem and dominated convergence. The same proof applies for $1 \leq p < 2$ if B is type p-stable. ∎

4.2 Remark: We do not have a definitive result for $p = 1$. It is relatively easy to prove that $X \in CLT_1$ and $E|\xi|^{1+\delta} < \infty$ for some $\delta > 0$ implies $\xi X \in CLT_1$.

4.3 Theorem. *Let $1 < p < 2$. Let ξ be a real valued random variable and let X be a Banach valued random variable (independent of ξ). If $X \in CLT_p$ and if $\xi \in L_{p,1}$, i.e. if*

$$(4.7) \qquad \int_0^\infty [P\{|\xi| > t\}]^{1/p} dt < \infty,$$

then $\xi X \in CLT_p$. Moreover, there exist B-valued random variables X in some Banach spaces B such that $X \in CLT_p$ but $\xi X \notin CLT_p$ unless $\xi \in L_{p,1}$.

We will only sketch the proof of this theorem because it is very similar to the case $p = 2$. For the first part, the obvious changes in pp. 938–939 of [6] give the inequality

$$E\|\|\sum_{j=1}^n \varepsilon_j \xi_j X_j / n^{1/p}\|\| \leq (\int_0^\infty [P\{|\xi| > t\}]^{1/p} dt) \max_{k \leq n} E\|\|\sum_{j=1}^k \varepsilon_j X_j / k^{1/p}\|\|$$

for any pseudonorm $\|\|\cdot\|\|$. We can take for instance the pseudonorms $\|\|\cdot\|\|$ to be distances to finite dimensional subspaces and obtain flat concentration of $\{\sum_{j=1}^n \varepsilon_j \xi_j X_j / n^{1/p}\}$ from flat concentration of $\{\sum_{j=1}^n \varepsilon_j X_j / n^{1/p}\}$, and likewise for stochastic boundedness. For the second part, one can make obvious modifications to the proof of the same result for $p = 2$ in [11]. For instance, the norm $\|x\|_F$ on top of page 918 in [11] must now be replaced by

$$\|(x_1, \ldots, x_q)\|_F = \sup\{\sum_{j=1}^n |x_{i_j}|/j^{1/p'} : \{i_1, \ldots, i_m\} \subset \{1, \ldots, q\}\}$$

where $1/p + 1/p' = 1$. then, the construction in [11] gives a Banach space B and a bounded B-valued random variable X such that $\sum_{j=1}^n \varepsilon_j X_j / n^{1/p} \longrightarrow 0$ but $\{\sum_{j=1}^n \varepsilon_j \xi_j X_j / n^{1/p}\}$ is not stochastically bounded unless $\xi \in L_{p,1}$. To obtain a random variable satisfying this property, but in the domain of normal attraction of a nondegenerate p-stable law Z, add to X and independent copy of Z.

References

[1] K. Alexander, *The non-existence of a universal multiplier moment for the central limit theorem*, pp. 15–16 in "Probability in Banach Spaces V", Lecture Notes in Math. **1153**. Springer Verlag 1985.

[2] A. Araujo and E. Giné, "the Central Limit Theorem for real and Banach valued random variables," Wiley, New York, 1980.

[3] K.L. Chung, "A course in Probability theory," Harcourt, Brace and World, New York, 1968.

[4] W. Feller, "An Introduction to Probability Theory and its Appreciations," Vol. II, Wiley, New York, 1966.

[5] E. Giné and J. Zinn, *Central limit theorems and weak laws of large numbers in certain Banach spaces*, Zeits. Wahrs. v. Geb. **62** (1983), 323–354.

[6] —————————, *Some limit theorems for empirical processes*, Ann. Probability **12** (1984), 929–989.

[7] J. Hoffmann-Jørgensen, *Sums of independent Banach space valued random variables*, Studia Math. **52** (1974), 159–184.

[8] N. Jain and M.B. Marcus, *Central limit theorems for C(S)-valued random variables*, J. Funct. Anal. **19** (1975), 216–231.

[9] —————————, *Integrability of infinite sums of independent vector valued random variables*, Trans. Amer. Math. Soc. **212** (1975), 1–36.

[10] J.P. Kahane, "Some Random Series of Functions," Heath, Lexington, Mass., 1968.

[11] M. Ledoux and M. Talagrand, *Conditions d'integrabilité pour le TLC Banachique*, Ann. Probability **14** (1986), 916–921.

[12] —————————, *Characterization of the law of iterated logarithm for Banach space valued random variables*, Ann. Probability **16** (1988), 1242–1264.

[13] R. LePage, M. Woodroofe and J. Zinn, *Convergence to a stable distribution via order statistics*, Ann. Probability **9** (1981), 624–632.

[14] V. Mandrekar and J. Zinn, *Central limit problem for symmetric case: Convergence to non-Gaussian laws*, Studia Math. **67** (1980), 279–296.

[15] M.B. Marcus, *Weak convergence of the empirical characteristic function*, Ann. Probability **9** (1981), 194–201.

[16] —————————, *ξ-radial processes and random Fourier series*, Memoirs of the Amer. Math. Soc. **368** (1987).

[17] M.B. Marcus and G. Pisier, "Random Fourier Series with Applications to Harmonic Analysis," Princeton U. Press, Princeton, N.J., 1978.

[18] —————————, *Characterization of almost surely continuous p-stable random Fourier series and strongly stationary processes*, Acta Math. **152** (1984), 245–301.

[19] B. Maurey and G. Pisier, *Séries de variables aleatoires vectorielles in-dépendantes et propriétés géométriques des spaces de Banach*, Studia Math. **58** (1976), 45–90.

[20] E. Seneta, "Regularly varying functions," Lecture Notes in Math., Vol. 508, Springer-Verlag, 1976.

Texas A & M University, Department of Mathematics, College Station, TX 77843

City College, CUNY, Department of Mathematics, New York, NY 10031

Texas A & M University, Department of Mathematics, College Station, TX 77843

Some Probability and Entropy Estimates for Gaussian Measures

V. Goodman

1. Introduction

We compare two estimates for the measure of Banach neighborhoods of Hilbert balls in the reproducing kernel space. Borell's estimate [1] is quite general and is known to be sharp for certain cases which involve small probabilities. However, Talagrand [7] and Goodman [4] use the openness of certain sets to obtain alternative estimates for cases in which the probability is near one.

Other probability estimates provide bounds on the number of sets with small Banach diameter necessary to cover a Hilbert ball in the reproducing kernel space. This may be viewed as an adjoint operator question for a special case of the covering problem considered in Dudley [2].

2. Gaussian measures

Let μ denote a centered Gaussian measure on the Borel σ-algebra of a real separable Banach space, B with norm, $\| \cdot \|$. It is well known (see Goodman, Kuelbs, Zinn [3]) that the covariance operator $S : B^* \to B$ exists as a compact operator and defines an inner product $(,)$ on its range given by

$$(S(y), S(z)) = \langle y, S(z) \rangle \forall y, z \in B^*.$$

The completion of Range(S) in the associated norm is the *reproducing kernel Hilbert space*, denoted by H. The centered *unit Hilbert ball* of H will be denoted by K. It is known that K is compact in B.

If $C \subset B$ then for $\varepsilon > 0$ the ε-*metric entropy* of C is denoted by $H(\varepsilon, C)$. This quantity is the logarithm of the minimal cardinality of coverings of C by sets of diameter not exceeding 2ε (see Dudley [2]). A Banach ball with radius r centered at $x \in B$ will be denoted by $B_r(x)$.

3. Estimates for $\mu(\lambda K + E)$.

Borell's estimate, [1] implies that if $E \subset B$ is any Borel set, the inequality

$$(3.1) \qquad \mu(\lambda K + E) \geq \Phi(\lambda + \alpha)$$

holds for all $\lambda > 0$, where Φ denotes the standard normal distribution function and α is given by

$$\mu(E) = \Phi(\alpha).$$

For the case $E = B_\varepsilon(0)$, the left hand expression in (3.1) is greater than or equal to

$$\mu(\psi_\varepsilon < \lambda^2)$$

where ψ_ε is the function

$$(3.2) \qquad \psi_\varepsilon(x) = \inf_{h \in H \cap B_\varepsilon(x)} |h|^2.$$

The proof of Lemma 3 in Talagrand [7] implicitly shows that

$$E[\exp(\tfrac{1}{2}\psi_\varepsilon)] < \infty,$$

and Lemma 3.1 in Goodman [4] gives further details concerning this result. One applies the Chebyshev inequality to obtain

$$\mu(\psi_\varepsilon \geq \lambda^2) \leq c \exp(-\tfrac{1}{2}\lambda^2), \quad \lambda > 0$$

where $c < \infty$ depends only on ε and μ. This gives the estimate

$$(3.3) \qquad \mu(\lambda K + B_\varepsilon(0)) \geq 1 - c \exp(-\tfrac{1}{2}\lambda^2), \quad \lambda > 0.$$

Fix $\theta > 1$. Then for λ sufficiently large,

$$1 - \Phi(\lambda - \theta\lambda^{-1}\ln\lambda) \approx (2\pi)^{-\frac{1}{2}}\lambda^{-1}\exp\{-\tfrac{1}{2}(\lambda - \theta\ln\lambda/\lambda)^2\}$$
$$= (2\pi)^{-\frac{1}{2}}\lambda^{\theta-1}\exp\{-\tfrac{1}{2}\lambda^2 - \tfrac{1}{2}\theta^2(\ln\lambda/\lambda)^2\}$$
$$\geq c\exp(-\tfrac{1}{2}\lambda^2).$$

We combine this inequality, holding for $\lambda \geq \lambda_{\epsilon,\mu}$, with (3.3) to obtain

$$(3.4) \qquad \mu(\lambda K + B_\epsilon(0)) \geq \Phi(\lambda - \theta\lambda^{-1}\ln\lambda) \quad \forall\lambda \geq \lambda_{\epsilon,\mu}.$$

In (3.4) the probability estimate is of the form

$$\Phi(\lambda + \alpha_\lambda)$$

where, although α_λ is negative $\alpha_\lambda \to 0$ as $\lambda \to \infty$. This contrasts with (3.1) where α is fixed and α is negative if ϵ is sufficiently small. However, more is known concerning the distribution of ψ_ϵ and we may improve (3.4).

Theorem 1. *Let $\epsilon > 0$. There is a constant $\lambda_{\epsilon,\mu} < \infty$ such that*

$$(3.5) \qquad \mu(\lambda K + B_\epsilon(0)) \geq \Phi(\lambda) \quad \forall\lambda \geq \lambda_{\epsilon,\mu}.$$

Proof: Proposition 4.2 in Goodman [4] states that the function ψ_ϵ in (3.2) satisfies

$$\psi_\epsilon(x) \leq \psi'(x) + \psi''(x)$$

where the functions ψ', ψ'' are independent as μ random variables. Moreover, the proof of Corollary 4.2 in [4] shows that ψ' is dominated in tail distribution by a random variable of the form

$$(\chi_1 - \delta)^2$$

where $\delta > 0$ and χ_1^2 has a chi-square distribution. Following the proof of Lemma 3.1 in [4], one may choose the number of degrees of freedom so that the function ψ'' is also dominated in tail distribution by a random variable

$$(\chi_2 - \delta)^2$$

where χ_2^2 is chi-square with one degree of freedom.

Since ψ', ψ'' are independent, we may choose a probability space on which χ_1, χ_2 are independent, and for $\lambda \geq 0$,

$$\mu(\psi_\epsilon \geq \lambda^2) \leq P((\chi_1 - \delta)^2 + (\chi_2 - \delta)^2 \geq \lambda^2).$$

Since $\chi_i \geq 0$,

$$(\chi_1 - \delta)^2 + (\chi_2 - \delta)^2 \leq \chi^2 - 2\delta\chi + 2\delta^2,$$

where $\chi^2 = \chi_1^2 + \chi_2^2$ has a chi-square distribution. To complete the proof it suffices to show that for λ sufficiently large,

$$(3.6) \qquad P((\chi - \delta)^2 + \delta^2 \geq \lambda^2) \leq 1 - \Phi(\lambda).$$

Since

$$E[\exp(\tfrac{1}{2}(\chi - \tfrac{1}{2}\delta)^2)] = M < \infty,$$

we have

$$P(\chi - \tfrac{\delta}{2} \geq \alpha) \leq M \exp(-\tfrac{1}{2}\alpha^2) \quad \forall \alpha \geq 0.$$

This may be written as

$$P(\chi \geq \alpha) \leq M \exp(-\tfrac{1}{2}(\alpha - \tfrac{\delta}{2})^2).$$

Now if $\alpha = \sqrt{\lambda^2 - \delta^2} + \delta$, $P(\chi \geq \alpha)$ is the probability in (3.6), and so we obtain the upper bound

$$M \exp\left(-\tfrac{1}{2}\lambda^2 - \tfrac{1}{2}\delta\sqrt{\lambda^2 - \delta^2} + \frac{3\delta^2}{8}\right).$$

One easily sees that this quantity is $\sigma(1 - \Phi(\lambda))$ as $\lambda \to \infty$.

4. Entropy Estimates for K.

Theorem 2. *Suppose that for some $\alpha > 0$,*

$$(4.1) \qquad \underline{\lim_{\varepsilon \downarrow 0}} \varepsilon^{\alpha} ln(\mu(B_{\varepsilon}(0))) > -\infty.$$

Then

$$(4.2) \qquad \overline{\lim_{\varepsilon \downarrow 0}} \varepsilon^{2\alpha/(2+\alpha)} H(\varepsilon, K) < \infty.$$

Proof: Suppose $\lambda > 0$, $h \in \lambda K$, and $h \neq 0$. Lemma 3.2 in Goodman [4] states that

$$(4.3) \quad \mu(B_{\varepsilon}(h)) \geq \mu(B_{\varepsilon/2}(0)) \left\{ \Phi\left(|h| + \frac{\varepsilon|h|}{2\|h\|} \right) - \Phi\left(|h| - \frac{\varepsilon|h|}{2\|h\|} \right) \right\}.$$

Since

$$c = \inf_{h \in H_{\mu}} \frac{|h|}{2\|h\|} = \tfrac{1}{2} \sup_{k \in K} \|k\| = \tfrac{1}{2}\sigma > 0,$$

a lower bound in (4.3) is obtained by substituting $\Phi(|h| \pm c\varepsilon)$ for the terms $\Phi(|h| \pm \frac{\varepsilon|h|}{2\|h\|})$. The resulting difference is minimized by taking $|h|$ large, and from (4.3) we have

$$(4.4) \qquad
\begin{aligned}
\mu(B_{\varepsilon}(h)) &\geq \mu(B_{\varepsilon/2}(0))\{\Phi(\lambda + c\varepsilon) - \Phi(\lambda - c\varepsilon)\} \\
&= \mu(B_{\varepsilon/2}(0)) \frac{1}{\sqrt{2\pi}} \int_{-c\varepsilon}^{c\varepsilon} \exp(-\tfrac{1}{2}(\lambda + x)^2)dx \\
&> \mu(B_{\varepsilon/2}(0)) \frac{c\varepsilon}{\sqrt{2\pi}} \exp(-\tfrac{1}{2}\lambda^2).
\end{aligned}$$

Let $\varepsilon = \lambda^{-2/\alpha}$. Then $\varepsilon \downarrow 0$ as $\lambda \to \infty$. By assumption, there is a constant $C \geq 0$ such that

$$ln(\mu(B_{\varepsilon}(0)) \geq -C\varepsilon^{-\alpha}, \quad \lambda \to \infty.$$

It follows from (4.4) that

$$ln(\mu(B_{\varepsilon}(h))) \geq -C\lambda^2 - \tfrac{2}{\alpha}ln(\lambda) - \tfrac{1}{2}\lambda^2 + ln\left(\frac{c}{\sqrt{2\pi}} \right)$$

and we obtain

(4.5)
$$\varliminf_{\lambda \to \infty} \lambda^{-2} ln(\mu(B_{\lambda^{-2/\alpha}}(h))) \geq -(C + \tfrac{1}{2}).$$

Now suppose that for each λ the set

$$\{h_{n,\lambda} \in \lambda K : n \leq N_\lambda\}$$

is such that $\|h_{i,\lambda} - h_{j,\lambda}\| \geq 2\varepsilon$ for all $i \neq j$. Then

$$N_\lambda \cdot \min_{n \leq N_\lambda} \{\mu(B_\varepsilon(h_{n,\lambda}))\} \leq 1$$

and the estimate (4.5) implies

(4.6)
$$\varlimsup_{\lambda \to \infty} \lambda^{-2} ln(N_\lambda) \leq C + \tfrac{1}{2}.$$

Since the set λK is compact for each $\lambda > 0$, there exist sets of the form

$$\{h_{n,\lambda} : \|h_{i,\lambda} - h_{j,\lambda}\| \geq 2\varepsilon \quad \text{for} \quad i \neq j, \ i, \ j \leq N_\lambda\}$$

where N_λ is maximal. It follows from the maximality of N_λ that for any $h \in \lambda K$, $\|h - h_{n,\lambda}\| < 2\varepsilon$ for some n. That is, the collection $\{B_{2\varepsilon}(h_{n,\lambda}) : n \leq N_\lambda\}$ is a covering of λK. By the definition of ε-entropy, we have

$$H(2\varepsilon, \lambda K) \leq ln(N_\lambda)$$

and (4.6) gives the inequality

(4.7)
$$\varlimsup_{\lambda \to \infty} \lambda^{-2} H(2\lambda^{-2/\alpha}, \lambda K) < \infty.$$

Since $H(2\lambda^{-2/\alpha}, \lambda K) = H(2\lambda^{-1-2/\alpha}, K)$, we may reparametrize the $\varlimsup$ in (4.7) with $\varepsilon = 2\lambda^{-\frac{\alpha+2}{\alpha}}$ to obtain (4.2).

Remark: Hoffmann-Jørgensen, Shepp, Dudley [5] have shown there are infinite dimensional examples where (4.1) holds for various values of α. Also, the set K may be viewed as the image of the unit ball of H under the imbedding $T \colon H \to B$, where $S = T \circ T^*$. The ε-metric entropy of T^*U where U is the unit ball in B^* is a particular instance of the question considered in Dudley [2]. Some results concerning equivalence of entropy estimates for these two questions appear in Pajor, Tomczak-Jaegermann [6]. However, the equivalence of

$$\overline{\lim_{\varepsilon \downarrow 0}} \varepsilon^p H(\varepsilon, K) < \infty$$

and

$$\overline{\lim_{\varepsilon \downarrow 0}} \varepsilon^p H(\varepsilon, T^*U) < \infty$$

has not been shown for $p \leq 2$. The result

$$\overline{\lim_{\varepsilon \downarrow 0}} \varepsilon^2 H(\varepsilon, K) = 0$$

was obtained in [4].

References

[1] C. Borell, *The Brunn-Minkowski inequality in Gauss space*, Inventiones Math. **30** (1975), 205–216.

[2] R.M. Dudley, *The sizes of compact subsets of Hilbert space and continuity of Gaussian processes*, J. of Functional Anal. **1** (1967), 290–330.

[3] V. Goodman, J. Kuelbs and J. Zinn, *Some results on the LIL in Banach space with applications to weighted empirical processes*, Ann. Prob. **9** (1981), 713–752.

[4] V. Goodman, *Characteristics of normal samples*, Ann. Prob. **16** (1988), 1281–1290.

[5] J. Hoffman-Jørgensen, L.A. Shepp and R.M. Dudley, *On the lower tail of Gaussian seminorms*, Ann. Prob. **7** (1979), 319–342.

[6] A. Pajor and N. Tomczak-Jaegermann, *Remarques sur les nombres d'entropie d'un opérateur et de son transposé*, C.R. Acad. Sci. Paris Sér. 1 Math. **301** No. 15 (1985), 743–746.

[7] M. Talagrand, *Sur l'intégrabilité des vecteurs gaussiens*, Z. Wahrsch. verw. Gebiete **68** (1984), 1–8.

Department of Mathematics, Indiana University, Bloomington, Indiana 47405

Isometries of $L^p(X)$
and Vector Valued Ergodic Theorems

This is an outline of a joint work with Y. Raynaud.

Abstract. Let X be a Banach space. We prove that surjective isometries of $L^p(X)$, $1 < p < \infty$, satisfy a vector valued maximal inequality and therefore (when X is reflexive) the vector valued ergodic theorem (the Cesaro means $(\frac{1}{n} \sum_{i=1}^{n} T^i f)_{n \in \mathbb{N}}$ converge a.s. in norm).

When X is a Banach lattice and T a positive isometry from $L^p(X)$ into itself, we give a representation of T, under appropriate convexity and concavity properties on X. This representation implies the same vector valued maximal inequality.

When $X = L^q$, we give such a representation for all isometries of $L^p(X)$.

We also prove that the dilation theorem of Akcoglu does not extend in a natural way to positive contractions of $L^p(L^q)$.

Given an operator T on L^p ($1 < p < \infty$) the (scalar) ergodic theory, proves under various hypotheses on T the a.s. convergence of Cesaro means $(\frac{1}{n} \sum_{i=1}^{n} T^i f)_{n \in \mathbb{N}}$ where $f \in L^p$. We say that T verifies the *ergodic theorem* if this convergence holds.

The first results in that direction are due to G.H. Hardy & J.E. Littlewood [9], N. Dunford & J.T. Schwartz [6] and G. Birkhoff [3]. The more recent ones belong to M. Akcoglu and L. Sucheston [1]. If X is a Banach space and T is an operator on $L^p(X)$ ($1 < p < \infty$), we say that T verifies the *vector valued ergodic theorem* if for all $f \in L^p(X)$, the Cesaro means $(\frac{1}{n} \sum_{i=1}^{n} T^i f)_{n \in \mathbb{N}}$ converge a.s. in norm to a fixed point of T.

Such a vector valued extension of N. Dunford and J.T. Schwartz' theorem [6] (when T is a contraction in L^1 and in L^∞) was obtained by R. Chacon [4].

We are interested in a vector valued extension of the following results due to A. Ionescu-Tulcea [10] and R. Chacon & S.A. Mc Grath [5]: "If

$1 < p < \infty$ and T is a positive isometry of L^p or if $1 < p \neq 2 < +\infty$ and T is a surjective isometry of L^p, then T verifies the ergodic theorem".

In the case where X is a reflexive B-space, we know (cf. [11]) that a vector valued ergodic theorem, can be deduced from a *vector valued maximal inequality*; namely:

$$\exists C = C(p, X) \text{ such that } \forall f \in L^p(X), \left| \sup_n \left| \frac{1}{n} \sum_{i=1}^{n} T^i f \right|_X \right|_p \leq C \|f\|_{L^p(X)}.$$

Definition: Let T be a contraction in $L^p(X)$. We say that $\widetilde{T}$ is an L_p-*majorizing contraction* of T if:

$$\forall f \in L^p(X), \quad |Tf|_X \leq \widetilde{T}|f|_X \text{ a.e.}$$

The following result is a consequence of Ackoglu's theorem [1] for positive contractions of L^p and is the key of all our results:

Proposition 1. *Let $1 < p < \infty$, X a B-space and T a contraction in $L^p(X)$. If T has an L^p-majorizing contraction $\widetilde{T}$, then T verifies the vector valued maximal inequality:*

$$\forall f \in L^p(X), \left| \sup_n \left| \frac{1}{n} \sum_{i=1}^{n} T^i f \right|_X \right|_p \leq \frac{p}{p-1} \|f\|_{L^p(X)}.$$

1. Surjective Isometries of $L^p(X)$.

The starting point of these results is the description of surjective isometries of $L^p(X)$, when X has no p-projections and $p \neq 2$ which appears in [2] and [7]. We obtain the following result:

Theorem 1. *Let X be a B-space, $1 < p < \infty$ and T a surjective isometry of $L^p(X)$.*

In the following cases:

(i) $p \neq 2$

(ii) $p = 2$, X is a lattice with no 2-projections and T has the property that if $|f|_X \wedge |g|_X = 0$, then $|Tf|_X \wedge |Tg|_X = 0$ then T verifies the vector valued maximal inequality with constant $\frac{p}{p-1}$.

2. Positive Isometries of $L^p(X)$.

Definition: (i) We say that a Banach lattice X is $1-r$-*concave on disjoint vectors (resp. $1-r$-convex on disjoint vectors)* if for all positive x and y in X such that $x \wedge y = 0$, then:

$$|x+y|x \geq (|x|_X^r + |y|_X^r)^{1/r}$$
$$(\text{resp: } |x+y|x \leq (|x|_X^r + |y|_X^r)^{1/r}$$

(ii) We say that X is *strictly monotone* if:

$$\left.\begin{array}{l} 0 \leq x \leq y \\ |x|x = |y|x \end{array}\right\} \Rightarrow x = y$$

Theorem 2. *Let $1 \leq p < +\infty$ and $X = L^q(Y)$, where Y is a strictly monotone Banach lattice. Suppose $q \geq r > p$ and Y to be $1-q$-concave and $1-r$-convex on disjoint vectors; or $q \leq 1 < p$ and Y to be $1-q$-convex and $1-r$-concave; then every positive isometry of $L^p(X)$ into itself has an L^p-majorizing contraction.*

3. Isometries of $L^p(L^q)$.

When $X = L^q$, it is possible to characterize general isometries of $L^p(L^q)$ as well as positive or surjective ones, for certain p and q:

Theorem 3. *Let $1 \leq p < +\infty$. If we are not in the two cases: $p \leq q \leq 2$ or $q = 2$ (where L^q embeds in L^p) then all isometries of $L^p(L^q)$ has an L_p-majorizing contraction.*

4. A Remark on Positive Contractions of $L^p(L^q)$.

The Dilation theorem of Ackoglu says that if T is a positive contraction of $\dot{L}^p$, then there exists a bigger space $\tilde{L}^p$ and a positive isometry $\widetilde{T}$ on $\widetilde{L}^p$ such that the following commutative diagram holds

$$\begin{array}{ccc}
L^p & \xrightarrow{\ T\ } & L^p \\
{\scriptstyle P}\big\uparrow & & {\scriptstyle P}\big\uparrow \\
\widetilde{L}^p & \xrightarrow{\ T\ } & \widetilde{L}^p
\end{array}$$

where P is a positive projection.

On the contrary, we can construct a positive contraction T of $L^p(L^q)$ $(p \neq q)$ such that no diagram of the following kind:

$$\begin{array}{ccc}
L^p(L^q) & \xrightarrow{\ T\ } & L^p(L^q) \\
{\scriptstyle P}\big\uparrow & & {\scriptstyle P}\big\uparrow \\
\widetilde{L}^p(\widetilde{L}^q) & \xrightarrow{\ \widetilde{T}\ } & \widetilde{L}^p(\widetilde{L}^q)
\end{array}$$

commutes, where P is a projection and $\widetilde{T}$ a positive isometry.

This shows that our results on positive isometries of $L^p(X)$ cannot extend in a natural way to positive contractions.

References

[1] M. Akcoglu and L. Sucheston, *Dilations of positive contractions on L^p-spaces*, Canad. Math. Bull. **20** (1977), 285–292.

[2] E. Behrends et al., "L^p-structure in Real Banach Spaces," Lecture Notes in Math., Vol. 613. Springer Verlag 1977.

[3] G. Birkhoff, *Proof of ergodic theorem*, Proc. Nat. Acad. Sci. U.S.A. **17** (1931), 656–660.

[4] R. Chacon, *An ergodic theorem for operators satisfying norm conditions*, J. Math. Mech. **11** (1962), 165–172.

[5] R. Chacon and S.A. McGrath, *Estimates of positive contractions*, Pacific J. of Math. **30** (1969), 609–620.

[6] N. Dunford and J.T. Schwartz, *Convergence a.e. of operators avarages*, J. Rat. Mech. Anal. **5** (1956), 129–178.

[7] P. Greim, *Isometries and L^p-structures of separably valued Bochner L^p-spaces*, pp. 209–218 in "Measure Theory and Its Applications", Lecture Notes in Math. **1033**. Springer Verlag 1983.

[8] S. Guerre and Y. Raynaud, *Sur les isometries de $L^p(X)$ et le théoréme ergodique vectoriels*, Can. J. Math. **40** (1988), 360–391.

[9] G.H. Hardy and J.E. Littlewood, *A maximal theorem with function theoretic applications*, Arkiv für Math. **19, Band 3** (1955), 239–244.

[10] A. Ionescu-Tulcea, *Ergodic properties of isometries in L^p-spaces*, $1 < p < +\infty$, Bull. AMS **70** (1964), 366–371.

[11] U. Krengel, "Ergodic theorems," Studies in Mathematics, De Gruyter, 1985.

Equipe d'Analyse, U.A. No 754 au C.N.R.S., Université Paris VI, Tour 46 – 4éme Etage, 4, Place Jussieu, 75252 – Paris Cedex 05

Some Exponential Inequalities with Applications to the Central Limit Theorem in $C[0,1]$

Bernard Heinkel

Exponential inequalities are a very useful tool in many topics in probability theory and statistics. According to the problem to study, one form or another one of such inequalities is the most convenient to use: Bernstein's, Prohorov's, Bennett's, Hoeffding's ... . Among these results, the one of Hoeffding has the simplest statement and is of course preferred if it applies. Let's recall that result:

Theorem 0.1. (Hoeffding [11]). *Let* $X_1, \ldots, X_n$ *be independent r.v. obeying the restrictions* $a_i \leq X_i \leq b_i$, $i = 1, 2, \ldots, n$. *Then for* $t > 0$:

$$(1) \quad P\left(\sum_{k=1}^{n}(X_k - E(X_k)) \geq nt\right) \leq \exp(-(2n^2t^2)/\sum_{k=1}^{n}(b_k - a_k)^2).$$

One notices that the right hand side of inequality (1) can also be written $\exp(-(2n^2t^2)/\|(b_k - a_k)\|_2^2)$, where $\|\cdot\|_2$ denotes the usual norm in the space ℓ_2. Many other classical exponential inequalities are of the same ℓ_2 spirit. The goal of the present paper is to study exponential inequalities in which the bound is sharpened by replacing that ℓ_2-norm by a "weak ℓ_2" one, and more generally by a "weak ℓ_p" $(1 < p \leq 2)$ one.

In the two first sections, we will prove such exponential inequalities; in the last one we will test their sharpness in applying them for finding sufficient conditions for the central limit theorem (CLT) in the space $C[0,1]$.

Let's begin by recalling some facts about weak ℓ_p spaces.

Let $0 < p < +\infty$ be given and denote by $\ell_{p,\infty}$ the space of all sequences (a_n) of real numbers such that:

$$\sup_{t>0} t^p \, card(n \colon |a_n| > t) < +\infty.$$

That space $\ell_{p,\infty}$ is called the weak ℓ_p space. Furthermore, let's define:

$$\|(a_n)\|_{p,\infty} = \left(\sup_{t>0} t^p \; card(n: |a_n| > t)\right)^{1/p}.$$

If $p > 1$, the functional $\|\cdot\|_{p,\infty}$ is equivalent to a norm on $\ell_{p,\infty}$, and $\ell_{p,\infty}$ equipped with that norm is a Banach space. It is obvious that a sequence (a_n) belonging to $\ell_{p,\infty}$ is also in c_0, so the non-increasing rearrangement (a_n^*) or $(|a_n|)$ can be defined without difficulty; it is easy to check that:

$$\|(a_n)\|_{p,\infty} = \sup_{n\geq 1}(n^{1/p} a_n^*).$$

Now we state two very nice probabilistic results which will be used later:

Lemma 0.2. (M.B. Marcus and G. Pisier [17] Theorem 3.3). *Let (Z_n) be a sequence of independent, positive r.v. Then for any $0 < p < +\infty$ and all $c > 0$*

$$c^p P(\|(Z_n)\|_{p,\infty} > c) \leq \lambda \sup_{t>0} t^p \sum_{n=1}^{\infty} P(Z_n > t),$$

where $\lambda = 262$.

For stating the second lemma, we need some more notations. Let $q > 2$ be given and denote by Ψ_q the following function:

$$\forall x \in \mathbf{R}, \; \Psi_q(x) = \exp|x|^q - 1.$$

For any probability space $(\Omega, \mathcal{F}, P)$ one denotes by $L^{\Psi_q}(dP)$ the Orlicz space associated to Ψ_q and P:

$$L^{\Psi_q}(dP) = \{f: (\Omega, \mathcal{F}) \to (\mathbf{R}, \mathcal{B}(\mathbf{R})): \exists c > 0: E\Psi_q(|f|/c) < +\infty\};$$

that space will be equipped with the following norm:

$$\|f\|_{\Psi_q} = \inf(c > 0: E\Psi_q(|f|/c) \leq 1).$$

With these notations, one has:

Lemma 0.3. ([17] Lemma 3.1). *Let (α_i) belonging to $\ell_{p,\infty}$, $1 < p < 2$, be given, Consider (ε_k) a sequence of independent Rademacher r.v. and define:*

$$S = \sum_{k=1}^{\infty} \alpha_k \varepsilon_k.$$

Then $S \in L^{\Psi_q}(dP)$ — where $1/p + 1/q = 1$ — and one has:

$$(1/k_p)\|(\alpha_i)\|_{p,\infty} \le \|S\|_{\Psi_q} \le k_p\|(\alpha_i)\|_{p,\infty},$$

where k_p is a constant depending only on p.

We are now ready to turn our attention to exponential inequalities.

1. Weak ℓ_2 exponential inequalities.

We begin by proving a Hoeffding type result:

Theorem 1.1. *Let $X_1, \ldots, X_n$ be independent, centered, r.v. such that there exist positive constants $a_1, \ldots, a_n$ with:*

$$\forall i = 1, \ldots, n \quad |X_i| \le a_i \ a.s.$$

Let's denote by γ the weak ℓ_2 norm of the sequence $(a_i)_{i \le n}$ and by σ^2 the variance of $X_1 + \cdots + X_n$. Then:

1) If the X_i are symmetrically distributed, one has:

$$(2) \quad \forall t > 0, P\left(\sum_{k=1}^{n} X_k > t\right) \le 2\exp\left(-\frac{t^2}{36}\inf\left(\frac{1}{\gamma^2}; \frac{1}{e\sigma^2}\right)\right).$$

2) Without any assumption of symmetry, the following holds:

$$(3) \quad \forall t > 2\sqrt{2}\sigma, P\left(\sum_{k=1}^{n} X_k > t\right) \le 4\exp\left(-\frac{t^2}{144}\inf\left(\frac{1}{4\gamma^2}; \frac{1}{2e\sigma^2}\right)\right).$$

Proof: We consider first the symmetric case. Suppose that the X_i are defined on $(\Omega, \mathcal{F}, P)$ and let $(\Omega', \mathcal{F}', P')$ be another probability space on which one has defined a sequence $(\varepsilon_1, \ldots, \varepsilon_n)$ of independent Rademacher r.v.

By symmetry, one has:

$$\forall t > 0, P\left(\sum_{k=1}^{n} X_k > t\right) = P \otimes P'\left(\sum_{k=1}^{n} \varepsilon_k X_k > t\right).$$

An easy computation shows that:

$$\forall u \geq \sqrt{n}, P \otimes P'\left(\sum_{k=1}^{n} \varepsilon_k X_k > 3\gamma u\right) = 0.$$

Now, we will bound the preceding probability for $u \in]0, \sqrt{n}[$. We start by studying, for every $w \in \Omega$, the following quantity:

$$a(w) = P'\left(\sum_{k=1}^{n} \varepsilon_k X_k(w) > 3\gamma u\right).$$

Such an w being fixed, $(X_k^*(w))$ will be the non-increasing rearrangement of the sequence $(|X_k(w)|)_{k \leq n}$.

The following inequality is obvious:

$$a(w) \leq P'\left(\sum_{k=[u^2]+1}^{n} \varepsilon_k X_k^*(w) > \gamma u\right),$$

where [ ] denotes the integer part of a real number. Now for every $k = 1, \ldots, n$, one defines Y_k^* as the indicator function of the event $(X_k^* \leq \gamma([u^2] + 1)^{-1/2})$ and by Y_k the one of $(X_k \leq \gamma([u^2] + 1)^{-1/2})$.

From the preceding inequality one derives:

$$a(w) \leq P'\left(\sum_{k=[u^2]+1}^{n} \varepsilon_k X_k^*(w) Y_k^*(w) > \gamma u\right),$$

and by applying Lévy's inequality with respect to P', one obtains:

$$a(w) \leq 2P' \left(\sum_{k=1}^{n} \varepsilon_k X_k(w) Y_k(w) > \gamma u \right).$$

From this inequality, one gets:

$$\forall u \in]0, \sqrt{n}[, P \otimes P' \left(\sum_{k=1}^{n} \varepsilon_k X_k > 3\gamma u \right) \leq 2P \otimes P' \left(\sum_{k=1}^{n} \varepsilon_k X_k Y_k > \gamma u \right).$$

By Tchebychev's exponential inequality:

$$\forall h > 0, \forall u \in]0, \sqrt{n}[,$$

$$P \otimes P' \left(\sum_{k=1}^{n} \varepsilon_k X_k > 3\gamma u \right) \leq 2 \exp(-\gamma h u) E \exp \left(\frac{h^2}{2} \sum_{k=1}^{n} X_k^2 Y_k^2 \right).$$

Now we need to distinguish two cases. On the one hand, if $\gamma^2 \leq e\sigma^2$, one chooses $h = \gamma u/(2e\sigma^2)$, and one gets after a short computation:

$$P \otimes P' \left(\sum_{k=1}^{n} \varepsilon_k X_k > 3\gamma u \right) \leq 2 \exp \left(-\frac{\gamma^2 u^2}{4e\sigma^2} \right);$$

on the other hand, if $\gamma^2 > e\sigma^2$, then one chooses $h = u/(2\gamma)$, and one obtains:

$$P \otimes P' \left(\sum_{k=1}^{n} \varepsilon_k X_k > 3\gamma u \right) \leq 2 \exp \left(-\frac{u^2}{4} \right).$$

So finally:

$$P \otimes P' \left(\sum_{k=1}^{n} \varepsilon_k X_k > 3\gamma u \right) \leq 2 \exp \left(-\frac{u^2}{4} \inf \left(1, \frac{\gamma^2}{e\sigma^2} \right) \right),$$

and:

$$\forall t > 0, P \left(\sum_{k=1}^{n} X_k > t \right) \leq 2 \exp \left(-\frac{t^2}{36} \inf \left(\frac{1}{\gamma^2}, \frac{1}{e\sigma^2} \right) \right),$$

which is the claimed result in the symmetrical setting.

The general case follows easily by applying an elementary symmetrization argument.

Remarks:

1) From Theorem 1.1 one can easily derive the following result:

"Let $X_1, \ldots, X_n$ be independent, centered, and square integrable r.v.; then:

$$\forall t > 2\sqrt{2}\sigma, \quad P\left(\sum_{k=1}^{n} X_k > t\right) \leq 4 \inf_{\mathbf{a}} \left\{ \sum_{k=1}^{n} P(|X_k| > a_k) + g(t, \mathbf{a}) \right\},$$

where

$$g(t, \mathbf{a}) = \exp\left(-\frac{t^2}{144} \inf(1/(4\|(a_k)\|_{2,\infty}^2), 1/(2e\sigma^2))\right),$$

and the infimum is taken over all $\mathbf{a} = (a_1, \ldots, a_n) \in \mathbf{R}^{+n}$."

This result can be compared with those of [20], §1.

2) It is easy to construct an example for which Theorem 1.1 gives a better bound than Hoeffding's result. Let $\xi_1, \ldots, \xi_n$ be independent, symmetrically distributed r.v. such that:

$$\forall k = 1, \ldots, n \quad P(\xi_k = 1) = P(\xi_k = -1) = \frac{1}{2k},$$
$$P(\xi_k = 0) = 1 - \frac{1}{k}.$$

Define:

$$\forall k = 1, \ldots, n \quad X_k = \frac{1}{\sqrt{k}}\xi_k.$$

It is clear that in that case the "best possible" sequence (a_k) which bounds the r.v. $|X_k|$ is: $a_k = \frac{1}{\sqrt{k}}$; so Hoeffding's result gives:

$$\forall t > 0 \quad P\left(\sum_{k=1}^{n} X_k > t\right) \leq \exp\left(-\frac{t^2}{2(1 + Log\, n)}\right);$$

by applying Theorem 1.1, one obtains the following bound which is much better for large n and t:

$$\forall t > 0 \quad P\left(\sum_{k=1}^{n} X_k > t\right) \le 2\exp\left(-\frac{t^2}{6e\pi^2}\right).$$

The method used for proving Theorem 1.1 gives also a Bernstein-Yurinskii exponential inequality; its statement is as follows:

Theorem 1.2. *Let $X_1, \ldots, X_n$ be independent, centered, square integrable r.v. Let's denote by σ^2 the variance of the sum $X_1 + \cdots + X_n$. Then*

1) If the X_i are symmetrically distributed, one has:

(4) $\forall t > 0$,

$$P\left(|\sum_{k=1}^{n} X_k| > t\right) \le \inf_{c>0}\{P(\|(X_k)\|_{2,\infty} > c) + 2\exp\left(\frac{-\frac{t^2}{72} + \frac{tc}{12}}{\sigma^2 + \frac{c^2}{2}}\right)\}.$$

2) Without any symmetry assumption, one has:

(5) $\forall t > 2\sqrt{2}\sigma$,

$$P\left(|\sum_{k=1}^{n} X_k| > t\right) \le 4\inf_{c>0}\{P(\|(X_k)\|_{2,\infty} > \frac{c}{4}) + \exp\left(\frac{-\frac{t^2}{288} + \frac{tc\sqrt{2}}{24}}{2\sigma^2 + \frac{c^2}{2}}\right)\}.$$

Proof: Here again we need only to prove statement 1), the second one being an easy consequence of the symmetrization inequalities. Suppose as we did before that the X_i are defined on $(\Omega, \mathcal{F}, P)$ and that $(\Omega', \mathcal{F}', P')$ is a probability space on which a sequence $(\varepsilon_1, \ldots, \varepsilon_n)$ of independent Rademacher r.v. has been defined.

For every $c > 0$, one notes:

$$A = (\|(X_k)\|_{2,\infty} \le c).$$

Such a positive c being fixed, one has for every $t > 0$:

$$P\left(|\sum_{k=1}^{n} X_k| > t\right) = P \otimes P'\left(|\sum_{k=1}^{n} \varepsilon_k X_k| > t\right)$$

$$\leq P(A^c) + P \otimes P'\left(|\sum_{k=1}^{n} \varepsilon_k X_k|I_A > t\right).$$

Note $t = 3cu$.

If $u \geq \sqrt{n}$, the second term of the right hand side in the above inequality vanishes. So we suppose from now that $u < \sqrt{n}$. By the same arguments as in the proof of Theorem 1.1 one obtains:

$$\forall w \in \Omega, a(w) = P'\left(|\sum_{k=1}^{n} \varepsilon_k X_k(w)|I_A(w) > 3cu\right)$$

$$\leq 2P'\left(|\sum_{k=1}^{n} \varepsilon_k X_k Y_k| > cu\right),$$

where Y_k denotes the indicator function of the event $(|X_k| \leq c/u)$. Let's now recall Yurinskii's exponential inequality [22]:

Lemma 1.3. *Let $\lambda_1, \ldots, \lambda_n$ be independent r.v. taking their values in a real separable Banach space $(B, \|\cdot\|)$. Suppose that these r.v. are strongly square integrable and that there exists a constant $L > 0$ such that for every $i = 1, \ldots, n$ and every $m \geq 2$:*

$$E\|\lambda_i\|^m \leq (m!/2)E\|\lambda_i\|^2 L^{m-2}.$$

Define: $S = \lambda_1 + \cdots + \lambda_n$ and $\Lambda^2 = \sum_{i=1}^{n} E\|\lambda_i\|^2$. Then:

$$\forall r > 0, P(\|S\| \geq r) \leq \exp\left(\frac{\frac{-r^2}{8} + \frac{r}{4}E\|S\|}{\Lambda^2 + (rL/2)}\right).$$

By applying this lemma one obtains

$$P\left(|\sum_{k=1}^{n} X_k| > 3cu\right) \leq P(A^c) + 2\exp\left(\frac{-\frac{c^2u^2}{8} + \frac{cu\sigma}{4}}{\sigma^2 + \frac{c^2}{2}}\right).$$

Relation (4) is then easily derived from the preceding one.

Remarks:

1) By applying Lemma 0.2, one obtains easily the following corollary of Theorem 1.2:

Let $X_1, \ldots, X_n$ be independent, centered, square integrable r.v. Then:

$\forall t > 2\sqrt{2}\sigma, \forall c > 0,$

$$P\left(|\sum_{k=1}^{n} X_k| > t\right) \leq \frac{16,768}{c^2}\left(\sup_{x>0} x^2 \sum_{k=1}^{n} P(|X_k| > x)\right)$$
$$+ 4\exp\left(\frac{-\frac{t^2}{288} + \frac{t\sigma\sqrt{2}}{24}}{2\sigma^2 + \frac{c^2}{2}}\right).$$

This result also can be compared with those stated in [20] §1, which are of a close spirit.

2) From the above proof one can take out the following partial result which will be used later:

Proposition 1.4. *Let $X_1, \ldots, X_n$ be independent, symmetrically distributed, square integrable r.v. Let's denote by σ^2 the variance of the sum $X_1 + \cdots + X_n$, and by $\varepsilon_1, \ldots, \varepsilon_n$ a sequence of independent Rademacher r.v., which are independent of the X_k.*

Furthermore, for every $c > 0$ we denote by $A(c)$ the following event:

$$A(c) = (\|(X_k)\|_{2,\infty} \leq c).$$

Then for every B which is $\sigma(X_1, \ldots, X_n)$-measurable and such that $B \subset A(c)$ one has:

$$\forall t > 0, P\left(|\sum_{k=1}^{n} \varepsilon_k X_k|I_B > t\right) \leq 2\exp\left(\frac{-\frac{t^2}{72} + \frac{t\sigma}{12}}{\sigma^2 + \frac{c^2}{2}}\right).$$

3) One sees easily that Theorem 1.2 has the following extension to Banach space valued r.v., which can be compared to Lemma 1.3:

Proposition 1.5. *Let $X_1, \ldots, X_n$ be independent r.v. with values in a real separable Banach space $(B, \|\cdot\|)$, which are symmetrically distributed and strongly square integrable. Define: $S = X_1 + \cdots + X_n$ and $\Lambda^2 = \sum_{k=1}^n E\|X_k\|^2$. Then:*

$$\forall t > 0, \quad P\left(\|\sum_{k=1}^n X_k\| > t\right)$$

$$\leq \inf_{c>0}\{P(\|(\|X_k\|)\|_{2,\infty} > c) + 2\exp\left(\frac{-\frac{t^2}{72} + \frac{t}{12}E\|S\|}{\Lambda^2 + (c^2/2)}\right)\}.$$

For conclusion of this section, we will give a new proof of a more specialized exponential inequality which is due to G. Morrow [19]. His proof is very nice, but much more combinatorial than probabilistic; A. Zoglat [23] has found a completely probabilistic proof of that inequality using "classical" exponential inequalities and some ideas of A. de Acosta [1]. In the papers of G. Morrow and A. Zoglat that specialized inequality is used to study the CLT for r.v. which increments, suitably normalized, are exponentially integrable.

Here we will give a proof of Morrow's result which reduces to an exercise of application of the above Theorem 1.2. G. Morrow's result ([19] Lemma 2.2) can be formulated as follows:

Proposition 1.6. *Let $X_1, \ldots, X_n$ be independent copies of a centered r.v. X such that there exists $\alpha \in]0, 2[$ with:*

$$E\exp(|X|^\alpha) \leq \exp(1/8).$$

Then there exists two positive constants $\gamma(\alpha)$ and $t(\alpha)$, depending on α, but independent of the law of X and of the size n of the sample, such that:

$$\forall t \geq t(\alpha), \quad P\left(|\sum_{k=1}^n \frac{X_k}{\sqrt{n}}| > t\right) \leq \exp(-\gamma(\alpha)t^\alpha).$$

Proof: It suffices to prove Proposition 1.6 for symmetrically distributed r.v., the general case following by an easy symmetrization argument. So let's suppose that X is symmetrically distributed.

a) We begin by showing the asserted property when $t < n^\beta$, where:

$$\beta = \inf\left(\frac{1}{2\alpha}, \frac{1}{4}\right).$$

Then:

$$nP(|X| > n^\beta) \le 2n\exp(-n^{\alpha\beta}) \le K(\alpha)\exp\left(-\frac{t^\alpha}{2}\right).$$

So it will be sufficient to find the claimed bound for $P(|\sum_{k=1}^n Y_k|/\sqrt{n} > t)$, where for every $k = 1, 2, \ldots, n$:

$$Y_k = X_k I_{(|X_k| \le n^\beta)}.$$

One has obviously:

$$P(\|(\frac{Y_k}{\sqrt{n}})\|_{2,\infty} > t^{1-\alpha/2}) \le P\left(\sum_{k=1}^n \frac{Y_k^2}{n} > t^{2-\alpha}\right).$$

Under the hypotheses of Proposition 1.6, there exists a constant $g(\alpha)$ which doesn't depend on the law of X and such that:

$$EX^2 \le g(\alpha).$$

If we suppose now that $t > \sup(1, (2eg(\alpha))^{1/(2-\alpha)})$, then an easy application of Tchebychev's exponential inequality yields:

$$P\left(\sum_{k=1}^n \frac{Y_k^2}{n} > t^{2-\alpha}\right) \le \exp\left(-\frac{n^{1-2\beta}t^{2-\alpha}}{2}\right) \le \exp\left(-\frac{t^\alpha}{2}\right).$$

Applying now Theorem 1.2, one obtains for t large enough:

$$P\left(|\sum_{k=1}^n \frac{X_k}{\sqrt{n}}| > t\right) \le (K(\alpha) + 1)\exp\left(-\frac{t^\alpha}{2}\right) + 2\exp\left(-\frac{t^\alpha}{144}\right).$$

This is the claimed result for this first class of values of t.

b) Let's now consider the case: $t \in [n^\beta, n^{1/\alpha}]$. Then:

$$nP(|X| > t) \leq K'(\alpha) \exp\left(-\frac{t^\alpha}{2}\right).$$

Defining that time $Y_k = X_k I_{(|X_k|\leq t)}$, we will again bound the quantity $P(|(\sum_{k=1}^n Y_k)/\sqrt{n}| > t)$. For doing this, we notice the following chain of inequalities:

$$\forall u > 0, \ P(\|(Y_k/\sqrt{n})\|_{2,\infty} > u) = P\left(\sup_{k=1}^n (k/n)^{1/\alpha}(Y_k^*)^{2/\alpha} > u^{2/\alpha}\right)$$

$$\leq P\left(\sup_{k=1}^n (k/n)^{1/\alpha}Y_k^* > u^{2/\alpha}t^{1-2/\alpha}\right) \leq P\left(\sum_{k=1}^n \frac{|X_k|^\alpha}{n} > u^2 t^{\alpha-2}\right)$$

$$\leq \exp((n/8) - nu^2 t^{\alpha-2}).$$

By choosing as we did before $u = t^{1-\alpha/2}$ we get:

$$P(\|(Y_k/\sqrt{n})\|_{2,\infty} > t^{1-\alpha/2}) \leq \exp(-7t^\alpha/8).$$

By applying Theorem 1.2 in the same way as in the first step, one obtains the announced bound also for $t \in [n^\beta, n^{1/\alpha}]$.

c) Let's finally consider the remaining case $t > n^{1/\alpha}$. Keeping the notations of step 2 and arguing as we did in the step 2, one obtains easily:

$$P(\|(Y_k/\sqrt{n})\|_{2,\infty} > t/(2\sqrt{n})) \leq \exp(-t^\alpha/8).$$

From the following easy implication

$$\|(Y_k(w)/\sqrt{n})\|_{2,\infty} \leq t/(2\sqrt{n}) \Rightarrow |\sum_{k=1}^n \frac{Y_k(w)}{\sqrt{n}}| \leq t,$$

one obtains, for t large enough:

$$P\left(|\sum_{k=1}^n \frac{X_k}{\sqrt{n}}| > t\right) \leq K'(\alpha)\exp(-t^\alpha/2) + \exp(-t^\alpha/8),$$

and this ends the proof of Proposition 1.6.

2. Weak ℓ_p exponential inequalities $(1 < p < 2)$

Exponential inequalities similar to that one given by Theorem 1.2, but with weak ℓ_p $(1 < p < 2)$ hypotheses, are not very difficult to guess from the statement of Lemma 0.3! For that reason, we don't treat in details the case $p \in]1,2[$. We will only give a direct proof — which doesn't need Lemma 0.3 — of the following result:

Theorem 2.1. *Let $X_1, \ldots, X_n$ be independent, symmetrically distributed r.v. such that there exists $p \in]1,2[$ with:*

$$\lambda(p) = \sup_{t>0} t^p \sum_{k=1}^{n} P(|X_k| > t) < +\infty.$$

Let's denote by q the conjugate exponent of p and by $\beta(p)$ the numerical constant

$$4(q+1)(4\lambda(p)/(2-p))^{1/p};$$

for every $x \geq \beta(p)$, $I(p,x)$ will be the interval

$$[\beta(p)/(4(q+1)), x/(4(q+1))].$$

Then the following inequality holds: $\forall x \geq \beta(p)$,

$$P\left(\sum_{k=1}^{n} X_k > x\right) \leq \inf_{c \in I(p,x)} \{P(\|(X_k)\|_{p,\infty} > c)$$

$$+ 2\exp\left(-\frac{1}{16}\left(\frac{x}{(q+1)c}\right)^q\right)\}.$$

Sketch of the proof: Let c be positive; by the same arguments as in the proof of Theorem 1.1 one obtains:

$$\forall u > 0, \quad P\left(\sum_{k=1}^{n} X_k > (q+1)cu\right)$$

$$\leq 2P\left(\sum_{k=1}^{n} X_k I_{(|X_k| \leq cu - (1/(p-1)))} > cu\right)$$

$$+ P(\|(X_k)\|_{p,\infty} > c).$$

An elementary computation shows that:

$$\sum_{k=1}^{n} E(X_k^2 I_{(|X_k| \leq cu^{-(1/(p-1))})}) \leq \frac{2\lambda(p)}{(2-p)} c^{2-p} u^{\frac{p-2}{p-1}}.$$

If now c is chosen larger or equal than $\beta(q)/(4(q+1))$ and $u \geq 4$, then, by Yurinskii's inequality:

$$P\left(\sum_{k=1}^{n} X_k I_{(|X_k| \leq cu^{-(1/(p-1))})} > cu\right) \leq \exp\left(-\frac{u^q}{16}\right),$$

and the inequality stated in Theorem 2.1 easily follows.

The following Hoeffding type result can be derived from Theorem 2.1:

Corollary 2.2. *Let $X_1, \ldots, X_n$ be independent, symmetrically distributed r.v. such that there exist positive constants $a_1, \ldots, a_n$ with:*

$$\forall i = 1, \ldots, n \quad |X_i| \leq a_i \quad a.s.$$

Then, if $p \in]1, 2[$ and q is its conjugate, one has:

$$\forall x \geq 4(q+1)(4/(2-p))^{1/p}\|a_i\|_{p,\infty}$$

$$P\left(\sum_{k=1}^{n} X_k > x\right) \leq 2\exp\left(-\frac{1}{16}\left\{\frac{x}{q+1}\left(\frac{(2-p)}{4\|a_i\|_{p,\infty}^p}\right)^{1/p}\right\}^q\right).$$

Now we will derive some sufficient conditions for the CLT in $C[0,1]$ from the exponential inequalities of sections 1 and 2.

3. Some applications to the central limit theorem in $C[0,1]$

Throughout this whole section, we will consider r.v. taking their values in $(C[0,1], C)$, C being the Borel σ-field associated to the usual norm $\|\cdot\|_\infty$ on $C[0,1]$.

Consider X such a r.v. which is weakly square integrable and centered:

(6) $$\forall s \in [0,1], EX(s) = 0; EX^2(s) < +\infty.$$

Let (X_n) denote a sequence of independent copies of X; one says that X satisfies the central limit theorem $(X \in CLT)$ if the sequence of r.v. $Sn/\sqrt{n} = (X_1 + \cdots + X_n)/\sqrt{n}$ converges in law. From the finite dimensional CLT it is easy to see that if $X \in CLT$, then the limiting r.v. G of the sequence $Sn/\sqrt{n})$ is a gaussian r.v. It is well known that the CLT holds if the following property is true

$$(7) \qquad \forall \varepsilon > 0, \exists \delta > 0: \sup_n P \left(\sup_{|s-t| \leq \delta} \left| \frac{S_n(s) - S_n(t)}{\sqrt{n}} \right| > \varepsilon \right) \leq \varepsilon.$$

This property (7), and the fact that the limiting r.v. G is gaussian, make it not surprising that for checking the CLT property one uses techniques which were introduced for studying the regularity of gaussian random functions. In particular, the following continuity lemma — which extends a famous result of A. Garsia, E. Rodemich and J. Rumsey Jr. [5] — is very useful for proving the CLT property for many classes of r.v.:

Theorem 3.1. [8] *Let:*

i) $\varphi: \mathbf{R} \to \mathbf{R}^+$ *be a function which is continuous, strictly increasing on $\mathbf{R}^+$, even, convex and such that $\varphi(0) \leq 1$; φ^{-1} will denote its $(\mathbf{R}^+$-valued) inverse function.*

ii) ρ *be a pseudo–metric on $[0,1]$;*

iii) μ *be a probability measure on $[0,1]$ (equipped with its ρ-Borel σ-field $\mathcal{B}_\rho$) such that:*

$$\lim_{\varepsilon \to 0} \sup_{x \in [0,1]} \int_0^\varepsilon \varphi^{-1}(1/\mu^2(y:\rho(x,y)<u))du = 0;$$

iv) $f: [0,1] \to \mathbf{R}$ *be a ρ-continuous function.*

If one denotes by $\tilde{f}$ the following real valued function:

$$\forall (s,t) \in [0,1]^2, \quad \tilde{f}(s,t) = \frac{f(s) - f(t)}{\rho(s,t)} I_{(\rho \neq 0)}(s,t),$$

and by c the following integral:

$$c = \int_{[0,1]^2} \varphi(\tilde{f}(s,t))d\mu \otimes \mu(s,t),$$

then one has:

(8) $\forall(s,t) \in [0,1]^2,$

$$|f(s) - f(t)| \le 20 \sup_{x \in [0,1]} \int_0^{\frac{\rho(s,t)}{2}} \varphi^{-1}(c/\mu^2(y : \rho(x,y) < u))du.$$

For proving that (7) holds, one applies relation (8) to the continuous functions $\frac{S_n}{\sqrt{n}}(w, \cdot)$ — for suitable φ, ρ and μ — and one checks that:

$$\forall \varepsilon > 0, \exists \delta > 0:$$

$$P\left(w: \sup_{x \in [0,1]} \int_0^\delta \varphi^{-1}(c_n(w)/\mu^2(y : \rho(x,y) < u))du > \frac{\varepsilon}{20}\right) \le \varepsilon.$$

For several important classes of r.v. X this last step can be done by using Theorem 1.2 or one of its corollaries; this is in particular the case for "strongly lipschitzian" r.v. as we will see. Let's suppose from now that X is a $C[0,1]$–valued r.v. such that there exist a continuous pseudo-metric ρ on $[0,1]$ and a positive r.v. M with:

$$\forall(s,t) \in [0,1]^2, \quad |X(s) - X(t)| \le \rho(s,t)M.$$

In the sequel such a r.v. will be called (ρ, M)–lipschitzian.

Following the pioneering work of R.M. Dudley and V. Strassen [4], many probabilists have studied the CLT for such r.v. The following result — which extends a well known theorem of N.C. Jain and M.B. Marcus [12] — seemed for a long time to be the optimal sufficient condition for the CLT for (ρ, M)–lipschitzian r.v.:

Proposition 3.2. [7]. *Let X be a $C[0,1]$–valued, centered and weakly square integrable (ρ, M)–lipschitzian r.v. If M is square integrable and if there exists a probability measure μ on $([0,1], \mathcal{B}_\rho)$ such that:*

$$(9) \qquad \lim_{\varepsilon \to 0} \sup_{x \in [0,1]} \int_0^\varepsilon \left(Log(1/\mu(y : \rho(x,y) < u)) \right)^{1/2} du = 0,$$

then $X \in CLT$.

The sharpest necessary condition for $X \in CLT$ in terms of integrability of the norm of X being:

$$\lim_{t \to +\infty} t^2 P(\|X\|_\infty > t) = 0,$$

the hypothesis made on M in Proposition 3.2 appears somewhat unnatural. In searching how to make more satisfactory hypotheses on M, A. Zoglat [23] obtained the following result:

Proposition 3.3.   *Let X be a $C[0,1]$–valued, centered and weakly square integrable, (ρ, M)–lipschitzian r.v. Suppose that the following hold:*

  i) $\sup_{t>0} t^2 P\{M > t\} < +\infty;$
  ii) $\sup_{\rho(s,t) \neq 0} \dfrac{E(X(s) - X(t))^2}{\rho^2(s,t)} < +\infty;$
  iii) *there exists a probability measure μ on $([0,1], \mathcal{B}_\rho)$ such that:*

$$\lim_{\varepsilon \to 0} \sup_{x \in [0,1]} \int_0^\varepsilon \Psi(1/_{\mu(y:\rho(x,y)<u)}) du = 0,$$

  where:

$$\forall x \in [0, e^e] : \quad \Psi(x) = Log\, x, \quad and$$
$$\forall x > e^e : \quad \Psi(x) = (Log\, x)/(Log\, Log\, x)$$

  Then $X \in CLT$.

The proof of this result uses Theorem 3.1 and the sharpest of the "classical" exponential inequalities; so it seemed hopeless to obtain an analogue of Proposition 3.2 needing only restriction i) on M. This pessimism was unjustified, because in December 1985, N.T. Andersen, E. Giné, M. Ossiander and J. Zinn [2] obtained the following statement:

Theorem 3.4. *Let X be a $C[0, 1]$–valued, centered and weakly square integrable, (ρ, M)–lipschitzian r.v., such that:*

a) $\sup_{t>0} t^2 P(M > t) < +\infty;$

b) $\lim_{x \to +\infty} x^2 P(\|X\|_\infty > x) = 0,$

c) ρ *and the pseudo–metric τ induced by the covariance of X are both pregaussian.*

Then $X \in CLT$.

Remark: According to M. Talagrand's [21] Theorem, assumption c) implies that there exists a probability measure μ on $[0,1]$ such that (9) holds for ρ. So Theorem 3.4 appears clearly as the optimal extension of Proposition 3.2.

In [2] Theorem 3.4 is obtained as a corollary of a more general result, which proof uses delicate empirical processes technique. Here we will show that Theorem 3.4 can also be obtained directly, as a simple corollary of Proposition 1.4.

Proof of Theorem 3.4: As usual, it suffices to prove the result for symmetrically distributed r.v. So let's suppose that X is symmetrically distributed. By replacing if necessary ρ by $\tilde{\rho} = (\tau^2 + \rho^2)^{1/2}$ there is of course no loss of generality in supposing that:

$$\sup_{\rho(s,t) \neq 0} \frac{\tau(s,t)}{\rho(s,t)} \leq 1.$$

It follows from Talagrand's Theorem [21] that there exists a probability measure μ on $([0,1], \mathcal{B}_\rho)$ such that (9) holds for μ and ρ; so relation (9) will also hold for μ and every pseudo–metric $\gamma = \theta\rho$, where θ is a positive constant which will be specified later.

Suppose now that a sequence (X_k) of independent copies of X is defined on $(\Omega, \mathcal{F}, P)$; suppose furthermore that (M_k) is a sequence of independent copies of M such that:

$$\forall k \in \mathbf{N},\ \forall (s,t) \in [0,1]^2,\ |X_k(s) - X_k(t)| \leq \rho(s,t)M_k.$$

Finally, (ε_k) will be a sequence of independent Rademacher r.v. defined on another probability space $(\Omega', \mathcal{F}', P')$. We will use the following notations:

$$\forall n \in \mathbf{N},\ \forall (s,t) \in [0,1]^2,\ G_n(s,t) = \sum_{k=1}^{n} \frac{\varepsilon_k(X_k(s) - X_k(t))}{\sqrt{n}\gamma(s,t)} I_{(\gamma \neq 0)}(s,t),$$

and

$$c_n = \int_{[0,1]^2} \exp G_n^2(s,t)d\mu \otimes \mu(s,t).$$

According to the method for proving the CLT property which was described at the beginning of this section, one has to show:

(10) $\forall \varepsilon > 0, \exists \delta > 0 :$

$$P \otimes P' \left(\sup_{x \in [0,1]} \int_0^\delta (\mathrm{Log}(c_n/\mu^2(y \colon \gamma(x,y) < u)))^{1/2}du > \frac{\varepsilon}{20} \right) < \varepsilon.$$

Let $\varepsilon > 0$ be fixed.

First we choose $c > 1$ such that:

$$\sup_n P(\| \left(\frac{M_k}{\sqrt{n}}\right)_{k=1,\dots,n} \|_{2,\infty} > c) < \frac{\varepsilon}{2};$$

such a choice is of course possible by hypothesis a) and Lemma 0.2. This c being fixed, we define for every integer n:

$$B_n = (\| \left(\frac{M_k}{\sqrt{n}}\right)_{k=1,\dots,n} \|_{2,\infty} \leq c).$$

Now we fix $\theta = 20c$.

Relation (10) will be true if we find $\delta > 0$ such that for every n:

$$u_n = P \otimes P' \left(\sup_{x \in [0,1]} \int_0^\delta (I_{B_n} Log(c_n/\mu^2(y \colon \gamma(x,y) < u)))^{1/2} du > \frac{\varepsilon}{20} \right) < \frac{\varepsilon}{2}.$$

For finding such a δ, we first notice that for every $(s,t) \in [0,1]^2$ such that $\rho(s,t) \neq 0$, one has by applying Proposition 1.4:

$$\forall x > 1, P \otimes P'(I_{B_n} |G_n(s,t)| > x) \leq 2 \exp \left(\frac{-\frac{x^2}{72} + \frac{x}{240}}{\frac{1}{400} + \frac{1}{800}} \right)$$

$$\leq 2 \exp(-2x^2).$$

From this inequality — which applies for every n — it follows that:

$$K = \sup_n EI_{B_n}(Log c_n)^{1/2} < +\infty.$$

Tchebychev's inequality leads to:

(11) $\quad \forall n \in \mathbf{N}$

$$u_n \leq \frac{20}{\varepsilon}(\delta K + \sup_{x \in [0,1]} \int_0^\delta (Log(1/\mu^2(y \colon \gamma(x,y) < u)))^{1/2} du).$$

It is then easy to choose δ in the right hand side of (11) small enough in order to imply $\sup_n u_n < \varepsilon/2$, and this ends the proof of Theorem 3.4.

Remarks:

1) It can be noticed that hypothesis b) (which is necessary in any case for the CLT) is not needed in the above proof of Theorem 3.4.

2) A natural question at that stage is the following one: "Are the exponential inequalities of sections 1 and 2 also able to give other limit theorems for (ρ, M)–lipschitzian r.v.?" The answer to this question is positive. For instance, the reader will check easily that in a similar way as for Theorem 3.4 — by using, instead of Proposition 1.4 a lemma of the same spirit, obtained as a corollary of Theorem 2.1 — one can prove:

Theorem 3.5. *Let X be a $C[0,1]$–valued, centered (ρ, M)–lipschitzian r.v. Let $p \in]1,2[$ be given and denote by q its conjugate. Suppose that the following hold:*

  a) $\sup_{t>0} t^p P(M > t) < +\infty$.
  b) *The n–dimensional finite joint distributions of $(X(t), t \in [0,1])$ are in the domain of normal attraction of a p–stable measure on $\mathbf{R}^n$, for every integer n.*
  c) *There exists a probability measure μ on $([0,1], B_\rho)$ such that:*

$$\lim_{\epsilon \to 0} \sup_{x \in [0,1]} \int_0^\epsilon (Log(1/\mu(y: \rho(x,y) < u)))^{1/q} du = 0.$$

Then X is in the domain of normal attraction of a p–stable measure on $C[0,1]$.

Theorem 3.5 was obtained by M.B. Marcus and G. Pisier [18] with an entropy hypothesis instead of c), and was extended to the preceding statement by D. Juknevičiené [13].

Several authors [14], [15], [10], [2], [3] have shown that the techniques which are efficient for studying the CLT for (ρ, M)–lipschitzian r.v. are also efficient in studying the law of the iterated logarithm (LIL) for such r.v. So it is not surprising that Proposition 1.4 is also an efficient tool for studying the LIL. We will not develop this remark, but only give an example of a LIL-statement which can be obtained by applying Proposition 1.4: by a similar proof as for Theorem 3.4 and by applying the necessary and sufficient condition for the LIL in Banach spaces discovered recently by M. Ledoux and M. Talagrand [16], one obtains the following result, which precises in the lipschitzian setting the general relation theorem between CLT and LIL [9], [6]:

Theorem 3.6. ([3], §6). *Let X be a $C[0,1]$–valued, centered and weakly square integrable, (ρ, M)–lipschitzian r.v. such that:*

  i) $\sup_{t>0} t^2 P\left\{\frac{M}{\sqrt{L_2 M}} > t\right\} < +\infty;$

ii) $E\left(\frac{\|X\|^2}{L_2\|X\|_\infty}\right) < +\infty;$

iii) ρ *and the pseudo–metric* τ *induced by the covariance of* X *are pregaussian.*

Then X *satisfies the compact LIL in* $C[0,1]$.

Acknowledgement

I am indebted to Michel Talagrand for several useful comments on the first draft of this paper.

References

[1] A. De Acosta, *Strong exponential integrability of sums of independent B-valued random vectors*, Prob. Math. Stat. 1, fasc. 2 (1980), 133–150.

[2] N.T. Andersen, E. Giné, M. Ossiander and J. Zinn, *The central limit theorem and the law of the iterated logarithm for empirical processes under local conditions*, Probab. Theory Relat. Fields 77 (1988), 271–305.

[3] N.T. Andersen, E. Giné and J. Zinn, *The central limit theorem for empirical processes under local conditions: the case of Radon infinitely divisible limits without Gaussian component*, Trans. Amer. Math. Soc. 308 (1988), 603–635.

[4] R.M. Dudley and V. Strassen, *The central limit theorem and ε–entropy*, Lecture Notes in Math. 89 (1969), 224–231.

[5] A. Garsia, E. Rodemich and J. Rumsey jr., *A real variable lemma and the continuity of paths of some gaussian processes*, Indiana Univ. Math. J. 20 (1970), 565–578.

[6] V. Goodman, J. Kuelbs and J. Zinn, *Some results on the law of the iterated logarithm in Banach space with applications to weighted empirical processes*, Ann. Prob. 9 (1981), 713–752.

[7] B. Heinkel, *Mesures majorantes et théorème de la limite centrale dans* $C(S)$, Z. Wahr. verw. Geb. 38 (1977), 339–351.

[8] ————, *Quelques remarques relatives au théorème central-limite dans* $C(S)$, pp. 204–211 in "Vector space measures and applications I", Lecture Notes in Math. 644. Springer Verlag 1978.

[9] ————, *Relation entre théorème central-limite et loi du logarithme itéré dans les espaces de Banach*, Z. Wahr. verw. Geb. 49 (1979), 211–220.

[10] ————, *Mesures majorantes et loi du logarithme itéré pour les variables aléatoires sous-gaussiennes*, Journal of Multivariate Analysis 13 (1983), 353–360.

[11] W. Hoeffding, *Probability inequalities for sums of bounded random variables*, Ann. Stat. Assoc. 58 (1963), 13–29.

[12] N.C. Jain and M.B. Marcus, *Central limit theorems for C(S)-valued random variables*, J. Funct. Anal. 19 (1975), 216–231.

[13] D. Jukneviciené, *Sur la condition de mesure majorante pour le théorème central limite dans C[0,1]*. Preprint 1985. A paraître dans Lietuvos Matematikos Rinkinys.

[14] J. Kuelbs, *The law of the iterated logarithm in C[0,1]*, Z. Wahr. Verw. Geb. 33 (1976), 221–235.

[15] M. Ledoux, *Loi du logarithme itéré dans C(S) et fonction caractéristique empirique*, Z. Wahr. verw. Geb. 60 (1982), 425–435.

[16] M. Ledoux and M. Talagrand, *La loi du logarithme itéré dans les espaces de Banach*, C.R. Acad. Sc. Paris Série I, Tome 303 (1986), 57–60.

[17] M.B. Marcus and G. Pisier, *Characterizations of almost surely continuous p-stable random Fourier series and strongly stationary processes*, Acta Math. 152 (1984), 245–301.

[18] —————————, *Some results on the continuity of stable processes and the domain of attraction of continuous stable processes*, Ann. Inst. Henri Poincaré 20 (1984), 177–199.

[19] G.J. Morrow, *On a central limit theorem motivated by some random Fourier series with dependent coefficients*. Preprint 1984.

[20] S.V. Nagaev, *Large deviations of sums of independent random variables*, Ann. Prob. 7 (1979), 745–789.

[21] M. Talagrand, *Regularity of gaussian processes*, Acta Math. 159 (1987), 99–149.

[22] V.V. Yurinskii, *Exponential bounds for large deviations*, Theor. Prob. Appl. 19 (1974), 154–155.

[23] A. Zoglat, *Mesures majorantes et propriété de limite centrale dans C[0,1]*. (Thèse de 3e cycle, Strasbourg 1986).

Département de Mathématique, 7, rue René Descartes, 67084 Strasbourg Cédex (France)

Extreme Values and LIL Behavior

J. KUELBS AND M. LEDOUX

Abstract. The law of the iterated logarithm is examined when extreme values are deleted from the partial sums of an i.i.d. sequence in a variety of contexts. Results are included which cover random vectors in the domain of attraction of a stable law, or, more generally, whose partial sums can be centered and normalized to be stochastically compact.

1. Introduction

Let X be a random variable and suppose $S_n = X_1 + \cdots + X_n$ where $X_1, X_2, \ldots$ are independent copies of X. If $\{S_n/n^{1/p}\}$ is bounded in probability and $0 < p \leq 2$, then

$$(1.1) \qquad \overline{\lim_n} \frac{\|S_n\|}{n^{1/p}(\log n)^{p'}} = 0 \quad \text{w.p.l.}$$

for any $p' > 1/p$. This is an immediate corollary of Theorem 1, and is best possible in the sense that if $p' \leq 1/p$, then there are vector valued X such that $\{S_n/n^{1/p}\}$ is bounded in probability and yet

$$(1.2) \qquad \overline{\lim_n} \frac{\|S_n\|}{n^{1/p}(\log n)^{p'}} = +\infty \quad \text{w.p.l.}$$

If $p > 2$, the situation is without interest since $p > 2$ and $\{S_n/n^{1/p}\}$ bounded in probability together imply $P(X = 0) = 1$. In case $p = 2$ and $\{S_n/n^{1/2}\}$ is bounded in probability, then by slight modifications of [11, Theorem 4.1] it is possible to prove the more precise result

$$(1.3) \qquad \overline{\lim_n} \frac{\|S_n\|}{\sqrt{nL_2 n}} < \infty \quad \text{iff} \quad E\{\|X\|^2/L_2\|X\|\} < \infty$$

where $L_2 n = \max\{1, \log \log n\}$. Further, the limit in (1.3) is non-zero unless $P(X = 0) = 1$. However, for $0 < p < 2$, the situation is more complicated in the sense that if X is a symmetric stable law of index

$0 < p < 2$, then there are no regular normalization constants $\gamma_n \uparrow \infty$ such that

$$(1.4) \qquad\qquad \overline{\lim_n} \|S_n\|/\gamma_n = c \quad \text{w.p.1.}$$

where $0 < c < \infty$. That is, if $\{\gamma_n\}$ is a sufficiently regular sequence with $\gamma_n \uparrow \infty$, then for X symmetric stable of index $p < 2$,

$$(1.5) \qquad\qquad \overline{\lim_n} \|S_n\|/\gamma_n = +\infty \quad \text{or } 0 \quad \text{w.p.1.}$$

The main question investigated here is: can a portion of the partial sums $\{S_n\}$ be normalized to obtain non-trivial limits in (1.5) for X stable, and for far more general X as well?

Our main results relate the behavior of the maximal terms in the sample $\{\|X_1\|, \ldots, \|X_n\|\}$ and the behavior of the partial sums $\{S_n\}$. Results of this type were obtained in [16], and in more detail in [18], [19] when X was in the domain of attraction of a Gaussian law. This interplay between the maximal terms of the sample and the partial sums has been studied in a variety of contexts by a number of authors. For example, Feller's paper [10] deals with the law of the iterated logarithm (LIL), and those of Mori [21], [22] examine this relationship in the setting of the strong law of large numbers. A related central limit theorem with Gaussian limit was obtained in [6] and [20], and LIL results for symmetric stable processes were obtained in [12]. Now we turn to some notation.

Throughout B will denote a real separable Banach space with topological dual B^* and norm $\|\cdot\|$. We assume $X, X_1, X_2, \ldots$ is a sequence of independent identically distributed B-valued random variables on some probability space $(\Omega, \mathcal{F}, P)$ and as usual let $S_n = X_1 + \cdots + X_n$. We use Lt to denote the function $\max(1, \log_e t)$ and write $L_j t$ for the function L composed j times. $\mathcal{L}(X)$ denotes the law of X. A sequence of random variables $\{W_n\}$ is said to be tight if for each $\varepsilon > 0$ there is a compact set

K_ε such that $\inf_n P(W_n \in K_\varepsilon) > 1 - \varepsilon$. The sequence $\{\mathcal{L}(W_n)\}$ converges weakly to $\mathcal{L}(W)$, and we write

$$\mathcal{L}(W_n) \xrightarrow{w} \mathcal{L}(W),$$

if $\lim_n E(f(W_n)) = E(f(W))$ for all bounded continuous f on the range space of $\{W_n\}$. A random variable is degenerate if its law is concentrated at a single point. Otherwise, it is said to be non-degenerate. Finally, a sequence of random variables $\{W_n\}$ is said to be stochastically compact if $\{W_n\}$ is tight and all weak limits of subsequences of $\{W_n\}$ are non-degenerate. The stochastically compact laws on $\mathbf{R}^1$ arising from suitably normalized partial sums of an i.i.d. sequence were studied by Feller in [9], and more recently in some work by Pruitt [25] and in Griffin, Jain, Pruitt [13]. Of course, since B is complete and separable it is well known that for B-valued $\{W_n\}$, tightness of $\{W_n\}$ implies every subsequence of $\{W_n\}$ contains a weakly convergent subsequence.

The notation $a_n \approx b_n$ is used if there is a $c \in (1, \infty)$ such that for all n sufficiently large

$$1/c < a_n/b_n < c,$$

and $a_n \sim b_n$ if $\lim_n a_n/b_n = 1$. We write $fg(t)$ to denote the composition of f and g.

2. Statement of results

Our first result deals with limit results of the type in (1.1). It is quite elementary, but, along with the examples provided, and the remarks related to (1.4) and (1.5), it provides some motivation for the results which follow. We are grateful for some comments of R.M. Dudley which led us to consider Theorem 1.

Theorem 1. *Let $X, X_1, X_2, \ldots$ be i.i.d. B-valued such that for some p, $0 < p \leq 2$, $\{S_n/n^{1/p}\}$ is bounded in probability. Then, for each $p' > 1/p$ and integer $j \geq 0$,*

$$(2.1) \qquad \overline{\lim_n} \|S_n\| / (n \, Ln \, L_2 n \ldots L_j n)^{1/p} (L_{j+1} n)^{p'} = 0 \quad w.p.1.$$

Examples. The result in Theorem 1 is best possible in the sense that for each $o \in (0, 2]$ and integer j, there are examples with $\{S_n/n^{1/p}\}$ bounded in probability yet

$$(2.2) \qquad \overline{\lim_{n}}\|S_n\|/(n \, Ln \, L_2n \ldots L_{j+1}n)^{1/p} = \infty \quad \text{w.p.l.}$$

To obtain such examples we first consider the case $0 < p < 2$. In this situation simply take X to be a non-degenerate symmetric stable law of index p taking values in any Banach space B. Then $\mathcal{L}(S_n/n^{1/p}) = \mathcal{L}(X)$ and by [1] or [5],

$$(2.3) \qquad \lim_{t \to \infty} t^p P(\|X\| > t) = c_p$$

where $0 < c_p < \infty$. Hence for each $M < \infty$ and integer n

$$\sum_{k \leq n} P(\|X_k\| > M(kLk \ldots L_{j+1}k)^{1/p}) \approx c_p \sum_{k \leq n} M^{-p}(kLk \ldots L_{j+1}k)^{-1},$$

and since $\sum_{k \geq 1}(kLk \ldots L_{j+1}k)^{-1} = \infty$,

$$(2.4) \qquad \overline{\lim_{k}}\|X_k\|/(kLk \ldots L_{j+1}k)^{1/p} = \infty \quad \text{w.p.l.}$$

Thus (2.2) holds for $0 < p < 2$.

When $p = 2$ the examples are slightly more subtle, but are an elementary modification of an example of Pisier-Zinn [24]. That is, let ε be a Rademacher random variable and let

$$(2.5) \qquad X = \varepsilon \sum_{k \geq 1} e_k I(N^2 < k \leq N^2 + N)$$

where $\{e_k : k \geq 1\}$ is the standard basis in the sequence space ℓ_α and N is an integer valued random variable independent of ε. We assume $2 < \alpha < \infty$ and define N such that

$$(2.6) \qquad P(N = k) \approx (k^{1+\frac{2}{\alpha}} L_{j+2}k)^{-1} \qquad (k \geq 1).$$

Hence $P(N \geq k) \approx (k^{2/\alpha} L_{j+2} k)^{-1}$ and using the results of [24] it is easy to show that X satisfies the classical central limit theorem when $2 < \alpha < \infty$. Hence $\{S_n/n^{1/2}\}$ actually converges weakly to a Gaussian measure, but since $\|X\|_{\ell_\alpha} = N^{1/\alpha}$ we have for each $M > 0$ that

$$\sum_{k \leq n} P(\|X_k\|_{\ell_\alpha} \geq M(kLk \ldots L_{j+1}k)^{1/2})$$

$$= \sum_{k \leq n} P(N \geq M^\alpha (kLk \ldots L_{j+1}k)^{\alpha/2})$$

$$\approx \sum_{k \leq n} (kLk \ldots L_{j+2}k)^{-1}.$$

Hence

$$\overline{\lim_k} \|X_k\| / (kLk \ldots L_{j+1}k)^{1/2} = \infty,$$

and thus (2.2) holds when $p = 2$ and X is as in (2.5).

In order to state our main results we need some further notation.

For $n \geq 1$ and $1 \leq j \leq n$, let

$$F_n(j) = \#\{i \colon \|X_i\| > \|X_j\| \text{ for } 1 \leq i \leq n \text{ or } \|X_i\| = \|X_j\| \text{ for } 1 \leq i \leq j\};$$

here $\#D$ denotes the cardinality of the set D. If $F_n(j) = k$, set $X_n^{(k)} = X_j$. Then $\|X_j\|$ is the k'th largest element of the sample $\{\|X_1\|, \ldots, \|X_n\|\}$ when $F_n(j) = k$.

For each $\beta > 1$, set $n_0 = 0$, $n_k = [\beta^k]$, and $I(k) = (n_k, n_{k+1}]$ for $k \geq 1$ where $[\cdot]$ denotes the greatest integer function. Then for any $r > 0$, $n \geq r$, $\tau > 0$ and positive function $d(t)$ defined on $[0, \infty)$, define

$$(2.7) \qquad {}^{(r)}S_n(\beta, \tau) = S_n - \sum_{j=1}^{[r]} X_n^{(j)} I(\|X_n^{(j)}\| > \tau d\bar{\alpha}(n_k))$$

provided $n \in I(k)$, $\alpha(t) = t/L_2 t$, and $\bar{\alpha}(S) = [\alpha(S)]$. Here, of course, $\sum_{j=1}^{0} \alpha_j$ is taken to be zero. Hence ${}^{(r)}S_n(\beta, \tau)$ denotes the partial sum S_n with the $[r]$ largest terms of the sample $\{\|X_1\|, \ldots, \|X_n\|\}$ deleted, provided they exceed $\tau d\bar{\alpha}(n_k)$ in norm when $n \in I(k)$. It would be desirable to eliminate the block aspects of the definition of $^{(r)}S_n(\beta, \tau)$, and if one is willing to accept almost sure boundedness results which are less precise than those of Theorem 2, this can be done (see Remark III following Theorem 2). However, to obtain the more precise results, and the non-degeneracy results in (2.13) at the level of generality of Theorem 2, blocking seems to be required.

We also define the centerings

$$(2.8) \qquad \delta_n(\beta, \tau) = \sum_{j=1}^{n} E(X_j I(\|X_j\| \leq \tau d\bar{\alpha}(n_k)))$$

for $n \in I(k)$, $\beta > 1$, $\tau > 0$, and for each positive function $d(t)$ on $[0, \infty)$ the normalizations

$$\gamma_n = \gamma(n) = L_2 n \, d\alpha(n)$$

where $\alpha(t) = t/L_2 t$. If q is a semi-norm on B, we let $\|q\| = \sup_{\|x\| \leq 1} q(x)$. Then q continuous implies $\|q\| < \infty$.

Theorem 2. *Let $X, X_1, X_2, \ldots$ be independent, identically distributed, B-valued random variables such that*

(2.9) *for the centerings $\{\bar{\delta}_n\}$ and positive normalizing constants $\{\bar{d}_n\}$ the sequence*

$$\{(S_n - \bar{\delta}_n)/\bar{d}_n\}$$

is stochastically bounded, and

(2.10) *for some linear functional $h \in B^*$ the sequence*

$$\{h(S_n - \bar{\delta}_n)/\bar{d}_n\}$$

is stochastically compact.

Then there is an increasing continuous function $d(t)$ defined on $[0, \infty)$ such that

$$(2.11) \qquad\qquad d(n) \approx \bar{d}_n.$$

Further, for each $\tau > 0$ there is a positive constant ξ_0 such that if $r_n = \xi_n L_2 n$ where $\xi_n \geq \xi_0$ and $r_n \leq n$, and q is any continuous semi-norm, then for all $\beta \in (1, \Lambda]$,

$$(2.12) \quad \overline{\lim_n} q(^{(r_n)} S_n(\beta, \tau) - \delta_n(\beta, \tau))/\gamma_n \leq M(\tau)\|q\|(2\Lambda + 1) \quad w.p.l.$$

where $M(\tau)$ is a finite function of $\tau > 0$ which is independent of $\beta > 1$, the semi-norm q, and $\Lambda > 1$. In addition, there is a $\tau_0 > 0$ such that if $\tau \geq \max(\tau_0, 1)$ and $r_n = \xi_n L_2 n$ is as above, then

$$(2.13) \quad \overline{\lim_n} q(^{(r_n)} S_n(\beta, \tau) - \delta_n(\beta, \tau))/\gamma_n > 0 \quad w.p.l.$$

provided $q(x) \geq c|h(x)|$ for all $x \in B$ and some $c > 0$, and h is as in (2.10).

Remarks: (I) The left hand side of (2.12) is actually a constant with probability one. We prove this in Section 4.

(II) It is possible to show (2.13) for all $\tau > 0$ if (4.27) holds for all $\tau > 0$. If $B = \mathbf{R}^1$ this is possible using a result of Pruitt [25] which characterizes the limit laws arising from the partial sums of a stochastically compact i.i.d. sequence. From the proof of [20, Theorem 2] we also have (4.27) in the generality of Banach spaces provided X is in the domain of attraction of a stable law. Corollary 2 below contains a related result.

(III) Let $r_n = \xi_0 L_2 n$ where ξ_0 is as in Theorem 2, and set

$$^{(r_n)} S_n = S_n - \sum_{j=1}^{[r_n]} X_n^{(j)}.$$

Then for $\beta \in (1, \Lambda]$ and $\tau > 0$,

$$\overline{\lim_n} q(^{(r_n)} S_n - \delta_n(\beta, \tau))/\gamma_n \leq M(\tau)\|q\|(2\Lambda + 1) + \xi_0 \tau$$

with probability one. This follows from (2.12) since for $n \in I(k)$,

$$^{(r_n)}S_n - {}^{(r_n)}S_n(\beta, \tau) = \sum_{j=1}^{[r_n]} X_n^{(j)} I(\|X_n^{(j)}\| \leq \tau d\bar{\alpha}(n_k)),$$

and hence for $n \in I(k)$,

$$q\left(^{(r_n)}S_n - {}^{(r_n)}S_n(\beta, \tau)\right) / \gamma_n \leq (\xi_0 L_2 n_{k+1} \tau d\bar{\alpha}(n_k)) / \gamma_{n_k} - \xi_0 \tau.$$

Of course, when X is symmetric this implies

$$\varlimsup_n q\left(^{(r_n)}S_n / \gamma_n\right) < \infty \qquad \text{w.p.l.}$$

In dealing with non-symmetric X it is natural to question whether the centerings $\delta_n(\beta, \tau)$ could be replaced by $nE(X)$ when $E(X)$ exists. One instance when this can be carried out is contained in the following corollary. first, however, we need the following definition.

It is said that X is in the domain of attraction of a non-degenerate stable law Z of index $p \in (0, 2]$ with respect to the centerings $\{\bar{\delta}_n\}$ and normalizations $\{\bar{d}_n\}$, if

$$\mathcal{L}((S_n - \bar{\delta}_n) / \bar{d}_n) \xrightarrow{w} \mathcal{L}(Z).$$

Stable laws and some of their immediate properties are described in [5], [8] for the interested reader.

Corollary 1. *Let $X, X_1, X_2, \ldots$ be i.i.d. B-valued such that X is the domain of attraction of a nondegenerate stable law of index $p \in (1, 2]$. Let $d(t)$ be as in Theorem 2 and let $\beta \in (1, \Lambda]$. Then, for each $\tau > 0$ there exists $\xi_0 \in (0, \infty)$ independent of β such that if $r_n = \xi_n L_2 n$, where $\xi_n \geq \xi_0$ and $r_n \leq n$, then for all continuous semi-norms q,*

$$(2.14) \quad \varlimsup_n q\left(\left(^{(r_n)}S_n(\beta, \tau) - nE(X)\right) / \gamma_n\right)$$

$$\leq M(\tau)\|q\|(2\Lambda + 1) + \Lambda q\left(\int_{\|y\| \geq \tau} y \, d\mu(y)\right) \qquad \text{w.p.l.}$$

where $M(\tau)$ is as in Theorem 2 and μ is the Lévy measure of the stable limit law.

An improvement of (2.12) can be obtained if (2.9) is strengthened and B is a Hilbert space. In case B is not a Hilbert space one can formulate a result of the type in Theorem 3, but one needs some extra assumptions. These will be discussed at the end of Section 5.

If $\{x_n\}$ is a sequence in a metric space (M, d) with $A \subseteq M$, we write $\{x_n\} \twoheadrightarrow A$ if $\lim_n d(x_n, A) = 0$ and for each $x \in A$ there is a subsequence of $\{x_n\}$ which converges to x. $C(\{x_n\})$ denotes the set of all limit points of $\{x_n\}$.

Theorem 3. *Assume the conditions of Theorem 2, with (2.9) replaced by the condition*

$$(2.15) \qquad \{(S_n - \bar{\delta}_n)/\bar{d}_n\} \quad \text{is tight,}$$

and that B is a Hilbert space. Let $d(t)$ be as in Theorem 2 and let $\beta \in (1, \Lambda]$. Then, for each $\tau > 0$ there exists $\xi_0 \in (0, \infty)$ independent of β such that if $r_n = \xi_0 L_2 n$ where $\xi_n \geq \xi_0$ and $r_n \leq n$, then

$$(2.16) \qquad \left\{ \left(\overset{(r_n)}{S_n}(\beta, \tau) - \delta_n(\beta, \tau) \right) / \gamma_n \right\}$$

is conditionally compact in B with probability one. Further, for each $\beta \in (1, \Lambda]$ and $\tau > 0$ there is a non-random compact set $A(\beta, \tau)$ of B such that

$$(2.17) \qquad \left\{ \left(\overset{(r_n)}{S_n}(\beta, \tau) - \delta_n(\beta, \tau) \right) / \gamma_n \right\} \twoheadrightarrow A(\beta, \tau) \quad w.p.l.$$

Remark: If $\tau > 0$ is taken so that (2.13) also holds, then it follows that the limit set $A(\beta, \tau)$ in (2.17) must contain elements of B other than the zero vector.

If the condition (2.9) is strengthened beyond (2.15) it is sometimes possible to show $A(\beta, \tau)$ is more than the zero vector even when $\tau > 0$ is small. For the purposes of the following result let $\tau = 1$ (any $\tau > 0$ will do) and define

$$
\begin{aligned}
{}^{(r)}S_n(\beta) &= {}^{(r)}S_n(\beta, 1) \\
S_n(\beta) &= S_n(\beta, 1).
\end{aligned}
$$

(2.18)

The set $A \subseteq B$ is said to be non-degenerate if it contains elements of B other than the zero vector.

Corollary 2. *Let $X, X_1, X_2, \ldots$ be independent, identically distributed, Hilbert space valued random variables, and assume X is in the domain of attraction of a non-degenerate stable law Z of index $p \in (0, 2]$. Let $d(t)$ be as in Theorem 2, and let $\beta \in (1, \Lambda]$. Then there exists $\xi_0 \in (0, \infty)$ independent of β such that if $r_n = \xi_n L_2 n$ where $\xi_n \geq \xi_0$ and $r_n \leq n$, then*

(2.19) $$\left\{ \left({}^{(r_n)}S_n(\beta) - \delta_n(\beta) \right) / \gamma_n \right\} \twoheadrightarrow A_\beta \qquad w.p.l.$$

where A_β is a compact, non-degenerate, non-random subset of B.

3. Proof of Theorem 1

Since $\{S_n / n^{1/p}\}$ is bounded in probability with $X, X_1, X_2, \ldots$ i.i.d., the argument in [23, Proposition 2.1] easily shows that if X is symmetric, then

(3.1) $$\sup_n \sup_{c>0} c^p P(\|S_n / n^{1/p}\| > c) < \infty.$$

When X is not symmetric, the inequality

$$
\begin{aligned}
\frac{1}{2} P(\|S'_n / n^{1/p}\| > c) &\leq P(\|S'_n\| > cn^{1/p}, \|S_n\| < dn^{1/p}) \\
&\leq P(\|S_n - S'_n\| > (c - d)n^{1/p})
\end{aligned}
$$

for $\{S_n'\}$ an independent copy of $\{S_n\}$ and $d > 0$ such that $P(\|S_n\| \le dn^{1/p}) \ge \frac{1}{2}$, gives (3.1) in general.

To prove (2.1) fix $\varepsilon > 0$, and let $n_k = 2^k$ for $k \ge 1$. Further, let $p' > 1/p$ and fix an integer $j \ge 0$. then

$$(3.2) \qquad \frac{S_n}{(nLn \cdots L_j n)^{1/p}(L_{j+1}n)^{p'}} \xrightarrow{\text{prob}} 0,$$

and by a standard application of Ottaviani's inequality and the Borel Cantelli lemma, (2.1) follows provided

$$(3.3) \qquad \sum_{k \ge 1} P(\|S_{n_k}\| > \varepsilon(n_k Ln_k \cdots L_j n_k)^{1/p}(L_{j+1}n_k)^{p'}) < \infty$$

for arbitrary $\varepsilon > 0$. Of course, since $p' > \frac{1}{p}$ and $n_k = 2^k$, (3.1) easily implies (3.3) and Theorem 2 is proved.

4. Proof of Theorem 2

Let $\Gamma_q(\beta, \tau)$ be the left hand side of (2.12). To see that $\Gamma_q(\beta, \tau)$ is actually a constant with probability one is a simple application of Kolmogorov's zero-one law. That is, since $d(n) \approx \bar{d}_n$ (see below) and from (2.10) we must have $\lim_n \bar{d}_n = \infty$, it follows that for each finite integer J,

$$(4.1) \qquad \lim_n \max_{j \le J} q(X_j)/d\alpha(n) = 0,$$

and hence the event $E_t = \{\omega : \varlimsup_n q\left(\overset{(r_n)}{S_n(\beta, \tau)} - \delta_n(\beta, \tau)\right)/\gamma_n \le t\}$ is in the σ-field $\{\mathcal{F}(X_j) : j \ge J + 1\}$ for each $t < \infty$. Thus $\Gamma_q(\beta, \tau)$ is measurable with respect to the tail σ-field of $\{X_j\}$, and hence $\Gamma_q(\beta, \tau)$ is a constant as claimed. That $\sup_{1 < \beta \le \Lambda} \Gamma_q(\beta, \tau)$ is dominated by the right hand side of (2.12) will be shown next.

The next step of the proof of Theorem 2 is to define the function $d(t)$ and in the next two lemmas prove various properties of $d(t)$. There are two cases to consider: they are $Eh^2(X) < \infty$, and $Eh^2(X) = \infty$ where $h \in B^*$ is as in (2.10).

If $Eh^2(X) < \infty$, then standard symmetrization arguments and (2.10) imply that $\bar{d}_n \approx \sqrt{n}$, so in this case we define

$$(4.2) \qquad\qquad\qquad d(t) = \sqrt{t}.$$

Thus (2.11) holds in this case.

Now assume $Eh^2(X) = \infty$, To define $d(t)$ we proceed as in [20] where the approach of [13], [14], and [25] is followed. That is, let

$$(4.3) \qquad\qquad U(t) = E(h^2(X) \wedge t^2) \qquad (t \geq 0).$$

Then the function

$$(4.4) \qquad\qquad f(t) = \begin{cases} U(t)/t^2 & t > 0 \\ P(|h(X)| > 0) & t = 0 \end{cases}$$

is positive, continuous, decreasing, and $\lim_{t \to \infty} f(t) = 0$. Also, f is strictly decreasing on $[a, \infty)$ where

$$(4.5) \qquad\qquad a = \inf\{x : P(|h(X)| \leq x) > 0\}.$$

Thus $1/f(s)$ is strictly increasing on $[a, \infty)$ with range $[1/f(0), \infty)$, and the function $d(t)$ is defined to be the inverse of $1/f(s)$ on $[1/f(0), \infty)$. That is,

$$(4.6) \qquad d(t) = \begin{cases} \inf\{s > 0 : \dfrac{1}{t} = f(s)\} & t \geq 1/f(0) \\ a & 0 \leq t \leq 1/f(0) \end{cases}$$

then $d(t)$ is continuous, non-decreasing, and strictly increasing on $[1/f(0), \infty)$ with $\lim_{t \to \infty} d(t) = \infty$.

The remainder of the proof now proceeds via a sequence of lemmas. The first lemma contains some useful properties of the function $d(t)$ when $Eh^2(X) = \infty$.

Lemma 4.1.   *If $Eh^2(X) = \infty$, then the function $d(t)$ satisfies the following conditions:*

$$(i) \quad d(t) = \sqrt{t}\,U^{1/2}(d(t)) \quad \text{for all } t \geq 1/f(0).$$

$$(ii) \quad d(n) \approx \bar{d}_n \quad \text{and } d(n+1) \approx d(n).$$

(4.7) $(iii) \quad \alpha^{-1} d\alpha(n) \sim L_2 n\, d\alpha(n).$

$$(iv) \quad \text{If } \gamma_n = L_2 n\, d\alpha(n) \text{ and } n_k = [\beta^k] \text{ for } \beta > 1,$$

$$\text{then } \gamma_{n_{k+1}} \approx \gamma_{n_k} \text{ uniformly in } \beta \in (1, \Lambda].$$

Proof: The proof of (4.7-i, ii) is given in Lemma 3.1 of [20]. To prove (4.7-iii) we define

$$(4.8) \qquad\qquad \phi(t) = E(h^2(X)I(|h(X)| \leq t))$$

for $t \geq 0$. Since $\{h(S_n - \bar{\delta}_n)/\bar{d}_n\}$ is stochastically compact, [9, p. 387] implies

$$\varlimsup_{x \to \infty} x^2 P(|h(X)| > x)/\phi(x) < A_1 < \infty,$$

and for some constants $\gamma, \tau, A_2 > 0$,

$$\phi(tx)/\phi(t) \leq A_2 x^{2-\gamma}$$

for $x > 1$ and $t > T$. Now

$$S^2/U(S) = (s^2/\phi(S))\{1 + s^2 P(|h(x)| > s)/\phi(s)\}^{-1},$$

and hence for large s

$$s^2/U(s) \leq s^2/\phi(s) \leq (s^2/U(s))(1 + A_1).$$

Thus $U(s) \approx \phi(s)$ as $s \to \infty$ and there is a constant A_3 such that for $s \geq 2T$,

$$U(s) \leq A_2\phi(s) = A_3\phi(2Ts/2T)$$
$$\leq A_3 A_2\phi(2T)(s/2T)^{2-\gamma}.$$

Thus (4.7-i) implies that for $d(t) \geq 2T$

$$d(t) \leq t^{1/2}\{A_3 A_2 \phi(2T)/(2T)^{2-\gamma}\}^{1/2} d(t)^{(2-\gamma)/2},$$

and since $\gamma > 0$ there is a constant $c_2 < \infty$ such that

$$d(t) \leq c_2 t^{1/\gamma}$$

for all t sufficiently large. On the other hand, since $U(s)$ is non-decreasing and eventually positive, (4.7-i) implies there is a $c_1 > 0$ such that $d(t) \geq c_1 t^{1/2}$ for t large. Hence $L_2 d\alpha(t) \sim L_2 t$ and $t \to \infty$ and (4.7-iii) is proved.

Using the methods to prove (4.7-iii) and arguing as in Lemma 4.2 below we have

$$\lambda d(t) \geq d(\lambda^{\gamma/2} t)$$

for $\lambda \geq \lambda_1$ and $t \geq t_1$ where $\gamma > 0$ is as above. Hence for all $k \geq 1$ and uniformly in $\beta \in (1, \Lambda]$ we have a positive number θ depending on $\gamma, \lambda_1, t_1, \Lambda$ such that

$$
\begin{aligned}
1 \leq \gamma_{n_{k+1}}/\gamma_{n_k} &= \frac{L_2 n_{k+1} d\alpha(n_{k+1})}{L_2 n_k d\alpha(n_k)} \\
&\leq \theta^{1/2} d\alpha(n_{k+1})/d\alpha(n_k) \\
&= \theta^{1/2} d\left(\alpha(n_k)\frac{\alpha(n_{k+1})}{\alpha(n_k)}\right)/d\alpha(n_k) \\
&\leq \theta^{1/2} d\left((\alpha(n_k) \vee t_1)(\lambda_1^{\gamma/2} \vee \frac{\alpha(n_{k+1})}{\alpha(n_k)})\right)/d\alpha(n_k) \\
&\qquad\qquad\qquad\qquad\qquad \text{since } d \text{ increases} \\
&\leq \theta^{1/2}(\lambda_1^{\gamma/2} \vee \alpha(n_{k+1})/\alpha(n_k))^{2/\gamma} d(\alpha(n_k) \vee t_1)/d\alpha(n_k) \\
&\leq \theta^{1/2}(\lambda_1^{\gamma/2} \vee \alpha(n_{k+1})/\alpha(n_k))^{2/\gamma} d(t_1)/d(1) \\
&\leq \theta.
\end{aligned}
$$

This proves (4.7-iv).

Remark. Note that the constant $\gamma \in (0.2)$ in the proof of Lemma 4.1 does not depend on the semi-norm q, but only on $h(X)$. Condition (4.17-iii) is useful in comparing the results obtained here with those in [18] and [19].

We now have (2.11), and hence we replace $\bar{d}_n$ by $d(n)$ and turn to the proof of (2.12).

Lemma 4.2. *the condition (2.9) and $d(n) \approx \bar{d}_n$ together imply that for each $\tau > 0$ and continuous semi-norm q we have*

$$(4.9) \qquad \begin{aligned} &(i) \sup_{t>0} tP(q(X) > \tau d(t)) = c(\tau, q) < \infty, \text{ and} \\ &(ii) \lim_{\lambda \to \infty} \sup_{t>0} tP(q(X) > \lambda d(t)) = 0. \end{aligned}$$

Proof: Since q is assumed to be a continuous semi-norm it suffices to prove the result when $q(\cdot) = \| \cdot \|$. Further, since $\tau > 0$ with $d(t)$ independent of the norm, (4.9-i) will follow if we show $C(1, q_1) < \infty$ for the norm $q_1(x) = \|x\|/\tau$.

To establish (4.9-i) for $q = q_1$ and $\tau = 1$ consider for each integer n the triangular array $\{Y_{n,j} = X_j/d(n) : 1 \le j \le n\}$ with independent copy $\{Y'_{n,j} : 1 \le j \le n\}$. Then for $\delta, t > 0$,

$$\begin{aligned} nP(q_1(Y_{n,1}) > t + \delta) &= nP(q_1(Y_{n,1}) > t + \delta, \, q_1(Y'_{n,1}) \le \delta) \\ &\quad + nP(q_1(Y_{n,1}) > t + \delta, \, q_1(Y'_{n,1}) > \delta) \\ &\le nP(q_1(Y_{n,1} - Y'_{n,1}) > t) \\ &\quad + nP(q_1(Y_{n,1}) > t + \delta)P(q_1(Y'_{n,1}) > \delta), \end{aligned}$$

and since $d(n) \uparrow \infty$,

$$P(q_1(Y'_{n,1}) > \delta) \le \frac{1}{2}$$

for all $n \ge n_0(\delta)$. Hence for $n \ge n_0(\delta)$,

$$nP(q_1(Y_{n,1}) > t + \delta) \le 2nP(q_1(Y_{n,1} - Y'_{n,1}) > t).$$

Letting $T_n = \sum_{j=1}^{n} Y_{n,j}$ and $T'_n = \sum_{j=1}^{n} Y'_{n,j}$ we have

$$P(q_1(T_n - T'_n) > t) \geq \frac{1}{2} P\left(\max_{1 \leq j \leq n} q_1(Y_{n,j} - Y'_{n,j}) > t\right)$$

$$\geq \frac{1}{2}\{1 - [P(q_1(Y_{n,j} - Y'_{n,j}) \leq t)]^n\}$$

$$\geq \frac{1}{2}\{1 - \exp\{-nP(q_1(Y_{n,j} - Y'_{n,j}) > t)\}\}$$

since $x \leq \exp\{-(1-x)\}$ for $0 \leq x \leq 1$. Since $T_n = S_n/d(n)$ (2.9), with $d(n)$ replacing $\bar{d}_n$, implies there is a number t_0 such that for $t \geq t_0$ we have

$$\sup_n P(q_1(T_n - T'_n) \geq t) \leq \frac{1}{4},$$

and hence for $t \geq t_0$ it is necessary that

$$\overline{\lim_n} nP(q_1(Y_{n,1} - Y'_{n,1}) > t) < \infty.$$

As a result, for $t \geq t_0$, $\delta > 0$, $s = t + \delta$ we have

$$\sup_n nP(q_1(X) \geq sd(n)) < \infty.$$

If $s \leq 1$, then (4.9-i) holds by an easy interpolation. If $s > 1$, then (4.7-i) implies

$$sd(t) = s\sqrt{t}U^{1/2}(d(t))$$

$$\leq \sqrt{s^2 t}U^{1/2}(d(s^2 t))$$

$$= d(s^2 t),$$

and hence $\sup_n nP(q_1(X) > d(s^2 n)) < \infty$. Thus for $t \in [s^2 n, s^2(n+1)]$ we have

$$tP(q_1(X) > d(t)) \leq s^2(n+1)P(q_1(X) > d(s^2 n))$$

$$\leq 2s^2 nP(q_1(X) > d(s^2 n)),$$

and hence $\sup_{t>0} tP(q_1(X) > d(t)) < \infty$ as required. Thus (4.9-i) holds for $q_1(\cdot)$ and, as mentioned previously, it also holds for all continuous semi-norms $q(\cdot)$ on B.

To prove (4.9-ii) we let $g(s) = d^{-1}(s)$ for $s > a$ and $1/f(0)$ for $0 \leq s \leq a$. Then for $\gamma \in (0,2)$ as in the proof of Lemma 4.1 we have

$$\begin{aligned}
g(\lambda d(t)) &= \lambda^2 d^2(t)/U(\lambda d(t)) \\
&= \lambda^2 tU(d(t))/U(\lambda d(t)) \\
&= (\lambda^\gamma t)\lambda^{2-\gamma}U(d(t))/U(\lambda d(t)) \\
&> c\lambda^\gamma t
\end{aligned}$$

for $\lambda \geq 1$ and some constant $c > 0$ provided t is sufficiently large. Now $g(d(\lambda^{\gamma/2}t)) = \lambda^{\gamma/2}t$ for λ, t large, so

$$\lambda d(t) \geq d(\lambda^{\gamma/2}t)$$

for $\lambda \geq \lambda_1, t \geq t_1$. Thus

$$\begin{aligned}
\overline{\lim_{\lambda\to\infty}} \sup_{t>0} tP(q(X) > \lambda d(t)) \leq{}& \varepsilon + \overline{\lim_{\lambda\to\infty}} \sup_{\varepsilon \leq t \leq t_1} tP(q(X) > \lambda d(t)) \\
&+ \overline{\lim_{\lambda\to\infty}} \sup_{t \geq t_1} tP(q(X) > \lambda d(t)) \\
\leq{}& \varepsilon + \overline{\lim_{\lambda\to\infty}} t_1 P(q(X) > \lambda d(\varepsilon)) \\
&+ \overline{\lim_{\lambda\to\infty}} \sup_{t \geq t_1} tP(q(X) > d(\lambda^{\gamma/2}t)) \\
\leq{}& \varepsilon + \overline{\lim_{\lambda\to\infty}} \lambda^{-\gamma/2} \sup_{t \geq t_1} \lambda^{\gamma/2}tP(q(X) > d(\lambda^{\gamma/2}t)) \\
={}& \varepsilon
\end{aligned}$$

where (4.9-i) is applied at the last step. Since $\varepsilon > 0$ was arbitrary (4.9-ii) holds and Lemma 4.2 is proved.

Returning to the proof of Theorem 2 recall $\tau > 0$, $\beta \in (1, \Lambda]$, $n_0 = 0$, and $n_k = [\beta^k]$ with $I(k) = (n_k, n_{k+1}]$, and define

$$(4.10) \qquad \left.\begin{aligned} u_j &= u_j(k) = X_j I(\|X_j\| \leq \tau d\bar{a}(n_k)) \\ v_j &= X_j - u_j \end{aligned}\right\} \quad 1 \leq j \leq n_{k+1}.$$

For $n \in I(k)$ now set

$$U_n = \sum_{j=1}^{n} u_j$$

$$V_n = \sum_{j=1}^{n} v_j.$$

Then

$$S_n = U_n + V_n$$

and

(4.11) $$\max_{n \in I(k)} q^{(r_n)}\left(S_n(\beta, \tau) - \delta_n(\beta, \tau) \right) \leq \max_{n \in I(k)} q(U_n - EU_n)$$

$$+ \max_{n \in I(k)} q\left(V_n - \sum_{j=1}^{[r_n]} X_n^{(j)} I(\|X_n^{(j)}\| > \tau d\bar{\alpha}(n_k)) \right)$$

since $\delta_n(\beta, \tau) = EU_n$ when $n \in I(k)$.

Lemma 4.3. *Let $\tau > 0$ be fixed and q be a continuous semi-norm. Let $\{r_n\}$ satisfy $r_n = \xi_n L_2 n$ where $\xi_n \geq \xi_0$ and assume $r_n \leq n$. If $c_k = \sup_{t \geq \bar{\alpha}(n_k)} t P(\|X\| > \tau d(t))$ and $\xi_0 = \xi_0(\tau)$ is a positive number chosen such that*

(4.12) $$\varlimsup_k (4 \Lambda c_k e / \xi_0)^{\xi_0} \leq e^{-4},$$

then for each $\beta \in (1, \Lambda]$,

(4.13) $$\lim_k \max_{n \in I(k)} q\left(V_n - \sum_{j=1}^{[r_n]} X_n^{(j)} I(\|X_n^{(j)}\| > \tau d\bar{\alpha}(n_k)) \right) = 0 \quad w.p.l.$$

Remark. Since $c_k \leq c(\tau, \|\cdot\|) < \infty$ by (4.9-i), the existence of a positive number ξ_0 satisfying (4.12) is obvious. Further, this inequality provides ξ_0 uniformly for $\beta \in (1, \Lambda]$.

Proof: Fix $\varepsilon > 0$, and set $\lambda_k = [r_{n_k}] + 1$, $p_k = P(\|X\| > \tau d\bar{\alpha}(n_k))$. Then

$$I_k = P\left(\max_{n \in I(k)} q\left(V_n - \sum_{j=1}^{[r_n]} X_n^{(j)} I(\|X_n^{(j)}\| > \tau d\bar{\alpha}(n_k)) \right) > \varepsilon \right)$$

$$\leq P(\text{at least } \lambda_k \text{ of the } X_j's$$

$$(1 \leq j \leq n_{k+1}) \text{ satisfy } \|X_j\| > \tau d\bar{\alpha}(n_k))$$

$$= \sum_{j=\lambda_k}^{n_{k+1}} \binom{n_{k+1}}{j} p_k^j (1 - p_k)^{n_{k+1}-j},$$

and hence by [7, p. 173] and Stirling's formula for all k sufficiently large

$$I_k \leq n_{k+1} \binom{n_{k+1} - 1}{\lambda_k - 1} \int_0^{p_k} t^{\lambda_k - 1}(1 - t)^{n_{k+1}-\lambda_k} dt$$

$$\leq n_{k+1}(n_{k+1}p_k)^{[r_{n_k}]} \int_0^{p_k} (1 - t)^{n_{k+1}-\lambda_k} dt/(\lambda_k - 1)!$$

$$\leq \frac{n_{k+1}}{n_{k+1} - [r_{n_k}]} (n_{k+1}p_k e/[r_{n_k}])^{[r_{n_k}]} (\lambda_k - 1)^{-1/2}.$$

Now

$$p_k = P(\|X\| > \tau d\bar{\alpha}(n_k))$$

$$\leq c_k/\bar{\alpha}(n_k) \leq 2c_k L_2 n_k/n_k,$$

and hence for all k sufficiently large

$$I_k \leq (4\Lambda c_k e/\xi_0)^{(\xi_0 L_2 n_k)/2}$$

since $r_n = \xi_n L_2 n \geq \xi_0 L_2 n$ and $r_n \leq n$. Hence by (4.12)

$$I_k \leq \exp\{-2L_2 n_k\}$$

for all k sufficiently large and thus $\sum_k I_k < \infty$. The Borel Cantelli lemma and $\varepsilon > 0$ arbitrary now yield (4.13), so Lemma 4.3 is proved.

Remark. If $\lim_k c_k = 0$ in Lemma 4.3, then we can choose $\xi(k)$ converging to zero slowly enough so that both

$$(4.14) \qquad\qquad (4\Lambda c_k e/\xi(k))^{\xi(k)} \le e^{-4}$$

and

$$(4.15) \qquad\qquad \sum_k \exp\{-2Ln_k\}/(\xi(k)L_2 n_k)^{1/2} < \infty.$$

As a result, if $r_n = \xi_n L_2 n$ where $\xi_n = \xi(k)$ for $n \in I(k)$, and $\xi(k)$ is as above, then the conclusion of Lemma 4.3 holds. For example, if X is in the domain of attraction of a Gaussian law, then $\lim_k c_k = 0$ is well known.

We now turn to three lemmas which develop ideas from [4] for the case of stochastic boundedness rather than the shift compact case.

Lemma 4.4.   *Let $\{Y_{n,j} = X_j/d(n)\colon 1 \le j \le n\}$ and set $Y_{n,j,\tau} = Y_{n,j}I(\|Y_{n,j}\| \le \tau)$, $Y_{n,j}^\tau = Y_{n,j} - Y_{n,j,\tau}$, $S_n^\tau = \sum_{j=1}^n Y_{n,j}^\tau$, $S_{n,\tau} = S_n/d(n) - S_n^\tau$ for each $\tau > 0$. If (2.9) holds with $\bar d_n$ replaced by $d(n)$, then $\{S_n^\tau\colon n \ge 1\}$ is stochastically bounded for all $\tau > 0$.*

Remark. Since we have proved $\bar d_n \approx d(n)$, the conclusion of Lemma 4.4 holds under the assumptions in Theorem 2.

Proof: Let $c(\tau, q)$ be as in Lemma 4.2. Then $c(\tau, \|\cdot\|) < \infty$ for all $\tau > 0$, and we define

$$\varphi_n = \sum_{j=1}^n I(Y_{n,j}^\tau \ne 0).$$

Then

$$E\varphi_n = nP(\|X\| > \tau d(n)) \le c(\tau, \|\cdot\|) < \infty.$$

Hence

$$P(\|S_n^\tau\| > \lambda) = P(\|S_n^\tau\| > \lambda, \varphi_n > m) + P(\|S_n^\tau\| > \lambda, \varphi_n \le m)$$

$$\le P(\varphi_n > m) + P(\max_{1 \le j \le n} \|Y_{n,j}^\tau\| > \lambda/m)$$

$$\le E\varphi_n/m + nP(\|X\| \ge \frac{\lambda}{m} \cdot d(n))$$

$$\le c(\tau, \| \cdot \|)/m + nP(\|X\| \ge \frac{\lambda}{m} \cdot d(n)).$$

Fixing $\varepsilon > 0$ and taking $m = c(\tau, \| \cdot \|)/\varepsilon$ it follows that

$$P(\|S_n^\tau\| > \lambda) \le \varepsilon + nP(\|X\| > \lambda \varepsilon d(n)/c(\tau, \| \cdot \|)).$$

Hence by Lemma 4.2,

$$\varlimsup_{\lambda \to \infty} P(\|S_n^\tau\| > \lambda) \le \varepsilon,$$

and since $\varepsilon > 0$ was arbitrary, Lemma 4.4 is proved.

Lemma 4.5. *Let $\{Y_{n,j} : 1 \le j \le n\}$ be an arbitrary triangular array of independent random variables with $S_n = \sum_{j=1}^n Y_{n,j}$ and such that*

$$(4.16) \qquad \|Y_{n,j}\| \le A < \infty \quad w.p.l.$$

Further, assume $E(Y_{n,j}) = 0$ for all n, j. Then $\{S_n - \delta_n\}$ stochastically bounded for some sequence of centerings implies $\{S_n\}$ is stochastically bounded.

Proof: $\{S_n - \delta_n\}$ stochastically bounded implies $\{S_n - S_n'\}$ stochastically bounded when $\{S_n'\}$ is an independent copy of $\{S_n\}$. Thus by (4.16), and a well known technique developed by Hoffmann-Jørgensen (see Lemma 4.6 below for details on the application of these ideas), we have

$$\sup_n E\|S_n - S_n'\| < \infty,$$

and hence

$$\sup_n E\|S_n\| < \infty$$

as $E(S_n') = 0$. This completes the proof of Lemma 4.5.

Lemma 4.6. *Let $\{Y_{n,j} : 1 \leq j \leq n\}$ be the triangular array of Lemma 4.4 and $\{Y'_{n,j} : 1 \leq j \leq n\}$ be an independent copy. Assume $\{S_n/d(n) - x_n\}$ is stochastically bounded. then, for all $\tau > 0$*

(i) *$\{\frac{S_n}{d(n)} - E(S_{n,\tau})\}$ is stochastically bounded,*

(ii) *$\{S_{n,\tau} - E(S_{n,\tau})\}$ is stochastically bounded, and if*

$$(4.17) \qquad t_0 = \inf\{t : \sup_n P(\|S_{n,\tau} - S'_{n,\tau}\| \geq t) \leq \frac{1}{24}),$$

then

(iii) *$\sup_n E(\|S_{n,\tau} - S'_{n,\tau}\|) \leq 24t_0 + 12\tau < \infty$.*

Proof: First observe that

$$S_n/d(n) - x_n = (S_{n,\tau} - x_n) + S_n^\tau.$$

Then, given $\varepsilon > 0$, there is a number ρ such that

$$P(\|S_{n,\tau} - x_n\| > \rho) \leq P(\|S_n/d(n) - x_n\| > \frac{\rho}{2}) + P(\|S_n^\tau\| > \frac{\rho}{2}) < \varepsilon$$

by applying Lemma 4.4 since $\tau > 0$. By Lemma 4.5 we thus have $\{S_{n,\tau} - ES_{n,\tau}\}$ stochastically bounded for $\tau > 0$ and hence (4.17-ii) holds. Combining Lemma 4.4 with this, it follows that (4.17-i) holds for all $\tau > 0$.

To prove (4.17-iii) for $\tau > 0$ choose $t_0 = t_0(\tau)$ as indicated. Now $t_0 < \infty$ by (4.17-ii) and hence by Hoffmann-Jørgensen's inequality (see, for example, [5, p. 107]),

$$\int_0^A P(\|S_{n,\tau} - S'_{n,\tau}\| > t)dt = 3 \int_0^{A/3} P(\|S_{n,\tau} - S'_{n,\tau}\| > 3t)dt$$

$$\leq 3\left[4 \int_0^{A/3} P^2(\|S_{n,\tau} - S'_{n,\tau}\| > t)dt + \int_0^{2\tau} P(N_{n,\tau} > t)dt\right]$$

$$\text{since } N_{n,\tau} = \sup_{1 \leq j \leq n} \|Y_{n,j,\tau} - Y'_{n,j,\tau}\| \leq 2\tau$$

$$\leq 12t_0 + \frac{1}{2} \int_{t_0}^{A/3} P(\|S_{n,\tau} - S'_{n,\tau}\| > t)dt + 6\tau.$$

Thus by letting $A \uparrow \infty$ we have

(4.18) $$\sup_n E(\|S_{n,\tau} - S'_{n,\tau}\|) \leq 24t_0 + 12\tau < \infty$$

and the lemma is proved.

Lemma 4.7. *Fix $\tau > 0$ and let $\beta \in (1, \Lambda]$. Let $p_k = [L_2 n_k]$, $s_k = n_{k+1}/p_k$ and assume $\tilde{q}$ is a continuous semi-norm on B such that $\|\tilde{q}\| \leq 1$. Further, assume that for all k sufficiently large*

(4.19) $$E\left\{ \tilde{q}\left(\sum_{j=1}^{s_k+1} (u_j(k) - Eu_j(k)) \right) / d\alpha(n_k) \right\} \leq e^{-2(\tau+1)}.$$

Then

(4.20) $$\overline{\lim_k} \max_{n \in I(k)} \tilde{q}(U_n - EU_n)/\gamma_n \leq 4 \quad w.p.l.$$

Proof: Since $\{\tilde{q}(U_n - EU_n): n \in I(k)\}$ is a submartingale $\{\exp(\tilde{q}(U_n - EU_n)): n \in I(k)\}$ is also a submartingale, and hence, by the submartingale maximal inequality, we have

$$\begin{aligned}
J_k &= P(\max_{n \in I(k)} \tilde{q}(U_n - EU_n)/d\alpha(n_k) > AL_2 n_k) \\
&\leq P(\max_{n \in I(k)} \exp\{\tilde{q}(U_n - EU_n)/d\alpha(n_k)\} \geq \exp(AL_2 n_k)) \\
&\leq \exp\{-AL_2 n_k\} E(\exp\{\tilde{q}(U_{n_{k+1}} - EU_{n_{k+1}})/d\alpha(n_k)\}) \\
&\leq \exp\{-AL_2 n_k\} E\left(\prod_{m=1}^{p_k} e^{T_m} \right) \\
&\leq \exp\{-AL_2 n_k\} \prod_{m=1}^{p_k} E(e^{T_m})
\end{aligned}$$

where

$$T_m = \tilde{q}\left(\sum_{j=(m-1)s_k}^{m s_k} (u_j(k) - Eu_j(k))/d\alpha(n_k) \right)$$

and we define

$$\sum_{j=s}^{t} = \sum_{j=[s]+1}^{[t]}$$

whenever s or t are non-integer. Now by a result of A. de Acosta [2], for all $t > 0$:

$$E(e^{T_m}) \le e^{t+2\tau}$$

$$P\left(\sup_{\ell \le ms_k} \tilde{q}\left(\sum_{j=(m-1)s_k}^{\ell} (u_j(k) - Eu_j(k))/d\alpha(n_k) \right) > t \right) E(e^{T_m}) + e^t$$

since $d(t)$ is increasing and $\|\tilde{q}\| \le 1$ implies

$$\tilde{q}(u_j(k) - Eu_j(k))/d\alpha(n_k) \le 2\tau \|\tilde{q}\| \le 2\tau.$$

Setting $t = 2$ we have by the submartingale maximal inequality that for all large k

$$P\left(\sup_{\ell \le ms_k} \tilde{q}\left(\sum_{j=(m-1)s_k}^{\ell} (u_j(k) - Eu_j(k))/d\alpha(n_k) \right) > 2 \right)$$

$$\le \frac{1}{2} E\left(\tilde{q}\left(\sum_{j=(m-1)s_k}^{ms_k} (u_j(k) - Eu_j(k))/d\alpha(n_k) \right) \right)$$

$$\le \frac{1}{2} E\left(\tilde{q}\left(\sum_{j=1}^{s_k+1} (u_j(k) - Eu_j(k))/d\alpha(n_k) \right) \right)$$

by stationarity and Jensen's inequality

$$\le \frac{1}{2} e^{-2(\tau+1)} \quad \text{by (4.19)}.$$

Hence, for all large k, we have

$$E(e^{T_m}) \le e^{2(\tau+1)} \cdot \frac{1}{2} e^{-2(\tau+1)} E(e^{T_m}) + e^2$$

which implies $E(e^{T_m}) \leq 2e^2 \leq e^3$ for $1 \leq m \leq p_k$. Hence for all large k

$$J_k \leq \exp\{-AL_2 n_k\}(e^3)^{p_k} \leq \exp\{-(A-3)L_2 n_k\},$$

and choosing $A > 4$ we have $\sum_k J_k < \infty$. Thus Lemma 4.7 is proved.

To obtain the proof of (2.12) we let q be any continuous semi-norm and fix $\tau > 0$. Let $\beta \in (1, \Lambda]$ and let also $\theta(\tau) = e^{-2(\tau+1)}(2\Lambda+1)(24t_0 + 12\tau) + 1$ where $t_0 = t_0(\tau)$ is as in (4.17-iii). Next define the continuous semi-norm

$$\tilde{q}(x) = q(x)/(\theta(\tau)\|q\|) \quad (x \in B).$$

Then, for all k large enough,

$$E\Big(\tilde{q}\Big(\sum_{j=1}^{s_k+1}(u_j(k) - Eu_j(k))/d\alpha(n_k)\Big)\Big)$$

$$\leq E\Big(\|\sum_{j=1}^{s_k+1}(u_j(k) - Eu_j(k))/d\alpha(n_k)\|\Big)/\theta(\tau)$$

$$\leq (2\Lambda+1)(24t_0 + 12\tau)/\theta(\tau)$$

$$\leq e^{-(2\tau+1)}$$

by Lemma 4.6 where we have used the fact that $s_k + 1 \leq (2\Lambda+1)\bar{\alpha}(n_k)$ for large k. Now $\|\tilde{q}\| \leq 1$, so Lemma 4.7 implies

$$\lim_k \max_{n \in I(k)} \tilde{q}(U_n - EU_n)/\gamma_n \leq 4 \quad \text{w.p.l.}$$

Letting $M(\tau) = 4\theta(\tau)/(2\Lambda+1)$, yields

$$(4.21) \qquad \overline{\lim}_k \max_{n \in I(k)} q(U_n - EU_n)/\gamma_n \leq M(\tau)\|q\|(2\Lambda+1) \quad \text{w.p.l.}$$

where $M(\tau)$ is a finite function of $\tau > 0$ which is independent of $\beta > 1$, the semi-norm $q(\cdot)$, and $\Lambda > 1$. Combining (4.11), (4.13), and (4.21) now yields (2.12).

To complete the proof of Theorem 2 it suffices to verify (2.13). Since h is q-continuous this follows if

$$(4.22) \qquad \overline{\lim_n} \, h \left({}^{(r_n)} S_n(\beta, \tau) - \delta_n(\beta, \tau) \right) / \gamma_n > 0 \quad \text{w.p.1.}$$

Throughout the remainder of the proof we assume $\tau \geq 1$ and take $\beta \in (1, \Lambda]$. Let c_k be defined as in Lemma 4.3. Then c_k is actually a decreasing function of τ, so we choose ξ_0 satisfying (4.12) with $c_k = c_k(1)$. Then, for each $\tau \geq 1$ and $\beta \in (1, \Lambda]$ we have (4.13) holding and, since h is q-continuous, (4.22) holds if we show there is a τ_0 such that $\tau \geq \max(\tau_0, 1)$ implies

$$(4.23) \qquad \overline{\lim_n} \, |h(U_n - EU_n)| / \gamma_n > 0 \quad \text{w.p.1.}$$

Of course, in (4.23) U_n is defined for $n \in I(k)$ by $\sum_{j=1}^{n} u_j$ where $u_j = u_j(k)$ is defined as in (4.10). Further, from (4.13) and (2.12) there is a constant $A < \infty$ such that

$$(4.24) \qquad \overline{\lim_k} \, \max_{n \in I(k)} |h(U_n - EU_n)| / \gamma_n \leq A \quad \text{w.p.1.}$$

Let $m_k = [\beta^{k^2+1}]$ be a subsequence of $\{n_k\}$ and define

$$(4.25) \qquad
\begin{aligned}
E_k &= \left\{ \left| h \left(\sum_{j=m_{k-1}+1}^{m_k} (u_j(k^2) - Eu_j(k^2)) \right) \right| > 2c\gamma_{m_k} \right\} \\
F_k &= \left\{ \left| h \left(\sum_{j=1}^{m_{k-1}} (u_j(k^2) - Eu_j(k^2)) \right) \right| \leq c\gamma_{m_k} \right\}
\end{aligned}$$

for $k \geq 1$ and $c > 0$. Then the $\{E_k\}$ are independent and for all k sufficiently large we have by Markov's inequality and (4.8) that for

$\|h\| = 1$,

$$P(F_k^c) \leq m_{k-1}\phi(\tau d\bar{\alpha}(m_k))/(c^2\gamma_{m_k}^2)$$

$$= m_{k-1}\phi(\tau d\bar{\alpha}(m_k))/(c^2(L_2 m_k)^2\alpha(m_k)U(d\alpha(m_k)))$$

since $d(t) = \sqrt{t}U^{1/2}(d(t))$

$$\approx \frac{m_{k-1}\phi(d\bar{\alpha}(m_k))\max(1, A_2\tau^{2-\gamma})}{c^2(L_2 m_k)^2\alpha(m_k)U(d\alpha(m_k))}$$

(4.26)

where $0 < \gamma < 2$ is as in Lemma 4.1

$$\approx m_{k-1}/(m_k L_2 m_k)$$

since $U(t) \approx \phi(t)$ as in Lemma 4.1

and $\phi(d(n+1)) \approx \phi(d(n))$

$$\leq k^{-3}.$$

Further, by (3.19) of [20] we have a $\tau_0 > 0$ such that $\tau \geq \tau_0$ implies

(4.27) $$\lim_k \alpha(n_k)E(h^2(u_1(k) - Eu_1(k)))/(d\alpha(n_k))^2 > 0.$$

Now

$$\left\{\left|h\left(\sum_{j=1}^{m_k}(u_j(k^2) - Eu_j(k^2))\right)\right| > c\gamma_{m_k}\right\} \supseteq E_k \cap F_k,$$

and hence

(4.28) $$\left\{\left|h\left(\sum_{j=1}^{m_k}(u_j(k^2) - Eu_j(k^2))\right)\right| > c\gamma_{m_k} \text{ i.o. }\right\} \supseteq \{E_k \cap F_k \text{ i.o.}\}.$$

Further, (4.26) implies $P(F_k^c \text{ i.o.}) = 0$, and hence we have $P(E_k \cap F_k \text{ i.o.}) = P(E_k \text{ i.o.})$. Letting $\ell_k = m_k - m_{k-1}$ it follows from (4.27) and [26, p. 262] that for k sufficiently large and $\gamma > 0$

(4.29) $$P(E_k) = P\left(\left|\sum_{j=1}^{\ell_k}\frac{h(u_j(k^2) - Eu_j(k^2))}{\sqrt{L_2 m_k}d\alpha(m_k)}\right| > 2c\sqrt{L_2 m_k}\right)$$

$$\geq \exp\{-(1+\gamma)2c^2 L_2 m_k\}$$

for all $c > 0$ small enough. Thus for all k sufficiently large

$$(4.30) \qquad P(E_k) \geq k^{-6c^2(1+\gamma)},$$

and if $c > 0$ is sufficiently small we have $\sum_k P(E_k) = \infty$. Since $\{E_k\}$ is an independent sequence, the Borel Cantelli Lemma now implies

$$P(E_k \text{ i.o.}) = P(E_k \cap F_k \text{ i.o.}) = 1,$$

and hence (4.28) implies

$$(4.31) \qquad \overline{\lim_k} \left| h\left(\sum_{j=1}^{m_k} (u_j(k^2) - Eu_j(k^2)) \right) \right| / \gamma_{m_k} > 0 \quad \text{w.p.l.}$$

Thus (4.23) holds and the theorem is proved provided $\tau \geq \max(\tau_0, 1)$ where τ_0 is as in (4.27).

5. Proof of Corollaries 1 and 2 and Theorem 3

Applying Theorem 2, Corollary 1 can be proved by showing that

$$(5.1) \qquad \overline{\lim_n} q(nE(X) - \delta_n(\beta, \tau))/\gamma_n \leq \Lambda q\left(\int_{\|y\| \geq \tau} y \, d\mu(y) \right).$$

Now the left hand side of (5.1) is equivalent to

$$(5.2) \qquad \overline{\lim_n} q(nE(XI(\|X\| > \tau d\bar{\alpha}(n_k))I(n \in I(k)))/\gamma_n,$$

and, since $\alpha(n) = n/L_2 n$ and $\gamma_n = L_2 n d\alpha(n)$ is increasing, (5.1) follows once it is proved that

$$(5.3) \qquad \lim_n nE(XI\|X\| > \tau d(n)))/d(n) = \int_{\|y\| > \tau} y \, d\mu(y)$$

where μ is the Lévy measure of the non-degenerate stable limit law of index $p \in (1, 2]$. Of course, the right hand term of (5.3) makes sense as a Bochner integral in B since $1 < p \leq 2$. See [5] for details and additional references on the Lévy measure of a stable law in the Banach space setting.

To verify (5.3) consider the case $p = 2$ and then $1 < p < 2$. If $p = 2$, then μ is the zero measure, and (5.3) follows immediately from [18] (see the proof of Lemma 3.2). If $1 < p < 2$, then by Theorem 2.10 of [4] it follows that

$$(5.4) \qquad \mathcal{L}\left(\frac{S_n - nE(XI(\|X\| \leq \tau d(n)))}{d(n)}\right) \xrightarrow{w} c_\tau \operatorname{Pois}(\mu)$$

where $c_\tau \operatorname{Pois}(\mu)$ is defined as in [4]. Further, by Theorem 6.1 of [3] the sequence $\{\|S_n - nE(XI(\|X\| \leq \tau d(n)))\|/d(n)\}$ is uniformly integrable, and hence

$$(5.5) \qquad \lim_n E(S_n - nE(XI(\|X\| \leq \tau d(n))))/d(n) = E(W)$$

where $\mathcal{L}(W) = c_\tau \operatorname{Pois}(\mu)$. Now a standard calculation gives $E(W) = \int_{\|y\|>\tau} y\, d\mu(y)$, and hence by rewriting the left-hand side of (5.5) it follows that (5.3) is verified. Thus Corollary 1 is proved.

To prove Theorem 3 we again apply Theorem 2. Since $\{(S_n - \bar{\delta}_n)/\bar{d}_n\}$ is tight the results in [4] analogous to Lemma's 4.4–4.6 imply that for each $\tau > 0$ the sequence

$$(5.6) \qquad\qquad \{S_{n,\tau} - ES_{n,\tau}\} \quad \text{is tight in } B, \text{ and}$$

$$(5.7) \qquad\qquad \sup_n E\|S_{n,\tau} - ES_{n,\tau}\|^p < \infty \quad \text{for each } p > 0$$

where $S_{n,\tau}$ is as in Lemma 4.4. The claim in (5.7) follows immediately from (5.6), and the method of proof for (4.17-iii).

Now fix $\tau > 0$ and choose $\varepsilon > 0$. In view of Theorem 2, to prove (2.16) it suffices to show that there is a finite dimensional subspace F such that

$$(5.8) \qquad \varlimsup_n q_F\left(\left(\overset{(r_n)}{}S_n(\beta, \tau) - \delta_n(\beta, \tau)\right)/\gamma_n\right) \leq 2\varepsilon \quad \text{w.p.1.}$$

where $q_F(x) = \inf_{y \in F} \|x - y\|$. Of course, since B is a Hilbert space, $q_F(x) = \|Q_F(x)\|$ where Q_F is the orthogonal projection onto the orthogonal complement of F. Further, in view of (4.10) and Lemma 4.3, (5.8) will hold once it is shown that

$$(5.9) \qquad \overline{\lim_k} \max_{n \in I(k)} q_F(U_n - EU_n)/\gamma_n \leq 2\varepsilon \quad \text{w.p.l.}$$

For $n \in I(k)$, since q_F is an inner-product semi-norm and $\gamma_n = L_2 n d\alpha(n)$ increases, Jensen's inequality and stationarity implies that

$$
\begin{aligned}
E(q_F^2(U_n &- EU_n)/L_2 n(d\alpha(n))^2) \\
&\leq E(q_F^2(U_{n_{k+1}} - EU_{n_{k+1}})/L_2 n_k(d\bar{\alpha}(n_k))^2) \\
(5.10) \qquad &\leq 2\beta E q_F^2\left(\sum_{j=1}^{[L_2 n_k]} Z_{n_{k+1},j}/\sqrt{L_2 n_k}\right) \\
&\leq 2\beta E q_F^2(Z_{n_{k+1},1})
\end{aligned}
$$

where $Z_{n_{k+1},j} = \sum_{k=(j-1)\bar{\alpha}(n_k)+1}^{j\bar{\alpha}(n_k)}(u_j(k) - Eu_j(k))/d\bar{\alpha}(n_k)$. Now $\{Z_{n_{k+1},1}: k \geq 1\}$ is a subsequence of $\{S_{n,\tau} - ES_{n,\tau}\}$ and since (5.6) and (5.7) hold it follows by a standard argument that there is a finite dimensional subspace F such that

$$(5.11) \qquad \sup_k E(q_F^2(Z_{n_{k+1},1})) < \varepsilon^2/(64\Lambda e^{16\tau/\varepsilon}).$$

Combining (5.10) and (5.11) it follows that

$$(5.12) \qquad \sup_{n \in I(k)} E(q_F^2(U_n - EU_n)) \leq \varepsilon^2 L_2 n_k(d\bar{\alpha}(n_k))^2/(32 e^{16\tau/\varepsilon}).$$

Further, since $\gamma_n = L_2 n d\alpha(n)$, (5.10) and (5.11) combine to imply that

$$(5.13) \qquad \overline{\lim_k} \max_{n \in I(k)} E(q_F(U_n - EU_n)/\gamma_{n_k}) = 0,$$

and hence by Ottaviani's inequality for all k sufficiently large

(5.14)
$$P(\max_{n \in I(k)} q_F(U_n - EU_n)/\gamma_n > 2\varepsilon)$$
$$\leq 2P(q_F(U_{n_{k+1}} - EU_{n_{k+1}}) > \varepsilon \gamma_{n_k}).$$

Now (5.14) and the Borel Cantelli Lemma yield (5.9) if

(5.15)
$$\sum_k P(q_F(U_{n_{k+1}} - EU_{n_{k+1}}) > \varepsilon \gamma_{n_k}) < \infty.$$

To prove (5.15) define

(5.16)
$$\begin{array}{ll} \text{(i)} & b_{n_k} = \varepsilon \sqrt{L_2 n_k} \, d\bar{\alpha}(n_k)/4 \\ \text{(ii)} & \varepsilon_{n_k} = 2\sqrt{L_2 n_k} \\ \text{(iii)} & c_{n_k} = 8\tau/(\varepsilon \sqrt{L_2 n_k}). \end{array}$$

Since $q_F(x) \leq \|x\|$ it follows from (4.10) and (5.16) that

(5.17)
$$\begin{array}{ll} \text{(i)} & q_F(u_j(k) - Eu_j(k)) \leq 2\tau d\bar{\alpha}(n_k) \leq c_{n_k} b_{n_k} \\ \text{(ii)} & \varepsilon_{n_k} c_{n_k} = 16\tau/\varepsilon \\ \text{(iii)} & 2\varepsilon_{n_k} b_{n_k} = \varepsilon L_2 n_k d\bar{\alpha}(n_k) \leq \varepsilon \gamma_{n_k}, \end{array}$$

and by using the estimate $1 + x/3 + x^2/(4 \cdot 3) + x^3/(6 \cdot 5 \cdot 4) + \cdots \leq e^x$ the proof of Lemma 2.1 of [15] implies that

(5.18) $\quad P(q_F(U_{n_{k+1}} - EU_{n_{k+1}}) > \varepsilon \gamma_{n_k})$
$$\leq \exp\{-\varepsilon_{n_k}^2 [1 - \frac{1}{2} \sum_{j=1}^{n_{k+1}} E(q_F^2(u_j(k) - Eu_j(k)))e^{\varepsilon_{n_k} c_{n_k}}/b_{n_k}^2$$
$$- Eq_F(U_{n_{k+1}} - EU_{n_{k+1}})/2\varepsilon_{n_k} b_{n_k}]\}.$$

Now q_F being an inner product semi-norm, (5.12) and (5.16-i) together imply that

(5.19)
$$\frac{1}{2} e^{-16\tau/\varepsilon} \geq \sum_{j=1}^{n_{k+1}} E(q_F^2(u_j(k) - Eu_j(k)))/b_{n_k}^2,$$

and (5.13) and (5.17-iii) yield

$$(5.20) \qquad \overline{\lim_{k}} \, Eq_F(U_{n_{k+1}} - EU_{n_{k+1}})/(2\varepsilon_{n_k} b_{n_k}) = 0.$$

Thus (5.18) and (5.17-ii) imply that for k sufficiently large

$$(5.21) \qquad P(q_F(U_{n_{k+1}} - EU_{n_{k+1}}) > \varepsilon \gamma_{n_k}) \leq \exp\{-2L_2 n_k\},$$

and (5.21) yields (5.15), so (2.16) is proved.

Given (2.16), (2.17) follows as an elementary application of the proof of Lemma 1 of [17]. That is, one first proves that there is a non-random set $A(\beta, \tau)$ such that

$$(5.22) \qquad C(\{(^{(r_n)}S_n(\beta,\tau) - \delta_n(\beta,\tau))/\gamma_n)\}) = A(\beta,\tau) \quad \text{w.p.l.}$$

Then the compactness of $A(\beta, \tau)$ follows from (2.16), and to prove that

$$(5.23) \qquad \lim_{n} d((^{(r_n)}S_n(\beta,\tau) - \delta_n(\beta,\tau))/\gamma_n, A(\beta,\tau)) = 0 \quad \text{w.p.l.}$$

follows immediately from (2.16) and (5.22). Thus (2.17) holds.

   Remark. The important steps of the proof of Theorem 3 are that (5.10) holds, and that (5.12) implies something of the form (5.19). To prove (5.10) we used a type 2 assumption (which follows immediately from the inner-product structure), and to obtain an inequality like (5.19) from (5.12) it is convenient to have B of cotype 2. Since the only Banach spaces which are both of type and cotype 2 are isomorphic to Hilbert space, it is natural to assume B is a Hilbert space in Theorem 3. Of course, if (5.10) and (5.19) can be verified directly in some other way, then our result holds without the assumption B is a Hilbert space.

   Corollary 2 follows immediately from Theorem 3 by setting $\tau = 1$ and by showing A_β contains points other than the zero vector, i.e. A_β is non-degenerate. That is, recall from the proof of (2.13) that if (4.27) holds for $\tau > 0$, then (2.13) holds for such a τ. Letting $\tau = 1$ we have (4.27) from Lemma 4.2 of [20]. Indeed, Lemma 4.2 of [20] gives (4.27) for all $\tau > 0$, so (2.13) holds for all $\tau > 0$ when X is in the domain of attraction of a stable law. Now (2.13) and (2.16) holding for $\tau = 1$, gives A_β non-degenerate and hence the corollary is proved.

References

[1] A. de Acosta, *Asymptotic behavior of stable measures*, Ann. Probability 5 (1977), 494–499.

[2] —————, *Exponential moments of vector valued random series and triangular arrays*, Ann. Probability 8 (1980), 381–389.

[3] A. de Acosta and E. Giné, *Convergence of moments and related functionals in the general central limit theorem in Banach spaces*, Z. Wahrscheinlichkeitstheorie verw. Gebiete 48 (1979), 213–231.

[4] A. de Acosta, A. Araujo and E. Giné, "Poisson measures, Gaussian measures and the central limit theorem in Banach spaces," Advances in Probability, vol. IV, Dekker, New York, 1978, pp. 1–68.

[5] A. Araujo and E. Giné, "The Central Limit Theorem for Real and Banach Valued Random Variables," John Wiley and Sons, New York, 1980.

[6] S. Csörgo, L. Horváth and D. Mason, *What portion of the sample makes a partial sum asymptotically stable or normal?*, Probab. Th. Rel. Fields 72 (1984), 1–16.

[7] W. Feller, "An Introduction to Probability Theory and Its Applications," vol. I, 3rd edition, John Wiley and Sons, New York, 1968.

[8] —————, "An Introduction to Probability Theory and Its Applications," vol. II, 2nd edition, John Wiley and Sons, New York, 1971.

[9] —————, *On regular variation and local limit theorems*, Proc. Fifth Berkeley Symp. Math. Statist. Prob. II, Part 1, pp. 373–388. University of California Press, Berkeley, California 1967.

[10] —————, *An extension of the law of the iterated logarithm to variables without variance*, J. Math. and Mechanics 18 (1968), 343–355.

[11] V. Goodman, J. Kuelbs and J. Zinn, *Some results on the LIL in Banach space with applications to weighted empirical processes*, Ann. Probability 9 (1981), 713–752.

[12] P. Griffin, *Laws of the iterated logarithm for symmetric stable processes*, Z. Wahrscheinlichkeitstheorie verw. Geb. 68 (1985), 271–285.

[13] P. Griffin, N. Jain and W. Pruitt, *Approximate local limit theorems for laws outside the domain of attraction*, Ann. Probability 12 (1984), 45–63.

[14] N. Jain and S. Orey, *Domains of partial attraction and tightness conditions*, Ann. Probability **8** (1980), 584–599.

[15] J. Kuelbs, *Kolmogorov's Law of the iterated logarithm for Banach space valued random variables*, Illinois J. of Math. **21** (1977), 784–800.

[16] J. Kuelbs and J. Zinn, *Some results on LIL behavior*, Ann. of Probability **11** (1983), 506–557.

[17] J. Kuelbs, *When is the cluster set of $\{S_n/a_n\}$ empty?*, Ann. of Probability **9** (1981), 377–394.

[18] ————, *The LIL when X is in the domain of attraction of a Gaussian law*, Ann. of Probability **13** (1985), 825–859.

[19] J. Kuelbs abd M. Ledoux, *Extreme values and the law of the iterated logarithm*, Probab. Th. Rel. Fields **74** (1987), 319–340.

[20] ————————, *Extreme values and a Gaussian central limit theorem*, Probab. Th. Rel. Fields **74** (1987), 341–355.

[21] T. Mori, *The strong law of large numbers when extreme values are excluded from sums*, Z. Warsch. verw. Gebiete **36** (1976), 189–194.

[22] ————, *Stability for sums of i.i.d. random variables when extreme terms are excluded*, Z. Warsch. verw. Gebiete **41** (1977), 159–167.

[23] G. Pisier, *Le théorème limite centrale et la loi du logarithme itérée dans les espaces de Banach*, Seminaire Maurey-Schwartz (exposes III and IV), Paris 1975.

[24] G. Pisier and J. Zinn, *On the limit theorems for random variables with values in the spaces L_p ($1 \leq p < \infty$)*, Z. Wahrscheinlichkeitstheorie verw. Gebiete **41** (1978), 289–304.

[25] W.E. Pruitt, *The class of limit laws for stochastically compact normed sums*, Ann. of Probability **11** (1983), 962–969.

[26] W.F. Stout, "Almost Sure Convergence," Academic Press, New York, 1974.

Department of Mathematics, University of Wisconsin, Madison, Wisconsin 53706, U.S.A.

Département de Mathématique, Université de Strasbourg, 7, Rue René Descartes, 67084 Strasbourg, France

Entropy Numbers and Duality for Operators with Values in a Hilbert Space

H. König, V.D. Milman*, N.Tomczak-Jaegermann

1. Introduction and preliminaries

Let Y be a Banach space and let $T \subset Y$ be a compact body. Let $K \subset Y$ be a compact set. Recall that the covering number $N(K,T)$ is defined by

$$N(K,T) = \inf\left\{ N : \exists y_1, \ldots, y_N \text{ in } Y \text{ such that } K \subset \bigcup_1^N (y_i + T) \right\}.$$

Let X and Y be Banach spaces and let B_X and B_Y be the closed unit balls. For a compact operator $u \colon X \to Y$, the entropy numbers $e_k(u)$ are defined by

$$e_k(u) = \inf\{\varepsilon > 0 : N(u(B_X), \varepsilon B_Y) \le 2^{k-1}\},$$

(cf. [Pie]). In this note we shall study entropy numbers of finite rank operators whose domain or range is a Hilbert space. In particular, we shall address a duality problem for entropy numbers of such operators.

A weak version of a general duality problem for entropy numbers can be stated as follows. Do there exist constants $a > 1$ and $c > 1$ such that for every compact operator u and for every $k = 1, 2, \ldots,$

(1) $$c^{-1} e_{[ak]}(u^*) \le e_k(u) \le c e_{[a^{-1}k]}(u^*)?$$

This and related weaker questions were studied by several authors, e.g., [C], [G-K-S], [K-M], [P-T]. Let us state the result in this direction,

---

*Research supported in part by the Fund for Basic Research administered by the Israel Academy of Sciences.

due to König and Milman [K-M], which we shall use in the paper. It says that for every $b > 0$ there exist $a > 1$ and $c > 1$ (depending on b only) such that for every finite rank operator u, rank $u = n$ say, (1) holds for all $k \geq bn$.

In this note we consider the natural Euclidean structure on $\mathbf{R}^n$, that is the inner product $(\cdot, \cdot)$ and the Euclidean norm $\|x\|_2 = (x, x)^{1/2}$, for $x \in \mathbf{R}^n$. Let $D_n = \{x \in \mathbf{R}^n : \|x\|_2 \leq 1\}$ and let $\ell_2^n = (\mathbf{R}^n, \|\cdot\|_2)$.

Let us describe the asymptotic notions of a "random projection" and a "random embedding" which play an essential role in the sequel.

Given a family of properties $\mathcal{P} = \{\mathcal{P}_{n,k}\}$ we say that $\mathcal{P}$ is satisfied for a "random projection" if the measure of the set of orthogonal rank k projections in ℓ_2^n which do not satisfy $\mathcal{P}_{n,k}$ is "small". To make this more precise, for any n and $1 \leq k \leq n$, there exists $\alpha(n, k) > 0$ with $\alpha(n, k) \to 0$ as $n \to \infty$ and $k \to \infty$ such that the following holds. Fix an orthonormal projection $Q_{k,n} : \ell_n^2 \to \ell_n^2$ of rank k. Let a subset $\mathfrak{U}_{n,k}$ of the orthogonal group $O(n)$ consist of $U \in O(n)$ such that the projection $P_k = U^{-1} Q_{k,n} U$ satisfies $\mathcal{P}_{n,k}$. We then require that $\mu(\mathfrak{U}_{n,k}) \geq 1 - \alpha(n, k)$, where μ is the normalized Haar measure on $O(n)$. (In standard situations, $\alpha(n, k)$ is decreasing exponentially.) In an analogous way we define a "random k-dimensional subspace" $E_k \subset \mathbf{R}^n$ and a "random embedding" $J_k : E_k \to \ell_2^n$. In particular, $E_k = P_k(\ell_2^n)$ and $J_k = (P_k)^* |_{P_k(\ell_2^n)}$, where P_k is a "random projection" of rank k.

In this note we shall establish certain relations between entropy numbers $e_j(u)$ and $e_j(P_k u)$, for an operator $u : X \to \ell_2^n$ and a random projection P_k of rank k, and between $e_j(w)$ and $e_j(w J_k)$, for an operator $w : \ell_2^n \to X$ and a random embedding J_k. Combining with the König and Milman's theorem and results from [M-T] we obtain, under some additional assumptions, a duality result $e_j(u) \sim e_{[aj]}(u^*)$, for $X \to \ell_2^n$.

An investigation of entropy numbers requires an additional notation. Let $K \subset \mathbf{R}^n$ be a compact centrally symmetric convex body and let $\|\cdot\|_K$ be the corresponding norm (so that $K = \{x \in \mathbf{R}^n : \|x\|_K \leq 1\}$). The polar body K^0 is given by $K^0 = \{x \in \mathbf{R}^n : |(x,y)| \leq 1 \text{ for all } y \in K\}$. In particular, $\|x\|_{K^0} = \sup_{y \in K} |(x,y)|$. Set

$$M(k) = \int_{S^{n-1}} \|x\|_K d\lambda(x), \quad M(K^0) = \int_{S^{n-1}} \|x\|_{K^0} d\lambda(x),$$

where $S^{n-1} = \partial D_n$ is the Euclidean unit sphere and λ is the normalized rotation invariant measure on S^{n-1}.

In study of covering numbers $N(K, D_n)$ and $N(D_n, K)$, developed in [M-T], the crucial role is played by certain "interpolation" bodies K_t. For $t > 0$ consider the body

$$(2) \qquad\qquad K_t = K \cap tD_n.$$

Then $\|x\|_{K_t} = \max(\|x\|_K, t^{-1}\|x\|_2)$. Set

$$(3) \qquad M_t = M((K_t)^0) = M(\text{conv}(K^0 \cup t^{-1}D_n)).$$

We shall often use the easy fact that given a convex body $K \subset \mathbf{R}^n$ and a Euclidean ball D_n, for every $y \in \mathbf{R}^n$ there exists $x \in K$ such that

$$(4) \qquad K \cap (y + D_n) \subset K \cap (x + D_n).$$

It follows that if $K \subset \bigcup_1^N (y_i + D_n)$ for some $y_1, \ldots, y_N$ in $\mathbf{R}^n$ then $K \subset \bigcup_1^N (x_i + D_n)$ for some $x_1, \ldots, x_N$ in K.

2. Operators into a Hilbert space

In this section we study entropy numbers of finite rank operators acting into a Hilbert space, $u \colon X \to \ell_2$. We shall compare, for $1 \leq j \leq$ rank u, $e_j(u)$ and $e_j(P_k u)$, where P_k is a random projection of rank k, for some k. Clearly, our statements depend on the dimension of the range space and so we identify $u(X)$ with ℓ_2^n ($n = $ rank u) and consider always operators $u \colon X \to \ell_2^m$ which are onto.

Set $K = u(B_X) \subset \mathbf{R}^n$ and, for $t > 0$, let K_t and M_t be given by (2) and (3), respectively.

Proposition 1. *(a) There exist absolute constants $a > 1$ and $c > 0$ such that for every $aj \leq k \leq n$ and for a random projection P_k of rank k, we have*

$$c\sqrt{k/n}e_j(u) \leq e_j(P_k u).$$

(b) There exists an absolute constant $c > 0$ such that for any $A \geq 1$ there exists $C = C(A)$ such that for a random projection P_k of rank k we have

$$e_j(P_k u) \leq C\sqrt{k/n}e_j(u),$$

for $k = [A\,cn(M_\varepsilon/\varepsilon)^2]$, where $\varepsilon = e_j(u)$.

Proof: (a) We shall use the following lemma due to Johnson and Lindenstrauss.

Lemma. **[J-L].** *There exist absolute constants $a > 1$ and $c > 0$ with the following property. Let $1 \leq j \leq n$ and let $\varepsilon > 0$. Let $x_1, \ldots, x_{2^j}$ in $\mathbf{R}^n$ such that $\|x_i - x_\ell\|_2 > \varepsilon$, for $i \neq \ell$. Then, for every $aj \leq k \leq n$, and a random projection P_k of rank k we have*

$$\|P_k x_i - P_k x_\ell\|_2 > c\varepsilon\sqrt{k/n}, \quad \text{for } i \neq \ell.$$

Now fix $\varepsilon \leq e_j(u)$. There exist $x_1, \ldots, x_{2^j}$ in $K = u(B_X)$ such that $\|x_i - x_\ell\|_2 > \varepsilon$ for $i \neq \ell$. Thus $\|P_k x_i - P_k x_\ell\|_2 > c\varepsilon\sqrt{k/n}$ for $i \neq \ell$ and a random projection P_k, and hence $e_j(P_k u) \geq c\varepsilon\sqrt{k/n}$. This shows (a).

The proof of the lemma given in [J-L] is based on the isoperimetric inequality on the sphere. Since the argument involves only the distribution of the function $r_k(x) = x(1)^2 + \cdots + x(k)^2$ on the sphere, the use of the isoperimetric inequality can be replaced with a direct calculation.

(b) If $A = 1$ then the argument is based on the dual version of Dvoretzky theorem with the estimate on Euclidean sections from [M] (cf. also [F-L-M] or [M-Sch]). Given $t > 0$ we have, for every $k \leq [cn(M_t/t)^2]$ (c is an absolute constant) and a random projection P_k of rank k,

$$(5) \qquad \frac{2}{3}M_t(P_k D_n) \subset P_k K_t \subset 2M_t(P_k D_n).$$

In general, this is no longer true for arbitrary A and $k = [A\,cn(M_t/t)^2]$. However, in this case the right-hand side inclusion still holds (this is a consequence of an upper estimate in "Dvoretzky type" theorem which is valid for all A). Precisely, we have

$$(6) \qquad\qquad P_k K_t \subset C'(A) M_t (P_k D_n),$$

where $C'(A)$ depends on A only.

Indeed, let $f(x) = \|x\|_{K_t^0}$ be the norm induced by $(K_t)^0$. Applying 2.4 and 2.8 of [M-Sch] with $\varepsilon = \sqrt{A} M_t/t$ and $\theta = 1/2$, say, we obtain, for a random k-dimensional subspace E, a $1/2$-(Euclidean)-net $\mathcal{N}$ in $S_{n-1} \cap E$ such that $\|x\|_{K_t^0} \leq (1 + \sqrt{A}) M_t$ for $x \in \mathcal{N}$. This implies, by 4.1 of [M-Sch], $\|x\|_{K_t^0} \leq (2 + 2(1 + \sqrt{A})) M_t = C'(A) M_t$ for all $x \in S_{n-1} \cap E$. That is, $D_n \cap E \subset C'(A)((K_t)^0 \cap E)$, which is a dual form of (6). (One can alternatively use Proposition 2.3, Lemma 2.5 and Theorem 2.6 of [F-L-M].)

Now, set $\varepsilon = e_j(u)$. By (4), there exist $x_1, \ldots, x_{2^j}$ in K such that $K \subset \bigcup_1^{2^j} (x_i + \varepsilon D_n)$. Then

$$K \subset \bigcup_1^{2^j} (x_i + \varepsilon D_n) \cap K \subset \bigcup_1^{2^j} (x_i + 2(K \cap \varepsilon D_n)).$$

By (6) we have, for $k = [A\,cn(M_\varepsilon/\varepsilon)^2]$ and for a random projection P_k,

$$P_k K \subset \bigcup_1^{2^j} (P_k x_i + 2C'(A) M_\varepsilon (P_k D_n)).$$

So $e_j(P_k u) \leq 2C'(A) M_\varepsilon \leq C(A)\varepsilon \sqrt{k/n}$, completing the proof. ∎

Remark: Given A, a relation between j and k in Proposition 1(b) depends, of course, on M_ε. In general, if $j = o(n)$ then $k = o(n)$. Indeed, assume that $M_\varepsilon/\varepsilon > \alpha > 0$. Hence $M_\varepsilon > \alpha\varepsilon$. Since $M_s \leq s$, for all $s > 0$ then setting $\delta = \alpha\varepsilon/4$ and $t = 4/\alpha$ we have $M_{t\delta}/M_\delta > 4$. It was shown in [M-T], Corollary 2, that this condition implies the "inverse Sudakov inequality" and then we have $2^j = N(K, \varepsilon D_n) \geq \exp(c'n(M_\varepsilon/t\varepsilon)^2)$ with a universal constant $c' > 0$. Thus $j = O(n)$.

Corollary 2. *There exist absolute constants $c > 0$ and $\bar{c} > 0$ such that the following is true. Let $s \geq 1$ satisfy*

$$(7) \qquad\qquad e_{[j/2]}(u) \leq s e_j(u).$$

Let $\varepsilon = e_j(u)$ and $k = [cs^2 n(M_\varepsilon/\varepsilon)^2]$. Then $k \geq aj$, for an absolute constant $a \geq 1$ and there exists $C = C(s)$ such that

$$(8) \qquad\qquad \bar{c}\sqrt{k/n}e_j(u) \leq e_j(P_k u) \leq C\sqrt{k/n}e_j(u),$$

for a random projection P_k of rank k.

Proof: The assumption (7) is equivalent to

$$N(K, s\varepsilon D_n)^2 \leq N(K, \varepsilon D_n).$$

By Theorem 1 in [M-T], $2^j = N(K, \varepsilon D_n) \leq \exp(c's^2 n(M_\varepsilon/\varepsilon)^2)$, where $c' > 0$ is an absolute constant. Let $a > 1$ be the (absolute) constant as in Proposition 1(a). Clearly, we can choose $c > 0$ such that $[cs^2 n(M_\varepsilon/\varepsilon)^2] \geq aj$. Now the lower estimate follows from Proposition 1(a) and the upper estimate, from Proposition 1(b). ∎

The last corollary combined with results from [M-T], provide sufficient conditions for $k \sim j$. We state it in terms of mixed volumes.

For a convex body $K \subset \mathbf{R}^n$ set

$$A_k(K) = (\mathrm{Vol}_k D_k)^{-1/k} \left(\int_{G_{n,k}} \mathrm{Vol}_k(P_E K) d\sigma_k(D) \right)^{1/k},$$

where σ_k is the normalized Haar measure on the Grassman manifold $G_{n,k}$ of all k-dimensional subspaces E of $\mathbf{R}^n$, and P_E is the orthogonal projection onto E, for $E \in G_{n,k}$ and Vol_k denotes the k-dimensional Lebesgue measure on a k-dimensional subspace of $\mathbf{R}^n$ (cf. [M-Sch]). We shall use this formula in the following equivalent form

$$A_k(K) = (\mathrm{Vol}_k D_k)^{-1/k} \left(\int_{O(n)} \mathrm{Vol}_k(Q_{k,n} U K) d\mu(U) \right)^{1/k},$$

where $Q_{k,n}: \ell_2^m \to \ell_2^n$ is a fixed orthogonal projection of rank k.

Now, let $s \geq 1$, ε and k be as in Corollary 2. If additionally

$$(9) \qquad A_k(K)/M_\varepsilon \geq 2(1+\beta),$$

for some $\beta > 0$, then, by Theorem 1 in [M-T], $\exp(\beta k/2) \leq 2^j$. So $j \geq \beta k/2 \log 2$. It follows that (8) holds for $k \sim j$.

Observe that if for some $t > 1$ and $\beta > 0$

$$(10) \qquad M_{t\varepsilon}/M_\varepsilon \geq 6(1+\beta),$$

then (9) is satisfied for $k = [cs^2 n(M_\varepsilon/t\varepsilon)^2]$. Indeed, by (5) we have, for every $\ell \leq [cn(M_{t\varepsilon}/t\varepsilon)^2]$,

$$P_\ell(K_{t\varepsilon}) \supset \frac{2}{3} M_{t\varepsilon}(P_\ell D_n) = \frac{2}{3} M_{t\varepsilon} D_\ell,$$

for a random projection P_ℓ of rank ℓ. Since $k \leq [cn(M_{t\varepsilon}/t\varepsilon)^2]$ then

$$A_k(K) \geq \frac{1}{2}(\text{Vol}_k(P_k K_{t\varepsilon})/\text{Vol}_k D_k)^{1/k} \geq \frac{1}{3} M_{t\varepsilon}.$$

Combining with Corollary 2 we get that if $s \geq 1$ and $t > 1$ satisfy (7) and (10) then there exists k, with $aj \leq k \leq \bar{a}j$ such that for a random projection P_k we have

$$(11) \qquad \bar{c}\sqrt{k/n}\, e_j(u) \leq e_j(P_k u) \leq \bar{C}\sqrt{k/n}\, e_j(u).$$

Here $a > 1$ and $\bar{a} > 1$ and $\bar{c} > 0$ are absolute constants and $\bar{C} = \bar{C}(s,t)$ depends only on s and t.

3. Operators from a Hilbert space

Here we consider a problem dual to the one discussed in Section 2, and we study finite-rank operators acting from a Hilbert space, $w: \ell_2 \to Y$. In this context it is natural to assume that the dimension of the Hilbert space is equal to the rank of w, and so, w is one-to-one. Let n be a positive integer and let $1 \leq j \leq n$. For an operator $w \colon \ell_2^n \to Y$ we shall compare $e_j(w)$ and $e_j(wJ_k)$, where J_k is a random embedding, i.e., $J_k \colon E_k \to \ell_2^n$ is the canonical embedding of a random k-dimensional subspace $E_k \subset \mathbf{R}^n$ into ℓ_2^n.

Set $K = w^*(B_{Y^*})$ and, for $t > 0$, let K_t and M_t be as in (2) and (3). Observe that $K^0 = w^{-1}(B_Y)$. Moreover, $(K_t)^0 = \operatorname{conv}(K^0 \cup t^{-1}D_n)$, so $\frac{1}{2}(K^0 + t^{-1}D_n) \subset K_t^0 \subset (K^0 + t^{-1}D_n)$.

Proposition 3. *(a) There exist absolute constants $c > 0$ and $c' > 0$ such that for arbitrary $\varepsilon > 0$ and $k = [cn(M_\varepsilon/\varepsilon)^2]$ we have*

$$c'\varepsilon\sqrt{k/n} \leq e_k(wJ_k),$$

for a random embedding J_k.

(b) There exists an absolute constant $c > 0$ such that for any $A \geq 1$ there exists $C = C(A)$ such that

$$e_j(wJ_k) \leq C\sqrt{k/n}\,e_j(w),$$

for $k = [A\,cn(M_\varepsilon/\varepsilon)^2]$, where $\varepsilon = e_j(w)$.

In the proof we shall use important inequalities relating volumes of a body and of its polar. For arbitrary compact convex symmetric body $B \subset \mathbf{R}^k$ we have

$$(12) \qquad\qquad \delta^k \leq \frac{(\operatorname{Vol}_k B)(\operatorname{Vol}_k B^0)}{(\operatorname{Vol}_k D_k)^2} \leq 1,$$

where $\delta > 0$ is an absolute constant. The upper estimate is Santaló inequality [Sa]. The lower estimate was proved in [B-M.1,2]. The same method also shows the upper estimate with the constant 1 replaced by an absolute constant $C < \infty$ (which is still sufficient for our purpose) (cf. [B-M.1,2]). A much simplified version of these proofs was given in [P].

Proof: (a) First observe that since $P_k K \supset P_k K_\epsilon \supset \frac{2}{3} M_\epsilon D_k$ (cf. (5)), then

$$(13) \qquad (\mathrm{Vol}_k(P_k K)/\mathrm{Vol}_k D_k)^{1/k} \geq \frac{2}{3} M_\epsilon$$

for a random projection P_k.

Now, let E_k be a random k-dimensional subspace and fix $0 < \alpha < 1$ to be chosen later. Consider arbitrary covering $D_n \cap E_k \subset \cup_1^N (y_i + \alpha M_\epsilon K^0)$, for some $y_1, \ldots, y_N$ in Y. Then there exist $z_1, \ldots, z_N$ in $(D_n \cap E_k) \subset E_k$ such that

$$D_n \cap E_k \subset \bigcup_1^N (z_i + 2\alpha M_\epsilon K^0) \cap E_k = \bigcup_1^N (z_i + 2\alpha M_\epsilon (K^0 \cap E_k)).$$

Comparing volumes, using (12) and (13) and observing that $(K^0 \cap E_k)^0 = P_k K$, we get

$$N \geq (2\alpha M_\epsilon)^{-k} \mathrm{Vol}_k D_k / \mathrm{Vol}_k (K^0 \cap E_k)$$
$$\geq (\delta/2\alpha M_\epsilon)^k \mathrm{Vol}_k (P_k K)/\mathrm{Vol}_k D_k$$
$$\geq (\delta c/6\alpha)^k.$$

Setting $\alpha = \delta c/12$ we get $N \geq 2^k$. Thus $e_k(wJ_k) \geq \alpha M_\epsilon \geq c' \varepsilon \sqrt{k/n}$, where $c' = \delta \sqrt{c}/12$.

(b) Here the proof is based on the dual version of the argument used in Proposition 1(b). An upper estimate in "Dvoretzky type" theorem implies that for $k = [A \, cn(M_t/t)^2]$ we have

$$(14) \qquad K_t^0 \cap E_k \supset (C'(A)M_t)^{-1} D_n \cap E_k,$$

for a random k-dimensional subspace E_k, where $C'(A)$ is the same as in (6). (Indeed, (14) is a dual version of (6).)

From the definition of $\varepsilon = e_j(w)$ there exist $y_1, \ldots, y_{2^j}$ in Y such that

$$w(D_n) \subset \bigcup_1^{2^j}(y_i + \varepsilon B_Y).$$

Thus, for $x_i = w^{-1}(y_i)$, we have $D_n \subset \cup_1^{2^j}(x_i + \varepsilon K^0)$. Therefore,

$$\varepsilon K_\varepsilon^0 \subset D_n + \varepsilon K^0 \subset \cup_1^{2^j}(x_i + 2\varepsilon K^0).$$

By (14), for a random k-dimensional subspace E_k we have

$$\varepsilon (C'(A)M_\varepsilon)^{-1}(D_n \cap E_k) \subset \varepsilon(K_\varepsilon^0 \cap E_k) \subset \cup_1^{2^j}(x_i + 2\varepsilon K^0) \cap E_k.$$

Thus

$$(D_n \cap E_k) \subset \cup_1^{2^j}(x_i + 2C'(A)M_\varepsilon K^0),$$

and so $e_j(wJ_k) \le 2C'(A)M_\varepsilon \le C(A)\varepsilon\sqrt{k/n}$. ∎

Remarks: (a) The lower estimate (a) in Proposition 3 can be improved, under some regularity assumptions. Let $\sigma \ge 1$ satisfy, for some j,

$$(15) \qquad\qquad e_{[j/2]}(w) \le \sigma e_j(w).$$

Then, by [M-T], Th. 3, $2^j \le \exp(C n\sigma^2(M_\varepsilon/\varepsilon)^2)$, where C is an absolute constant and $\varepsilon = e_j(w)$. Thus $k = [cn(M_\varepsilon/\varepsilon)^2] \ge c''j/\sigma^2$ for an absolute constant $c'' > 0$. Thus, by Proposition 3(a),

$$c'\sqrt{k/n}e_j(w) \le e_{[c''j/\sigma^2]}(wJ_k).$$

(b) If additionally,

$$A_k(K)/M_\varepsilon \ge b,$$

for some fixed absolute constant $b > 2$, then, by [M-T] Theorem 3, $2^k \le 2^j$ and then $k \le j$.

In sections 2 and 3 we discussed some sufficient conditions for $k \sim j$. Under these assumptions, applying König-Milman's result stated in the introduction, we get a duality result for entropy numbers of finite rank operators acting to or from a Hilbert space.

If $u: X \rightarrow \ell_2$ has rank n, identify $u(X)$ with ℓ_2^n and consider $K = u(B_X)$ as a subset of ℓ_2^n. With this notation we have

Proposition 4. *Let $u: X \rightarrow \ell_2$ have a finite rank. Let $1 \leq j \leq \text{rank } u$. Let $s \geq 1$ and $\sigma \geq 1$ satisfy regularity conditions for entropy numbers*

$$e_{[j/2]}(u) \leq se_j(u) \quad \text{and} \quad e_{[j/2]}(u^*) \leq \sigma e_j(u^*).$$

Moreover, let $\varepsilon = e_j(u)$, $\delta = e_j(u^*)$ and let $k = [cs^2 n(M_\varepsilon/\varepsilon)^2]$, $k_1 = [c'n(M_\delta/\delta)^2]$, where $c > 0$ and $c' > 0$ are fixed absolute constants. Assume that

$$A_k(K)/M_\varepsilon \geq b \quad \text{and} \quad A_{k_1}(K)/M_\delta \geq b,$$

where $b > 0$ is a fixed absolute constant. Then there exist constants $\alpha = \alpha(s, \sigma)$ and $C = C(s, \sigma)$ such that

$$C^{-1} e_{[\alpha^{-1}j]}(u^*) \leq e_j(u) \leq C e_{[\alpha j]}(u^*).$$

Proof: In Corollary 2 and at the end of Section 2 it was shown that our regularity assumptions imply $aj \leq k = [cs^2 n(M_\varepsilon/\varepsilon)^2] \leq \bar{c}j$, where $a > 0$ and $\bar{c}$ are absolute constants. Similarly, from Remarks after Proposition 3, we have $c''j/\sigma^2 \leq k_1 = [cn(M_\delta/\delta)^2] \leq j$. Now to conclude the proof is enough to apply Proposition 1 and 3 together with König-Milman's result. ∎

Remark: The mixed volume assumptions in the above theorem are implied by the conditions

$$M_{t\varepsilon}/M_\varepsilon \geq 3b \quad \text{and} \quad M_{\tau\delta}/M_\delta \geq 3b,$$

for some $t > 1$ and $\tau > 1$. Then α and C depend on s, σ, t and τ.

4. The identity operator from ℓ_p^n to ℓ_2^n

In this section we investigate the formal identity operator $u: \ell_p^n \to \ell_2^n$, with $1 < p \le 2$. We shall discuss entropy numbers $e_j(P_k u)$ and $e_j(u^* J_k)$, for $j \sim k$ and for arbitrary and random projections P_k and embeddings J_k.

Fix $1 \le j \le n$.

Proposition 5. *(a) For every projection P_j of rank j and for every embedding J_j we have*

$$e_j(P_j u) \ge \frac{1}{2} n^{\frac{1}{2} - \frac{1}{p}} \quad and \quad e_j(u^* J_j) \ge \frac{1}{4} n^{\frac{1}{2} - \frac{1}{p}}.$$

(b) There exists an absolute constant $a \ge 1$ such that for a random projection P_j of rank j and for a random embedding J_j we have

$$e_{[aj]}(P_j u) \le n^{\frac{1}{2} - \frac{1}{p}} \quad and \quad e_{[aj]}(u^* J_j) \le n^{\frac{1}{2} - \frac{1}{p}}.$$

Proof: Set $p' = p/(p-1)$. By B_p^n (resp. $B_{p'}^n$) denote the unit ball in ℓ_p^n (resp. $\ell_{p'}^n$). Fix an arbitrary projection P_j of rank j. Set $E_j = P_j(\ell_2^n) \subset \mathbf{R}^n$, and let $J_j: E_j \to \ell_2^n$ be the natural embedding. Clearly, $(P_j B_p^n)^0 = B_{p'}^n \cap E_j$. We have

$$(16) \qquad n^{\frac{1}{2} - \frac{1}{p}} \le \left(\frac{\mathrm{Vol}_j(P_j B_p^n)}{\mathrm{Vol}_j D_j} \right)^{1/j} \le \left(\frac{\mathrm{Vol}_j D_j}{\mathrm{Vol}_j(B_{p'}^n \cap E_j)} \right)^{1/j}.$$

The first inequality is obvious since $B_p^n \supset n^{\frac{1}{2} - \frac{1}{p}} D_n$. The second follows from Santaló inequality (cf. (12)). A similar upper estimate is valid for a random projection P_j of rank j and a random subspace $E_j = P_j(\ell_2^n) \subset \mathbf{R}^n$. We have

$$(17) \qquad \left(\frac{\mathrm{Vol}_j D_j}{\mathrm{Vol}_j(B_{p'}^n \cap E_j)} \right)^{1/j} \le \left(\frac{1}{\delta} \right) \left(\frac{\mathrm{Vol}_j(P_j B_p^n)}{\mathrm{Vol}_j D_j} \right)^{1/j} \le C n^{\frac{1}{2} - \frac{1}{p}},$$

where $\delta > 0$ and C are absolute constants. The first inequality follows from the inverse Santaló inequality (cf. (12)). For the second, use the definition of mixed volumes, the concentration of measure on the orthogonal group and the Alexandroff inequality $A_j(K) \leq A_1(K) = M(K^0)$ (cf. [B-Z]) to get

$$\left(\frac{\mathrm{Vol}_j(P_j B_p^n)}{\mathrm{Vol}_j D_j}\right)^{1/j} \leq C' A_j(B_p^n) \leq C' M(B_{p'}^n) \leq C'' n^{\frac{1}{2}-\frac{1}{p}}.$$

Now we use a standard argument. Any covering

$$P_j B_p^n \subset \bigcup_1^N (x_i + \frac{1}{2} n^{\frac{1}{2}-\frac{1}{p}} D_n)$$

yields the covering $P_j B_p^n \subset \cup_1^N (P_j x_i + \frac{1}{2} n^{\frac{1}{2}-\frac{1}{p}} P_j D_n)$. Comparing volumes we get, by (16), $N \geq 2^j$. This implies the lower estimate for $e_j(P_j u)$.

Now set $\varepsilon = n^{\frac{1}{2}-\frac{1}{p}}$ and let P_j be a random projection of rank j. Let $x_1, \ldots, x_N$ be a maximal collection of vectors in $P_j B_p^n$ such that $\|x_i - x_\ell\|_2 > \varepsilon$ for $i \neq \ell$. Then the balls $x_i + \frac{\varepsilon}{2} P_j D_n$ are mutually disjoint and

$$\bigcup_1^N (x_i + \frac{\varepsilon}{2} P_j D_n) \subset \frac{3}{2} P_j B_p^n.$$

Thus, comparing volumes we get, by (17), $N \leq (3C)^j$, which yields $e_{[aj]}(P_j u) \leq \varepsilon$ for a suitable choice of an (absolute) constant $a \geq 1$.

The estimates for $e_j(u^* J_j)$ are proved by a very similar argument. ∎

Proposition 5 shows that, as j changes, $e_j(P_j u)$ and $e_j(u^* J_j)$ are essentially constant, for random projections P_j and random embeddings J_j. This behaviour strongly contrasts with the behaviour of entropy numbers of the original operators u and u^*. It was shown in [S] that

$$e_j(u) \sim e_j(u^*) \sim \begin{cases} 1 & \text{if } j \leq \log n \\ (\log(\frac{n}{j}+1)/j)^{\frac{1}{p}-\frac{1}{2}} & \text{if } \log n < j \leq n \\ 2^{-j/n} n^{\frac{1}{2}-\frac{1}{p}} & \text{if } j > n \end{cases}$$

It would be interesting to establish, given $1 \leq k \leq n$, the order of growth of $e_j(P_k u)$ and $e_j(u^* J_k)$, for all $1 \leq j \leq n$ and a random projection P_k and a random embedding J_k.

References

[B-M.1] J. Bourgain and V.D. Milman, *Sections euclidiennes et volume des corps symétriques convexes dans* $\mathbf{R}^n$, C.R. Acad. Sci. Paris 300 (1985), 435–438.

[B-M.2] —————————, *New volume ratio properties for convex symmetric bodies in* $\mathbf{R}^n$, Inventiones Math. 88 (1987), 319–340.

[B-Z] Y.D. Burago and V.A. Zalgaler, *Geometric inequalities*. in "Nauka", Leningrad, 1980 (Russian) and Springer Verlag 1987.

[C] B. Carl, *On Gelfand, Kolmogorov and entropy numbers of operators acting between special Banach spaces*, Journal of Approx. Theory.

[F-L-M] T. Figiel, J. Lindenstrauss and V.D. Milman, *The dimension of almost spherical sections of convex bodies*, Acta Math. 139 (1977), 53–94.

[G-K-S] Y. Gordon, H. König and C. Schütt, *Geometric and probabilistic estimates for entropy and approximation numbers of operators*, J. of Approx. Theory 49 (1987), 219–239.

[J-L] W.B. Johnson and J. Lindenstrauss, *Extensions of Lipschitz mappings into a Hilbert space*, Contemporary Mathematics 26 (1984), 189–206.

[K-M] H. König and V.D. Milman, *On the covering number of convex bodies*, pp. 82–95 in "Geometric Aspects of Functional Analysis", Israel Seminar 1985/86, Lecture Notes in Math. 1267. Springer Verlag 1987.

[M] V.D. Milman, *A new proof of the theorem of A. Dvoretzky on sections of convex bodies*, Funct. Anal. Appl. 5 (1971), 28–38. (translated from Russian).

[M-Sch] V.D. Milman and G. Schechtman, "Asymptotic theory of finite dimensional normed spaces," Lecture Notes in Math. 1200, Springer Verlag, 1986.

[M-T] V.D. Milman and N. Tomczak-Jaegermann, *Sudakov type inequalities for convex bodies in* $\mathbf{R}^n$, pp. 113–121 in "Geometric Aspects of functional Analysis", Israel Seminar 1985/86, Lecture Notes in Math. 1267. Springer Verlag, 1987.

[P] G. Pisier. private letter.

[Pie] A. Pietsch, "Operator ideals," VEB Deutscher Verlag, Berlin, North Holland, 1980.

[P-T] A. Pajor and N. Tomczak-Jaegermann, *Remarques sur les nombres d'entropie d'un operateur et son transposé*, C.R. Acad. Sci. Paris 301 (1985), 743–746.

[S] C. Schütt, *Entropy numbers of diagonal operators between symmetric spaces*, J. of Appr. Theory 40 (1983), 121–128.

[Sa] L.A. Santalo, *Un inveriant afin pasa los cuerpos convexos del espacio de n-dimensiones*, Portugal Math. 8 (1949), 155–161.

Mathematisches Seminar, der Universität Kiel, 2300 Kiel 1, West Germany

School of Mathematical Sciences, Raymond and Beverley Sackler, Faculty of Exact Sciences, Tel Aviv University, Tel Aviv, Israel

Department of Mathematics, University of Alberta, Edmonton, Alberta, Canada T6G 2G1

On the Convergence of Types
for Radon Probability Measures
in Banach Spaces

W. LINDE AND G. SIEGEL

Introduction

A Radon probability measure ν on a Banach space F belongs to the same type as a measure μ on a Banach space E provided that $\nu = T(\mu) * \delta_y$ for some linear operator T from E into F and some $y \in F$. Assume that the probability measures ν_n belong to the same type as the measures μ_n for each $n = 1, 2, \ldots$, and, moreover, $\{\mu_n\}$ and $\{\nu_n\}$ converge weakly to μ and ν, respectively. Then the present paper is concerned with the problem whether or not the limit ν belongs to the same type as the limit μ. A classical result due to Khinchin (cf. [6, Ch. VIII, §2, Lemma 1]) asserts that this is so on the real line provided that μ is non-degenerated. Similar results under weaker assumptions can be found in [12, Section 2.3]. Expansions of such results to n-dimensional spaces were studied in Billingsley [1] and Sharpe [13]. In arbitrary Banach or even more general spaces this problem was treated in Parthasarathy [11, p.58], and Csiszar/Rajput [3] in the special case $E = F$ and $\nu_n = (\alpha_n I)(\mu_n) * \delta_{y_n}$, where I denotes the identity operator of E and α_n are some real numbers. The case of arbitrary linear oparators T_n (instead of $\alpha_n I$) was investigated in Jouandet [7]; but these results are false. The basic aim of the present paper is to derive general convergence of types theorems in arbitrary Banach spaces. It turns out that one has to assume the boundedness of the sequence of operators in order to obtain the classical results in the infinite dimensional setting. Easy examples show that this boundedness restriction is in fact necessary. The main step of all former proofs is the verification of the compactness of the sequence of operators in a suitable topology. The crucial observation of our paper is that one has to replace the topology of uniform convergence by the strong operator topology in the infinite dimensional case. The main result of the present paper can be used for the investigation of operator-stable distributions in Banach spaces (cf. Siegel [14]).

1. Notation and definitions

E shall always be a real Banach space with (topological) dual space E', and $\mathcal{R}(E)$ denotes the set of Radon probability measures on the Borel subsets of E, i.e. we have

$$\sup\{\mu(K)\colon K \subseteq E,\ K \text{ compact}\} = 1$$

for each $\mu \in \mathcal{R}(E)$. The support of $\mu \in \mathcal{R}(E)$ is defined by

$$\operatorname{supp}(\mu)\colon\ = \{x \in E\colon \mu(U_\varepsilon(x)) > 0 \text{ for all } \varepsilon > 0\},$$

where $U_\varepsilon(x)$ denotes the open ball with radius $\varepsilon > 0$ and centre $x \in E$. Recall that $\sup p(\mu)$ is the smallest closed subset of E possessing the whole measure of μ. Sometimes it is necessary to investigate the closed linear subspace generated by $\operatorname{supp}(\mu)$, i.e., we set

$$\gamma(\mu)\colon\ = \overline{\operatorname{span}}(\operatorname{supp}(\mu)),$$

and call it linear support of μ. Notice that $\gamma(\mu)$ as well as $\operatorname{supp}(\mu)$ are always separable and, moreover,

$$\gamma(\mu) = \cap\{A \subseteq E\colon A \text{ closed subspace with } \mu(A) = 1\}.$$

A measure μ is said to be full if $\gamma(\mu) = E$. As usual δ_x denotes the probability measure concentrated at $x \in E$. If F is another real Banach space, $\mathcal{L}(E, F)$ stands for the set of linear continuous operators from E into F. For each $T \in \mathcal{L}(E, F)$ and each subspace $E_0 \subseteq E$ the operator $T|_{E_0} \in \mathcal{L}(E_0, F)$ denotes the restriction of T to E_0. For any $\mu \in \mathcal{R}(E)$ and any $T \in \mathcal{L}(E, F)$ the image measure $T(\mu)$ or $T\mu$ is defined by $T(\mu)(B) = \mu(T^{-1}(B))$, B any Borel subset of F. Observe that $T(\mu) \in \mathcal{R}(F)$ whenever $\mu \in \mathcal{R}(E)$ and $T \in \mathcal{L}(E, F)$. Let us now state a characterization of $\gamma(\mu)$ in terms of functionals. This is a well-known consequence of the Hahn-Banach extension theorem.

Lemma 1.1. *For each $\mu \in \mathcal{R}(E)$ we have*

$$\gamma(\mu) = \{a \in E' : a(\mu) = \delta_0\}^{\perp}.$$

Here as usual $A^{\perp} := \{x \in E : \langle x, a \rangle = 0 \text{ for all } a \in A\}$, $A \subseteq E'$. Analogously, if $B \subseteq E$, then $B^{\perp}$ is defined by

$$B^{\perp} = \{a \in E' : \langle x, a \rangle = 0 \text{ for all } x \in B\}.$$

Corollary 1.2. *A measure $\mu \in \mathcal{R}(E)$ is full iff*

$$\mu\{x \in E : \langle x, a \rangle = 0\} < 1 \text{ for each } a \in E' \smallsetminus \{0\}.$$

Corollary 1.3. *For any two measures $\mu, \nu \in \mathcal{R}(E)$ the inclusion $\gamma(\mu) \subseteq \gamma(\nu)$ is equivalent to*

$$\{a \in E' : a(\nu) = \delta_0\} \subseteq \{a \in E' : a(\mu) = \delta_0\}.$$

Given $\alpha \in \mathbf{R}$, the dilation $\alpha \circ \mu$, $\mu \in \mathcal{R}(E)$, is defined by

$$\alpha \circ \mu = (\alpha I)(\mu),$$

where I is the identity operator of E. Hence we have $(\alpha \circ \mu)(B) = \mu(\alpha^{-1}B)$, $\alpha \neq 0$, and $0 \circ \mu = \delta_0$. Observe that $\alpha \circ (T\mu) = T(\alpha \circ \mu) = (\alpha T)(\mu)$, $T \in \mathcal{L}(E, F)$, and, consequently, if the symmetrization μ^s of μ is defined by

$$\mu^s = \mu * ((-1) \circ \mu),$$

then $T(\mu^s) = (T(\mu))^s$ for any $T \in \mathcal{L}(E, F)$. Since

$$\mathrm{supp}(\mu^s) = \overline{\mathrm{supp}\, p(\mu) - \mathrm{supp}(\mu)},$$

in general neither $\mathrm{supp}(\mu^s) \subseteq \mathrm{supp}(\mu)$ nor the converse inclusion hold.

The following is a simple consequence of Lemma 1.1, Corollary 1.2 and 1.3.

Lemma 1.4. *For $\mu \in \mathcal{R}(E)$ and $x_0 \in \text{supp}(\mu)$ we have*

$$\gamma(\mu) = \{z + \lambda x_0 : z \in \gamma(\mu^s), \lambda \in \mathbf{R}\} = span(\gamma(\mu^s) + x_0),$$

that is, $\gamma(\mu)$ and $\gamma(\mu^s)$ differ at most in one dimension. In particular, it holds $\gamma(\mu^s) \subseteq \gamma(\mu)$; moreover these two sets coincide iff

$$\mu\{x \in E : \langle x, a\rangle = 1\} < 1 \text{ for each } a \in E',$$

i.e., the support of μ is not contained in a hyperplane $H \subseteq E$ with $0 \notin H$.

Remark: Let $Q : \gamma(\mu) \to \gamma(\mu^s)$ be the natural projection for some fixed $x_0 \in \text{supp}(\mu)$, i.e.,

$$Q(z + \lambda x_0) := z, \quad z \in \gamma(\mu^s), \quad \lambda \in \mathbf{R}.$$

Then Q is continuous (closed graph theorem) and it is easy to see that $Q(\mu) = \mu * \delta_{-x_0}$. In particular, each operator $T_0 \in \mathcal{L}(\gamma(\mu^s), F)$ admits an extension $T \in \mathcal{L}(\gamma(\mu), F)$ with $T(\mu) = T_0(\mu * \delta_{-x_0})$. Notice that we cannot write $T(\mu) = T_0(\mu) * \delta_{-T_0 x_0}$ because in general neither $T_0(\mu)$ nor $T_0 x_0$ make sense. This extension procedure turns out to be quite useful for our subsequent investigations.

2. Weak convergence of measures and strong operator topology

Here and hereafter we shall always endow $\mathcal{R}(E)$ or $\mathcal{R}(F)$ with the weak topology. A net $\{\mu_\alpha\}$ in $\mathcal{R}(E)$ converges weakly to μ ($\mu_\alpha \Rightarrow \mu$) iff

$$\int_E f(x)d\mu_\alpha(x) \to \int_E f(x)d\mu(x)$$

for each continuous bounded function $f : E \to \mathbf{R}$. It suffices that this convergence takes place for all uniformly continuous bounded functions (cf. [2]). The weak topology on $\mathcal{R}(E)$ is metrizable and one possible metric is the well-known Prokhorov metric ρ, for more information see [11] or [15].

For later use we also state the following easy property of ρ.

Lemma 2.1. *Let $\mu \in \mathcal{R}(E)$, $\nu \in \mathcal{R}(E)$, then*

$$\rho(T\mu, T\nu) \leq \max\{1, \|T\|\}\rho(\mu, \nu), \quad T \in \mathcal{L}(E, F).$$

From Lemma 2.1 we derive the next assertion.

Lemma 2.2. *If $\{\mu_n\} \subseteq \mathcal{R}(E)$ converges weakly to δ_0, then there are real numbers α_n with $\alpha_n \to \infty$ and $\alpha_n \circ \mu_n \Rightarrow \delta_0$ as well.*

Proof: Setting $\alpha_n = \rho(\mu_n, \delta_0)^{-1/2}$ and using Lemma 2.1 we obtain

$$\rho(\alpha_n \circ \mu_n, \delta_0) = \rho(\alpha_n \circ \mu_n, \alpha_n \circ \delta_0)$$
$$\leq \max\{1, \alpha_n\}\rho(\mu_n, \delta_0) \leq \rho(\mu_n, \delta_0)^{1/2}$$

which proves our assertion.

A subset $\mathcal{M} \subseteq \mathcal{R}(E)$ is called weakly relatively compact (w.r.c.), if its weak closure is compact in $\mathcal{R}(E)$. In view of the preceding remarks this happens iff $\mathcal{M}$ is sequentially compact. Prokhorov's theorem (cf. [11] or [15]) asserts that $\mathcal{M}$ is w.r.c. iff it is uniformly Radon, i.e., there are compact subsets $K_n \subseteq E$ with

$$\lim_{n \to \infty} \inf_{\mu \in \mathcal{M}} \mu(K_n) = 1.$$

A subset $\mathcal{M} \subseteq \mathcal{R}(E)$ is said to be relatively shift compact (r.s.c.) provided that there exist $x_\mu \in E$, $\mu \in \mathcal{M}$, such that $\{\mu * \delta_{x_\mu} : \mu \in \mathcal{M}\}$ forms a w.r.c. subset of $\mathcal{R}(E)$. Consider a sequence $\{\mu_n\} \subseteq \mathcal{R}(E)$ of measures with $\mu_n \Rightarrow \mu$, and operators T_n, $T \in \mathcal{L}(E, F)$ with $\|T_n - T\| \to 0$, i.e., $\{T_n\}$ converges uniformly to T. Then it is easy to check that $T_n(\mu_n) \Rightarrow T(\mu)$. But we shall see that the same conclusion is true only assuming the convergence of $\{T_n\}$ in the strong operator topology. Considering a net $\{T_\alpha\}$, $\{T_\alpha\}$ converges strongly to $T \in \mathcal{L}(E, F)$ ($T_\alpha \xrightarrow{s} T$) iff $T_\alpha x \to Tx$ for all $x \in E$. We write $\mathcal{L}_s(E, F)$ to indicate that $\mathcal{L}(E, F)$ is endowed with the strong operator topology.

Remark: Let $\{T_\alpha\} \subseteq \mathcal{L}(E, F)$ be a net with $\sup_\alpha \|T_\alpha\| < \infty$ and let $A \subseteq E$ be a subset with $\overline{\text{span}}(A) = E$.

(i) We have $T_\alpha \xrightarrow{s} T$ iff $T_\alpha x \to Tx$ for all $x \in A$.

(ii) If $T_\alpha \xrightarrow{s} T$, then even $T_\alpha x \to Tx$ uniformly on compact subsets of E.

A set $\mathcal{L} \subseteq \mathcal{L}(E, F)$ is called strongly relatively compact (s.r.c.) if its strong closure is strongly compact. The following characterization of s.r.c. sets is well known, see [5].

Lemma 2.3. *For any subset $\mathcal{L} \subseteq \mathcal{L}(E, F)$ the following are equivalent:*

(i) *$\mathcal{L}$ is s.r.c.*

(ii) *For all $x \in E$ the set $\{Tx : T \in \mathcal{L}\}$ is relatively compact (r.c.) in F.*

(iii) *We have $\sup\{\|T\| : T \in \mathcal{L}\} < \infty$ and $\{Tx : T \in \mathcal{L}\}$ is r.c. for all $x \in A$ where $\overline{\text{span}}(A) = E$.*

Moreover, if E is separable, then this is also equivalent to:

(iv) *Each sequence in $\mathcal{L}$ contains a strongly convergent subsequence.*

Notice that if E is arbitrary and $\dim(F) < \infty$, then $\mathcal{L} \subseteq \mathcal{L}(E, F)$ is s.r.c. iff it is bounded. Further, in the case $F = \mathbf{R}$ the strong operator topology coincides with the weak-*-topology $\sigma(E', E)$.

3. Sufficient conditions

In this section we study the convergence of the sequence $\{T_n(\mu_n)\}$ of image measures assuming the weak convergence of $\{\mu_n\} \subseteq \mathcal{R}(E)$ and the convergence of $\{T_n\} \subseteq \mathcal{L}(E, F)$ in a certain sense. To begin with we prove the following general result.

Theorem 3.1. *Let T_α, T be operators in $\mathcal{L}(E, F)$ and assume that $T_\alpha x \to Tx$ uniformly on compact subsets of E. If $\{\mu_\alpha\} \subseteq \mathcal{R}(E)$ converges weakly to $\mu \in \mathcal{R}(E)$, then $T_\alpha(\mu_\alpha) \Rightarrow T(\mu)$.*

Proof: Let $f : F \to \mathbf{R}$ be a bounded and uniformly continuous function. Given $\varepsilon > 0$ we choose a compact set $K \subseteq E$ with $\mu_\alpha(K) \geq 1 - \varepsilon$, $\alpha \in \mathcal{A}$.

Next we take a $\delta > 0$ such that $\|y_1 - y_2\| < \delta$ implies $|f(y_1) - f(y_2)| < \varepsilon$ and, finally, we choose α_0 satisfying

$$\left| \int_E f(Tx)d\mu_\alpha(x) - \int_E f(Tx)d\mu(x) \right| < \varepsilon, \quad \alpha > \alpha_0,$$

as well as

$$\|T_\alpha x - Tx\| < \delta, \quad x \in K, \quad \alpha > \alpha_0.$$

Hence, if $\alpha > \alpha_0$, we conclude

$$\left| \int_E f(T_\alpha x)d\mu_\alpha(x) - \int_E f(Tx)d\mu(x) \right| \leq$$
$$\left| \int_K (f(T_\alpha x) - f(Tx))d\mu_\alpha(x) \right| + 2\varepsilon \sup_{y \in F} |f(y)|$$
$$+ \left| \int_E f(Tx)d(\mu_\alpha - \mu)(x) \right|$$
$$\leq \mu_\alpha(K) + 2\varepsilon \sup_{y \in F} |f(y)| + \varepsilon \leq 2\varepsilon(1 + \sup_{y \in F} |f(y)|).$$

This being true for any uniformly continuous and bounded function f from F into $\mathbf{R}$ proves our assertion.

Corollary 3.2. *For any bounded subset $\mathcal{L} \subseteq \mathcal{L}_s(E, F)$ the mapping $(T, \mu) \to T(\mu)$ is continuous from $\mathcal{L} \times \mathcal{R}(E)$ into $\mathcal{R}(F)$.*

Remark: Observe that the mapping $(T, \mu) \to T(\mu)$ is never continuous from $\mathcal{L}_s(E, F) \times \mathcal{R}(E)$ into $\mathcal{R}(F)$ if $\dim(E) = \infty$. More general, for any fixed $\mu \in \mathcal{R}(E)$ the mapping $a \to a(\mu)$ from $[E', \sigma(E', E)]$ into $\mathcal{R}(\mathbf{R})$ is continuous iff there are finite dimensional subspaces $E_k \subseteq E$ with $\lim_{k \to \infty} \mu(E_k) = 1$ (cf. [4]).

Our next objective is to investigate special sets of measures. More precisely, we treat the image measures of a convergent sequence. By virtue of theorem 3.1 we have that $\mu_n \Rightarrow \mu$ and $T_n \xrightarrow{s} T$ imply $T_n(\mu_n) \Rightarrow T(\mu)$. This statement can also be derived from Th. 5.5, Ch. I in [2].

Proposition 3.3. *Assume $\mu_n \Rightarrow \mu$ in $\mathcal{R}(E)$ and let $\{T_n\} \subseteq \mathcal{L}(E, F)$ with $\sup_n \|T_n\| < \infty$.*

(i) *The sequence $\{T_n \mu_n\}$ converges weakly iff $\{T_n \mu\}$ does so. In either case, both sequences possess the same limit.*

(ii) *$\{T_n \mu_n\}$ is w.r.c. iff $\{T_n \mu\}$ is so.*

(iii) *$\{T_n \mu_n\}$ is r.s.c. iff $\{T_n \mu\}$ is so.*

Proof: (i) and (ii) follow from Lemma 2.1 by writing

$$\rho(T_n \mu_n, T_n \mu) \le \max\{1, c\} \rho(\mu_n, \mu), \quad c = \sup_n \|T_n\|;$$

(iii) can be similarly proved using

$$\rho(T_n \mu_n * \delta_{y_n}, T_n \mu * \delta_{y_n}) = \rho(T_n \mu_n, T \mu), \quad y_n \in F.$$

Remark: A little bit more than (iii) is valid, namely we have

$$(T_n \mu_n) * \delta_{y_n} \Rightarrow \nu \quad \text{iff} \quad (T_n \mu) * \delta_{y_n} \Rightarrow \nu, \quad \nu \in \mathcal{R}(F).$$

An immediate consequence of Proposition 3.3 is the following

Corollary 3.4. *Let $\{T_n\} \subseteq \mathcal{L}(E, F)$ be a bounded sequence and assume $\mu_n \Rightarrow \mu$ in $\mathcal{R}(E)$. Then*

$$T_n|_{\gamma(\mu)} \xrightarrow{s} T, \quad T \in \mathcal{L}(\gamma(\mu), F),$$

implies $T_n(\mu_n) \Rightarrow T(\mu)$ as well as $T_n(\mu) \Rightarrow T(\mu)$.

Note that in this case $T_n|_{\gamma(\mu)} \xrightarrow{s} T$ is equivalent to $T_n x \to T x$ for each $x \in \text{supp}(\mu)$. Indeed, $\{T_n\}$ is bounded.

4. Necessary conditions

We want to investigate the converse problem of the preceding section, i.e. we consider a sequence $\{\mu_n\}$ of probability measures on E and a sequence of operators $\{T_n\} \subseteq \mathcal{L}(E, F)$ such that $\{T_n(\mu_n)\}$ has some special properties. Then we ask for properties of the sequence $\{T_n\} \subseteq \mathcal{L}(E, F)$. Easy examples (even in $\mathbf{R}^m$, $m \geq 2$) show that we may have $\mu_n \Rightarrow \mu$, $T_n(\mu_n) \Rightarrow \nu$, yet $\{T_n\}$ is not convergent. Thus we will restrict ourselves to the question of boundedness or compactness of the sequence $\{T_n\}$. Let us start with the problem of boundedness. A first proposition tells us that the situation becomes completely different in the case $\dim(E) = \infty$.

Proposition 4.1. *Let E be infinite dimensional. Then for every $\mu \in \mathcal{R}(E)$ there are functionals $a_n \in E'$ with $\sup_n \|a_n\| = \infty$ and $a_n(\mu) \Rightarrow \delta_0$.*

Proof: This is trivial in the case $\gamma(\mu) \neq E$. Indeed, take some $a_0 \in \gamma(\mu)^\perp$, $a_0 \neq 0$, and set $a_n = na_0$, $n = 1, 2, \ldots$. Clearly, $a_n(\mu) = \delta_0$ and $\sup_n \|a_n\| = \infty$. Thus we may assume $\gamma(\mu) = E$ and, as a consequence, E is separable. Now take any sequence $\{a_n\} \subseteq E'$ $\sigma(E', E)$-converging to zero. Because of Cor. 3.2 we then have $a_n(\mu) \Rightarrow \delta_0$. If $a_n(\mu) \Rightarrow \delta_0$ always yields $\sup_n \|a_n\| < \infty$, in view of Lemma 2.2 we then even have $\|a_n\| \to 0$. Therefore the unit sphere of E' has to be norm compact (E is separable) which contradicts $\dim(E) = \infty$. The proof is complete.

Corollary 4.2. *Let E be infinite dimensional and let $\{\mu_n\} \subseteq \mathcal{R}(E)$ be an arbitrary sequence. Then there exist elements $a_n \in E'$ such that $a_n(\mu_n) \Rightarrow \delta_0$ and $\sup_n \|a_n\| = \infty$.*

Proof: We apply the preceding proposition to each measure μ_n. Then we find $a_n \in E'$ with $\rho(a_n(\mu_n), \delta_0) < 1/n$ and $\|a_n\| \geq n$, $n = 1, 2, \ldots$. Of course, the functionals a_n possess the desired properties.

Remark: One should compare Cor. 4.2 with the finite dimensional case treated by Sharpe [13]. Namely, if $\dim(E) = \infty$, F arbitrary, then for each sequence $\{\mu_n\} \subseteq \mathcal{R}(E)$ there are operators $T_n \in \mathcal{L}(E, F)$ such that $\sup_n \|T_n\| = \infty$ although $\{T_n(\mu_n)\}$ is w.r.c. We even may have $T_n(\mu_n) \Rightarrow \delta_0$.

In order to formulate the next result we need the following notation: If $\{a_n\} \subseteq E'$, then $\{a_n\}'$ denotes the set of all cluster points of $\{a_n\}$ with respect to the norm topology of E'. Further, we put $\frac{a}{\|a\|} = 0$ whenever $a = 0$.

Proposition 4.3. *Let $\{\mu_n\} \subseteq \mathcal{R}(E)$ be a sequence with $\mu_n \Rightarrow \mu$. If $\{a_n\} \subseteq E'$ satisfies*

(i) $\{a_n(\mu_n)\}$ *is w.r.c.,*

(ii) $\left\{ \dfrac{a_n}{\|a_n\|} \right\}$ *is r.c.,*

(iii) $\left\{ \dfrac{a_n}{\|a_n\|} \right\}' \cap \gamma(\mu)^\perp \subseteq \{0\}$,

then $\sup_n \|a_n\| < \infty$.

Proof: Let us assume $\sup_n \|a_n\| = \infty$. Then we may select a subsequence $\{n'\}$ with $\|a_{n'}\| \to \infty$. If $\alpha_{n'} := \|a_{n'}\|^{-1}$, because of (ii) there exists a further subsequence $\{n''\}$ for which $\alpha_{n''} a_{n''} \to a_0$, $a_0 \in E'$, $\|a_0\| = 1$ and, furthermore, $a_{n''}(\mu_{n''}) \Rightarrow \nu$ for some $\nu \in \mathcal{R}(\mathbf{R})$. Since $\alpha_{n''} \to 0$ it follows $(\alpha_{n''} a_{n''})(\mu_{n''}) \Rightarrow \delta_0$. On the other hand, in view of Cor. 3.2

$$(\alpha_{n''} a_{n''})(\mu_{n''}) \Rightarrow a_0(\mu).$$

Hence $a_0(\mu) = \delta_0$ and $a_0 \in \gamma(\mu)^\perp$ which contradicts (iii). This completes the proof.

Corollary 4.4. *Assume $\mu_n \Rightarrow \mu$ and let μ be full. If $\{a_n\} \subseteq E'$ satisfies*

(i) $\{a_n(\mu_n)\}$ *is w.r.c.,*

(ii) $\left\{ \dfrac{a_n}{\|a_n\|} \right\}$ *is r.c. in E',*

then $\sup_n \|a_n\| < \infty$.

The next corollary is an easy consequence of the principle of uniform boundedness applied to the sequence of dual operators.

Corollary 4.5.   *Assume $\mu_n \Rightarrow \mu$ and let $\{T_n\} \subseteq \mathcal{L}(E, F)$ with*

(i) *$\{T_n(\mu_n)\}$ is w.r.c. in F;*

(ii) *For every $b \in F'$ the set $\left\{ \dfrac{T_n' b}{\|T_n' b\|} \right\}$ is r.c. in E';*

(iii) *For every $b \in F'$ $\left\{ \dfrac{T_n' b}{\|T_n' b\|} \right\}' \cap \gamma(\mu)^\perp \subseteq \{0\}$.*

Then $\sup_n \|T_n\| < \infty$.

Remark: It is clear that (ii) in Cor. 4.5 is always satisfied for $\dim(E) < \infty$. Moreover, (iii) is always true provided that μ is full. Hence we find back one direction of the compactness lemma due to Sharpe [13].

The next lemma is the key for all subsequent investigations.

Lemma 4.6.   *Let $\{\mu_n\}$ be a w.r.c. sequence in $\mathcal{R}(E)$. If the elements $x_n \in E$ satisfy*

$$\varliminf_{n \to \infty} \mu_n(U_\varepsilon(x_n)) > 0, \quad \text{for all } \varepsilon > 0,$$

then $\{x_n\}$ has to be r.c.

Proof: For each $\varepsilon > 0$ we define $f(\varepsilon) > 0$ by

$$f(\varepsilon) = \varliminf_{n \to \infty} \mu_n(U_\varepsilon(x_n)).$$

Hence there are natural numbers $n(\varepsilon)$ such that

$$\mu_n(U_\varepsilon(x_n)) > f(\varepsilon)/2, \quad n \geq n(\varepsilon).$$

On the other hand, since $\{\mu_n\}$ is w.r.c. by Prokhorov's theorem there exist compact subsets $K_\varepsilon \subseteq E$ with

$$\mu_n(K_\varepsilon) \geq 1 - f(\varepsilon)/2, \quad n = 1, 2, \ldots.$$

Therefore, for each $n \geq n(\varepsilon)$ the two sets K_ε and $U_\varepsilon(x_n)$ cannot be disjoint, i.e.

$$x_n \in K_\varepsilon + U_\varepsilon(0), \quad n \geq n(\varepsilon).$$

This being true for each $\varepsilon > 0$ yields the precompactness of $\{x_n\}$ and since E is complete we are done.

Before we proceed to investigate our problem let us state two consequences of the preceding lemma which seem to be interesting in their own right.

Proposition 4.7. *Assume a r.s.c. subset $\{\mu_n\} \subseteq \mathcal{R}(E)$ satisfies*

$$\lim_{n \to \infty} \mu_n(U_\varepsilon(x_0)) > 0$$

for all $\varepsilon > 0$ and some $x_0 \in E$. Then $\{\mu_n\}$ has to be w.r.c.

Proof: Let $\{x_n\} \subseteq E$ be a sequence of shifts of $\{\mu_n\}$, i.e., $\{\mu_n * \delta_{x_n}\}$ is w.r.c. Because of

$$\mu_n(U_\varepsilon(x_0)) = (\mu_n * \delta_{x_n})(U_\varepsilon(x_0 + x_n)),$$

Lemma 4.6 lets us conclude the relative compactness of $\{x_0 + x_n\}$. Then so is $\{x_n\}$, and by virtue of Proposition 2.3.3 in [10] the sequence $\{\mu_n\}$ is w.r.c. as asserted.

Using exactly the same ideas as in the proofs of Lemma 4.6 and Proposition 4.2 we obtain a criterion for a r.s.c. set to be w.r.c.

Proposition 4.8. *Let $\{\mu_n\} \subseteq \mathcal{R}(E)$ be r.s.c. Then the following are equivalent:*

  (i) *$\{\mu_n\}$ is w.r.c.*
  (ii) *There exists a compact set $K \subseteq E$ with*

$$\lim_{n \to \infty} \mu_n(K) > 0.$$

We come back now to the investigation of sequences $\{T_n \mu_n\}$.

Theorem 4.9.   *Let* $\{T_n\} \subseteq \mathcal{L}(E, F)$ *be a bounded sequence and assume* $\mu_n \Rightarrow \mu$ *in* $\mathcal{R}(E)$. *Then the following are equivalent:*

   (i) $\{T_n\mu_n\}$ *is w.r.c.*

   (ii) $\{T_n\mu\}$ *is w.r.c'*

   (iii) *For each* $x \in \mathrm{supp}(\mu)$ *the sequence* $\{T_n x\}$ *is r.c.*

   (iv) *The restrictions* $\{T_n|_{\gamma(\mu)}\}$ *are s.r.c. in* $\mathcal{L}(\gamma(\mu), F)$.

Proof: We have (i) $\Leftrightarrow$ (ii) and (iii) $\Leftrightarrow$ (iv) because of proposition 3.3 and lemma 2.3, respectively. Cor. 3.4 lets us conclude (iv) $\Rightarrow$ (ii). Hence it remains to prove (ii) $\Rightarrow$ (iii). Setting $c = \sup_n \|T_n\|$, we obtain

$$\varlimsup_{n \to \infty} (T_n\mu)(U_\varepsilon(T_n x)) \geq \mu(U_{\varepsilon/c}(x)) > 0, \quad \varepsilon > 0.$$

In view of Lemma 4.6 the set $\{T_n x\} \subseteq F$ is r.c. ($\{T_n\mu\}$ is w.r.c.) proving (iii).

In view of cor. 4.5 we need not to assume the boundedness of $\{T_n\}$ in the case of finite dimensional spaces E and full μ. Hence we obtain the compactness lemma of Sharpe [13]:

Corollary 4.10.   *Let* E *be finite dimensional and assume* $\mu_n \Rightarrow \mu$ *and* μ *full. Then for any sequence* $\{T_n\} \subseteq \mathcal{L}(E, F)$ *the following are equivalent:*

   (i) $\{T_n\mu_n\}$ *is w.r.c.*

   (ii) $\{T_n\mu\}$ *is w.r.c.*

   (iii) $\{T_n\}$ *is s.r.c.*

Moreover, if $\dim(F) < \infty$ *as well, then this is also equivalent to*

   (iv) $\sup_n \|T_n\| < \infty$.

Finally we investigate operators $T_n \in \mathcal{L}(E, F)$ for which $\{T_n\mu_n\}$ is only r.s.c. and $\mu_n \Rightarrow \mu$. In order to obtain results similar to theorem 4.9 we have to replace $T_n|_{\gamma(\mu)}$ by $T_n|_{\gamma(\mu^*)}$. Further, we shall see that representations of shift elements of $\{T_n\mu_n\}$ depend on algebraic properties of $\mathrm{supp}(\mu)$.

Theorem 4.11. *Assume* $\mu_n \Rightarrow \mu$ *and* $\sup_n \|T_n\| < \infty$. *Then the following are equivalent:*

(i) $\{T_n \mu_n\}$ *is r.s.c.*

(ii) $\{T_n \mu\}$ *is r.s.c.*

(iii) *For some [each]* $x_0 \in \operatorname{supp}(\mu)$ *we have* $\{(T_n \mu_n) * \delta_{-T_n x_0}\}$ *w.r.c. or* $\{(T_n \mu) * \delta_{-T_n x_0}\}$ *w.r.c.*

(iv) *If* $x \in \operatorname{supp}(\mu) - \operatorname{supp}(\mu)$, *then* $\{T_n x\} \subseteq F$ *is r.c.*

(v) *The restrictions* $\{T_n|_{\gamma(\mu^s)}\}$ *are s.r.c. in* $\mathcal{L}(\gamma(\mu^s), F)$.

Proof: Because of proposition 3.3 we have (i) $\Leftrightarrow$ (ii). Since $(T_n(\mu))^s = T_n(\mu^s)$ the statements (ii), (iv), and (v) are equivalent in view of theorem 4.9. Indeed, $\{T_n \mu\}$ is r.s.c. iff $\{(T_n(\mu))^s\}$ is w.r.c. (cf. [10, Proposition 2.5.5]); moreover we have

$$\gamma(\mu^s) = \overline{span}(\operatorname{supp}(\mu) - \operatorname{supp}(\mu)).$$

Of course, if (iii) holds for some $x_0 \in \operatorname{supp}(\mu)$, then we have $\{T_n \mu_n\}$ r.s.c. or $\{T_n \mu\}$ r.s.c., respectively. Hence it suffices to prove (iv) $\Rightarrow$ (iii) for arbitrary elements of $\operatorname{supp}(\mu)$. To do so choose $x_0 \in \operatorname{supp}(\mu)$ and put $\sigma = \mu * \delta_{-x_0}$. Then

$$\operatorname{supp}(\sigma) = \operatorname{supp}(\mu) - x_0 \subseteq \operatorname{supp}(\mu) - \operatorname{supp}(\mu)$$

and, by (iv), $\{T_n x\}$ is r.c. for each $x \in \operatorname{supp}(\sigma)$. Theorem 4.9 lets us conclude $\{T_n \sigma\}$ w.r.c., i.e., $\{(T_n \mu) * \delta_{-T_n x_0}\}$ w.r.c. In view of proposition 3.3 the same is valid for the sequence $\{(T_n \mu_n) * \delta_{-T_n x_0}\}$ completing the proof.

Remarks: (1) If $\{\nu_n\} \subseteq \mathcal{R}(E)$ is r.s.c., then in general we have no information about centering sequences, i.e., about sequences $\{y_n\}$ for which $\{\nu_n * \delta_{y_n}\}$ is w.r.c. But the preceding theorem shows that we may take $y_n = -T_n x_0$, $x_0 \in \operatorname{supp}(\mu)$, in the case $\nu_n = T_n \mu_n$, T_n and μ_n as above.

(2) Let μ_n, μ, and T_n be defined as in theorem 4.11. Then the assumption

$$\mu\{x \in E : \langle x, a \rangle = 1\} < 1$$

for each $a \in E'$ implies $\gamma(\mu) = \gamma(\mu^s)$, see lemma 1.4. Hence in this case the assertions (i)-(v) of theorem 4.11 are also equivalent to

(vi) $\{T_n \mu_n\}$ w.r.c.

(vii) $\{T_n \mu\}$ w.r.c.

(viii) $\{T_n|_{\gamma(\mu)}\}$ s.r.c. in $\mathcal{L}(\gamma(\mu), F)$.

(3) Observe that the assumption

$$\mu\{x \in E : \langle x, a \rangle = 1\} < 1, \quad a \in E',$$

is necessary to obtain equivalence between (vi) and (i). Indeed, if $\mathrm{supp}(\mu)$ is contained in some hyperplane $H = \{x \in E : \langle x, a_0 \rangle = 1\}$, then we may define operators $T_n \in \mathcal{L}(E, l_2)$ by

$$T_n x = \langle x, a_0 \rangle e_n \quad (e_n \ n\text{-th unit vector in } l_2)$$

and $\{T_n \mu\}$ is r.s.c. because of $T_n(\mu^s) = \delta_0$. On the other hand, we have $T_n(\mu) = \delta_{e_n}$ and, consequently, $\{T_n(\mu)\}$ cannot be w.r.c.

(4) As we have seen the sequence $\{T_n \mu_n\}$ is r.s.c. iff it is w.r.c. provided that $\gamma(\mu) = \gamma(\mu^s)$. Especially this is valid in the case $\gamma(\mu^s) = E$, i.e., μ^s full. To verify this fact we had to assume $\sup_n \|T_n\| < \infty$. One might believe now that the boundedness of $\{T_n\}$ is not necessary; but it is because of the following general assertion:

For any $\mu \in \mathcal{R}(E)$ with $\dim \gamma(\mu) = \infty$ there are $\mu_n \in \mathcal{R}(E)$ and $a_n \in E'$ such that $\mu_n \Rightarrow \mu$ and $\{a_n(\mu_n)\}$ is r.s.c. but not w.r.c. To see this take $a_n \in E'$ with $a_n(\mu) \Rightarrow \delta_0$ and $\|a_n\| \to \infty$ (proposition 4.1). Hence there exist $x_n \in E$ with $\|x_n\| = 1$ and $\langle x_n, a_n \rangle \to \infty$. Now we get

$$z_n := x_n(\langle x_n, a_n \rangle)^{-1/2} \to 0$$

and thus

$$\mu_n := \mu * \delta_{z_n} \Rightarrow \mu.$$

We clearly have $a_n(\mu_n) = a_n(\mu) * \delta_{y_n}$ where — by construction —

$$y_n := \langle z_n, a_n \rangle = (\langle x_n, a_n \rangle)^{1/2} \to \infty.$$

So $\{a_n(\mu_n)\}$ is r.s.c. but it is not w.r.c.

Applying theorem 4.11 we are in a position to state and prove the following convergence of types theorem which turned out to be useful in the investigation of operator-stable probability measures on Banach spaces (cf. [14]).

Theorem 4.12. *Let μ_n, μ, and T_n as in theorem 4.11. Then any cluster point ν of shifts of $\{T_n \mu_n\}$ can be represented as $\nu = T(\mu) * \delta_y$ where $T \in \mathcal{L}(\gamma(\mu), F)$ and $y \in F$.*

Proof: Theorem 4.11 tells us that the restrictions $\{T_n|_{\gamma(\mu^s)}\}$ have to be s.r.c. in $\mathcal{L}(\gamma(\mu^s), F)$. Using lemma 4.2, (iv), there exist a subsequence $\{n'\}$ as well as an operator $T_0 \in \mathcal{L}(\gamma(\mu^s), F)$ so that

$$T_{n'}|_{\gamma(\mu^s)} \xrightarrow{s} T_0.$$

Setting $\sigma = \mu * \delta_{-x_0}$ for some fixed $x_0 \in \text{supp}(\mu)$, we obtain $\text{supp}(\sigma) \subseteq \gamma(\mu^s)$. From cor. 3.4 we now derive $T_{n'}(\sigma) \Rightarrow T_0(\sigma)$ and, accordingly,

$$T_{n'}(\mu) * \delta_{-T_n x_0} \Rightarrow T_0(\mu * \delta_{-x_0}).$$

On the other hand, theorem 4.11 implies that $\{T_n \mu_n\}$ is r.s.c. Therefore it is no loss of generality to assume that

$$T_{n'}(\mu) * \delta_{y_{n'}} \Rightarrow \nu.$$

Combining both weak convergences and applying cor. 2.3.5 in [10] we conclude that

$$T_{n'} x_0 + y_{n'} \to y \in F, \text{ say,}$$

and, moreover, $\nu = T_0(\mu * \delta_{-x_0}) * \delta_y$. To end the proof take the projection Q from $\gamma(\mu)$ onto $\gamma(\mu^s)$ constructed in the remark after lemma 1.4. Then

$$\mu * \delta_{-x_0} = Q(\mu),$$

and

$$\nu = (T_0 Q)(\mu) * \delta_y$$

which proves the assertion with $T = T_0 Q \in \mathcal{L}(\gamma(\mu), F)$.

The following example shows that the assumption $\sup_n \|T_n\| < \infty$ cannot be dropped in theorem 4.12.

Example: Let $\lambda_n \in \mathcal{R}(l_2)$ be defined by

$$\lambda_n(B) = \frac{n}{2} \lambda \{ t \in [-\frac{1}{n}, \frac{1}{n}] : t\, e_n \in B \},$$

λ Lebesgue measure on $[-1,1]$, B Borel subset of l_2. Furthermore, let $\varphi_n \in \mathcal{R}(l_2)$ be Gaussian with expectation $0 \in l_2$ and covariance operators given by

$$\langle C_n e_i, e_j \rangle = \begin{cases} \lambda^i \delta_{ij} & , \quad i \neq n \\ 0 & , \quad i = n, \end{cases}$$

for certain $\lambda^i > 0$, $i, j = 1, 2, \ldots$. It is clear that $\mu_n := \lambda_n * \varphi_n \Rightarrow \mu$ where μ is a full Gaussian measure with $\langle C e_i, e_j \rangle = \lambda^i \delta_{ij}$ for all i, j. On the other hand, we define $T_n \in \mathcal{L}(l_2, l_2)$ by writing

$$T_n x = n\, x^n e_1 + \sum_{j=2}^{\infty} x^j e_j$$

for each

$$x = \sum_{j=1}^{\infty} x^j e_j \in l_2.$$

Hence

$$T_n(\mu_n) \Rightarrow \nu := \lambda_1 * \varphi_1,$$

where λ_1, φ_1 are as above. Of course, a representation of ν as $T(\mu) * \delta_y = \nu$ cannot be true.

References

[1] P. Billingsley, *Convergence of types in k-spaces*, Z. Wahrscheinlichkeitsth. verw. Geb. 5 (1966), 175–179.

[2] —————, "Convergence of probability measures," Wiley, New York, 1968.

[3] I. Csiszar and B.S. Rajput, *A convergence of types theorem for probability measures on topological vector spaces with applications to stable laws,* Z. Wahrscheinlichkeitsth. verw. Geb. 36 (1976), 1–7.

[4] R.M. Dudley, *Random linear functionals,* Trans. Amer. Math. Soc. 136 (1969), 1–24.

[5] N. Dunford and J.T. Schwartz, "Linear operators I, general theory," Wiley, New York, 1964.

[6] W. Feller, "An Introduction to Probability Theory and Its Applications II," Wiley, New York, 1971.

[7] O. Jouandet, *Sur les convergence en type de variables aléatiores à valeurs dans des espaces d'Hilbert ou de Banach,* C.R. Acad. Sci. Paris, Sér. A 271 (1970), 1082–1085.

[8] J. Jurek, *On stability of probability measures in Euclidean spaces,* pp. 128–145 in "Probability theory on vector spaces II", Proc. Błazejewko, Poland 1979, Lecture Notes in Math. 818. Springer–Verlag, Berlin, 1980.

[9] G. Letta, *Eine Bemerkung zum Konvergenzsatz für Verteilungstypen,* Z. Wahrscheinlichkeitsth. verw. Geb. 2 (1964), 310–313.

[10] W. Linde, "Infinitely divisible and stable measures on Banach spaces," Wiley, New York, 1986.

[11] R.K. Parthasarathy, "Probability measures on metric spaces," Academic Press, New York, 1967.

[12] H.J. Rossberg, B. Jesiak and G. Siegel, "Analytic Methods of Probability Theory," Academie–Verlag, Berlin, 1985.

[13] M. Sharpe, *Operator–stable probability measures on vector groups,* Trans. Amer. Math. Soc. 136 (1969), 51–65.

[14] G. Siegel, *Operator–stable distributions in separable Banach spaces.*

[15] N.N. Vakhania, V.I. Tarieladze and S.A. Chobajan, "Probability distributions in Banach spaces (Russian)," Nauka, Moscow, 1985.

Friedrich–Schiller–Universität, Sektion Mathematik, UHH 17. OG, Jena 6900

Technische Universität Dresden, Sektion Mathematik, Mommenstr. 13, Dresden 8027

Gelfand Numbers and Euclidean Sections
of Large Dimensions

ALAIN PAJOR AND NICOLE TOMCZAK-JAEGERMANN*

Let $E = (\mathbf{R}^n, \|\cdot\|)$ be an n-dimensional Banach space and let B_E be the unit ball in E. We shall also consider a Euclidean structure on $\mathbf{R}^n$, and so, let $(\cdot,\cdot)$ denote an inner product and $\|\cdot\|_2$ the corresponding Euclidean norm. The dual space E^* is naturally identified to $(\mathbf{R}^n, \|\cdot\|_*)$, where

$$(1) \qquad \|x\|_* = \sup\{|(x,y)| \mid y \in B_E\}.$$

Let B_2^n be the Euclidean unit ball and let $S = S_{n-1}$ be the unit sphere. Let μ be the normalized rotation invariant measure on S. Set

$$(2) \qquad M(E) = \left(\int_S \|x\|^2 d\mu\right)^{1/2}, \quad M(E^*) = \left(\int_S \|x\|_*^2 d\mu\right)^{1/2}.$$

Let us recall the main result from [P-T], which was motivated by study of V. Milman (cf. [M.1], [M.2]).

Theorem 1. *Let* $E = (\mathbf{R}^n, \|\cdot\|)$ *and let* $\|\cdot\|_2$ *be a Euclidean norm on* $\mathbf{R}^n$. *For every integer* k, $1 \le k \le n$, *there exists a subspace* F *of* $\mathbf{R}^n$ *with* $\dim F = n - k$ *such that*

$$\|x\|_2 \le KM(E^*)(n/k)^{1/2}\|x\| \quad \text{for} \quad x \in F,$$

where K *is a universal constant.*

The main technical result of this note improves Theorem 1. In particular, it will enable us to obtain precise estimates in some cases of a

---

*Research partially supported by NSERC Grant A8854.

"non K-convex" space E, when a direct application of Theorem 1 gives a superfluous "logarithmic factor".

Before we go on, let us shortly discuss a natural interpretation of Theorem 1 in terms of operators. It requires some definitions. For an operator $u: X \to Y$ between Banach spaces X and Y, the Gelfand numbers are defined by

$$c_k(u) = \inf\{\||u|_Z\| \mid Z \subset X, \operatorname{codim} Z < k\},$$

and the Kolmogorov numbers are defined by

$$d_k(u) = \inf\{\|q_L u\| \mid L \subset Y, \dim L < k, q_L: Y \to Y/L \text{ the quotient map}\},$$

for $k = 1, 2, \ldots$. We have $d_k(u) = c_k(u^*)$ for $k = 1, 2, \ldots$. For an operator $u: \ell_2^n \to Y$ set

$$\ell(u) = \left(\int \|ux\|^2 d\gamma_n(x)\right)^{1/2},$$

where γ_n is the standard Gaussian probability measure on $\mathbf{R}^n$. In particular, $\ell(u) = n^{1/2} M(\widetilde{Y})$, where $\widetilde{Y} = (\mathbf{R}^n, |\cdot|)$ and $|x| = \|ux\|$ for $x \in \mathbf{R}^n$. Finally, for $u: \ell_2 \to Y$ set

$$\ell(u) = \sup\{\ell(uv) \mid v: \ell_2^n \to \ell_2, \|v\| \leq 1, n \in \mathbf{N}\}.$$

The operator version of Theorem 1 says

Theorem 2. *Let X be a Banach space and let $u: X \to \ell_2$ be a compact operator. Then*

$$\sup_{k \geq 1} k^{1/2} c_k(u) \leq K\ell(u^*),$$

where K is a universal constant.

The estimates obtained in Theorems 1 and 2 are, in general, the best possible. To see this let $1 < p < 2$ and consider the formal identity operator $i: \ell_p^n \to \ell_2^n$. It was proved by E.d. Gluskin [**G1**], [**G2**] that

Proposition 1. *Let* $1 < p < 2$. *Then*

$$c_k(i \colon \ell_p^n \to \ell_2^n) \geq c(p) \min(1, n^{1-1/p_k-1/2}),$$

where $c(p)$ depends only on p.

For $1 < p < \infty$ set $p' = p/(p-1)$. It is well-known that $\ell(i^* \colon \ell_2^n \to \ell_{p'}^n) = n^{1/2} M(\ell_{p'}^n) \geq c'(p) n^{1/p'}$ (cf. e.g. [M-Sch] or [T.2] XI.3). Combining with Proposition 1 we get, for $1 < p < 2$,

$$\sup_{k \geq 1} k^{1/2} c_k(i \colon \ell_p^n \to \ell_2^n) \geq c''(p) \ell(i^* \colon \ell_2^n \to \ell_{p'}^n),$$

where $c''(p) > 0$ depends only on p.

For a reader's convenience let us sketch the Gluskin's argument.

Proof of Proposition 1: Fix $0 < a(p) < 1$ to be defined later. Let $a(p) n^{2/p'} \leq k \leq a(p) n$. Fix $2 < s \leq p'$ such that $k = a(p) n^{2/s}$. Pick $L \subset \mathbf{R}^n$ with $\dim L = k$ such that $d_{k+1}(i^* \colon \ell_2^n \to \ell_{p'}^n) = \|q_L i^*\|$. For $j = 1, \ldots, n$ let e_j be the j-th unit vector in $\mathbf{R}^n$ and pick $y_j = (y_j(1), \ldots, y_j(n)) \in L$ such that

$$(3) \qquad \|e_j - y_j\|_{p'} = \|q_L i^* e_j\|_{p'} = \inf\{\|e_j - y\|_{p'} \mid y \in L\}.$$

We shall use the following inequality valid for all $t > 0$,

$$(4) \qquad \frac{1}{n} \sum_{j=1}^n |y_j(j)| \leq t^{-1/s}(k^{s/2} n^{-1})^{1/(s-1)} + t^{1-1/s} \frac{1}{n} \sum_{j=1}^n \|y_j\|_s^s.$$

We shall also use the following elementary inequality valid for all real numbers r,

$$|1 - r|^s \geq 1/2 - 2sr/3 + (2s)^{-s} |r|^s.$$

Combining these two inequalities we get

$$\frac{1}{n}\sum_{j=1}^{n}\|e_j - y_j\|_s^s \geq \frac{1}{2} - \frac{2s}{3n}\sum_{j=1}^{n}|y_j(j)| + (2s)^{-s}\frac{1}{n}\sum_{j=1}^{n}\|y_j\|_s^s$$

$$\geq \left(\frac{1}{2} - \frac{2s}{3}t^{-1/s}a(p)^{s/2(s-1)}\right)$$

$$+ \left((2s)^{-s} - \frac{2s}{3}t^{1-1/s}\right)\frac{1}{n}\sum_{j=1}^{n}\|y_j\|_s^s.$$

Setting $t^{1-1/s} = 3(2s)^{-s-1}$ and $a(p) = (3/16p')^2$ we easily check that $a(p) \leq a/4^4s^4 = (3t^{1/s}/8s)^{2-2/s}$ and the latter sum is greater than or equal to $1/4$. Thus there exists j_0, $1 \leq j_0 \leq n$, such that

$$\|e_{j_0} - y_{j_0}\|_{p'} \geq n^{1/p'-1/s}\|e_{j_0} - y_{j_0}\|_s \geq n^{1/p'-1/s}4^{-1/s}.$$

By (3) and the definition of s, we get

$$d_{k+1}(i^* : \ell_2^n \to \ell_{p'}^n) = \|q_L i^*\| \geq n^{1/p'-1/s}/2$$
$$= a(p)^{1/2}n^{1/p'}/2k^{1/2}.$$

Finally, for $k < n^{2/p'}$, the lower estimate for $d_{k+1}(i^*)$ is clearly satisfied, and for $k > a(p)n$ we have

$$d_{k+1}(i^* : \ell_2^n \to \ell_{p'}^n) \geq d_n(i^*) = n^{1/p'-1/2} \geq a(p)^{1/2}n^{1/p'}k^{-1/2}.$$

It remains to prove (4). Let $P : \ell_2^n \to \ell_2^n$ be the orthogonal projection onto L. Clearly, for the Hilbert-Schmidt norm of P we have $k^{1/2} = hs(P) = \left(\sum_{j=1}^{n}\|Pe_j\|_2^2\right)^{1/2}$. Thus

$$\frac{1}{n}\sum_{j=1}^{n}|y_j(j)| = \frac{1}{n}\sum_{j=1}^{n}|(y_j, Pe_j)| \leq \frac{1}{n}\sum_{j=1}^{n}\|y_j\|_2\|Pe_j\|_2$$

$$\leq \frac{1}{n}\left(\sum_{j=1}^{n}\|y_j\|_2^2\right)^{1/2}\left(\sum_{j=1}^{n}\|Pe_j\|_2^2\right)^{1/2}$$

$$\leq n^{-1/s}k^{1/2}\left(\frac{1}{n}\sum_{j=1}^{n}\|y_j\|_s^s\right)^{1/s}.$$

To conclude (4) it is enough to use the well-known inequality $a^{1-1/s}b^{1/s} \leq t^{-1/s}a + t^{1-1/s}b$, valid for all a, b and $t > 0$ and $s \geq 1$.   ∎

We pass now to the main technical result of this note. Given Banach space $E = (\mathbf{R}^n, \|\cdot\|)$ and a Euclidean norm $\|\cdot\|_2$ on $\mathbf{R}^n$, for every $\rho > 0$ set $\|x\|_\rho = \max(\|x\|, \rho^{-1}\|x\|_2)$ for $x \in \mathbf{R}^n$ and let $E_\rho = (\mathbf{R}^n, \|\cdot\|_\rho)$. With this notation we have

Theorem 3. *Let $E = (\mathbf{R}^n, \|\cdot\|)$ and let $\|\cdot\|_2$ be a Euclidean norm on $\mathbf{R}^n$. For every positive integer k, $1 \leq k \leq n$, there exists a subspace F of $\mathbf{R}^n$, with $\dim F = n - k$, such that for every $x \in F$, $\|x\|_2 \leq \rho_0\|x\|$, where*

$$\rho_0 = \inf\{\rho > 0 \ \mid \ M(E_\rho^*)/\rho \leq K(k/n)^{1/2}\},$$

and K is a universal constant.

Theorem 1 follows immediately from Theorem 3 as we observe that $M(E_\rho^*) \leq M(E^*)$.

An idea of improving Theorem 1 into Theorem 3 was suggested by a renorming approach of V. Milman and G. Pisier in [M-P]. The proof of Theorem 3 was given implicitly already in [P-T]. It is based on the Sudakov minoration theorem, which gives an upper estimate for a Euclidean entropy, and two lemmas which utilize direct calculations on the Euclidean sphere and Dvoretzky's theorem. (The proofs of these lemmas given in [P-T] are slightly different and based directly on the isoperimetric inequality on the sphere.)

In the proof which follows we may and we shall assume that $\|x\|_2 = \left(\sum_{j=1}^n |x(j)|^2\right)^{1/2}$, for $x = (x(1), \ldots, x(n)) \in \mathbf{R}^n$. For $k = 1, \ldots, n$, let $P_k: \ell_2^n \to \ell_2^n$ denote the orthogonal projection onto the first k coordinates. It can be checked by a direct computation that there exists $a > 0$ such that

$$(5) \quad \mu\{x \in S_{n-1} \ \mid \ \|P_k x\|_2 \leq a(k/n)^{1/2}\} < \min(1/2, \exp(-ak)).$$

Finally, let $\mathcal{P}$ denote the normalized Haar measure on the group $G(n)$

of all isometries of ℓ_2^n. Fix $x \in \mathbf{R}^n$, $x \neq 0$; then the measure μ on the sphere $S = \{\|x\|_2 = 1\}$ is the image of $\mathcal{P}$ by the map $U \to Ux/\|x\|_2$.

Lemma 1. *Let $1 \leq k \leq n$ and let $\Lambda \subset \mathbf{R}^n$ be a subset of cardinality smaller than $\exp(ak/2)$. Then*

$$\mathcal{P}\{U \in G(n) \mid \exists z \in \Lambda \ \|P_k U z\|_2 \leq a(k/n)^{1/2}\|z\|_2\} < 1/2.$$

Proof: We have, by (5),

$$\mathcal{P}\{U \in G(n) \mid \exists z \in \Lambda \ \|P_k U z\|_2 \leq a(k/n)^{1/2}\|z\|_2\}$$
$$\leq |\Lambda|\mu\{x \in S_{n-1} \mid \|P_k x\|_2 \leq a(k/n)^{1/2}\}$$
$$\leq \exp(-ak/2).$$

This concludes the lemma for k such that $\exp(-ak/2) \leq 1/2$. For the remaining k, $|\Lambda| = 1$ and the lemma follows directly from (5). ∎

Consider the Grassman manifold $G_{n,k}$ of all k-dimensional subspaces of $\mathbf{R}^n$ with the normalized measure μ_k. For $F \in G_{n,k}$, let P_F denote the orthogonal projection onto F. If $X = (\mathbf{R}^n, |\cdot|)$ is a Banach space with $\|x\|_2 \leq |x|$, for $x \in \mathbf{R}^n$, then Dvoretzky's theorem says (cf. e.g. [M-Sch]) that with a "high" probability in $G_{n,k}$, ℓ-dimensional sections of B_{X^*} are "close" to ℓ-dimensional Euclidean balls, if $\ell \leq [\eta n M(X^*)^2]$. Passing to the dual statement we have in particular that there exists a constant $\eta > 0$ such that for every $k \leq [\eta n M(X^*)^2]$ there exists a subset $\mathcal{A}$ of $G_{n,k}$, with $\mu_k(\mathcal{A}) \geq 1/2$ such that

$$(6) \qquad P_F B_X \subset 2M_*|P_F B_2^n| \subset 2\sqrt{2}M(X^*)(P_F B_2^n) \quad \text{for} \quad F \in \mathcal{A}.$$

(Here M_* denotes the median of $|\cdot|_*$ on the sphere S_{n-1} and so $M_* \leq \sqrt{2}M(X^*)$ (cf. [M-Sch], or [T.2] I.7.2).

Lemma 2. *Let $X = (\mathbf{R}^n, |\cdot|)$ and $\|x\|_2 \leq |x|$ for $x \in \mathbf{R}^n$. Let $1 \leq k \leq n$ and assume that $M(X^*) = (k/\eta n)^{1/2}$. Then*

$$\mathcal{P}\{U \in G(n) \mid \|P_k U : X \to \ell_2^n\| > 2(2k/\eta n)^{1/2}\} < 1/2.$$

Proof: Clearly,

$$\mathcal{P}\{U \in G(n) \mid \|P_k U : X \to \ell_2^n\| > 2(2k/\eta n)^{1/2}\}$$
$$= \mu_k\{F \in G_{n,k} \mid \|P_F : X \to \ell_2^n\| > 2(2k/\eta n)^{1/2}\}.$$

Since the latter subset is contained in the complement of $\mathcal{A}$, the conclusion follows from (6).   ■

Proof of Theorem 3: If $X = (\mathbf{R}^n, |\cdot|)$ and $\|\cdot\|_2$ is a Euclidean norm on $\mathbf{R}^n$ then, for $\varepsilon > 0$, there exists an ε-net in B_X in the Euclidean metric, say Λ, such that

(7) $$|\Lambda| \le \exp(9M(X^*)^2 n/\varepsilon^2).$$

This is a consequence of the Sudakov minoration theorem [**Su**]. Indeed, let $(g_j)_{j \le n}$ be independent standard Gaussian random variables. Consider a Gaussian process $Y_t = \sum_{j=1}^n t_j g_j$, $t = (t_1, \ldots, t_n) \in B_X$. Then we have

$$\|t - s\|_2 = (\mathbf{E}|Y_t - Y_s|)^{2 \ 1/2} \quad \text{for } t, s \in B_X.$$

By the Sudakov theorem there exists an ε-net in B_X, say Λ, such that

$$\varepsilon(log|\Lambda|)^{1/2} \le 3\mathbf{E}\sup\{|Y_t| \mid t \in B_X\}.$$

To conclude (7) it is enough to observe that

$$\sup\{|Y_t| \mid t \in B_X\} = |\sum_{j=1}^n g_i e_i|_*$$

and

$$\mathbf{E}|\sum_{j=1}^n g_j e_j|_* \le n^{1/2} M(X^*).$$

Now, let $a > 0$ and $\eta > 0$ be as in (5) and (6). Define $\rho_0 > 0$ by

$$(8) \qquad \rho_0 = \inf\{\rho > 0 \mid M(E_\rho^*)/\rho \le (a\sqrt{\eta a}/24)(k/n)^{1/2}\}.$$

Since $M(E^*)/\rho$ is equal to the average of the norm $\|\| \cdot \||_*$ dual to $\|\|x\|\| = \max(\rho\|x\|, \|x\|_2)$ for $x \in \mathbf{R}^n$, then it is a decreasing function on ρ and so, $M(E_{\rho_0}^*)/\rho_0 = (a\sqrt{\eta a}/24)(k/n)^{1/2}$. Let B_{ρ_0} denote the unit ball in E_{ρ_0}, i.e. $B_{\rho_0} = B_E \cap \rho_0 B_2^n$ and set $S_{\rho_0} = \{x \in B_{\rho_0} \mid \|x\|_2 = \rho_0\}$. Set $\varepsilon = \rho_0 a \sqrt{\eta}/4\sqrt{2}$. Let Λ be an ε-net in S_{ρ_0} with $|\Lambda| \le \exp(ak/2)$ (by (7)). For $\tau > 0$ let $W_\tau = \tau B_{\rho_0} \cap B_2^n$ and let $|\cdot|_\tau$ be the norm associated to W_τ and $X_\tau = (\mathbf{R}^n, |\cdot|_\tau)$. Clearly, $\|x\|_2 \le |x|_\tau$ for $x \in \mathbf{R}^n$. Let τ_0 such that $M(X_{\tau_0}^*) = (k/\eta n)^{1/2}$.

Let $x \in S_{\rho_0}$ and pick $z \in \Lambda$ such that $\|x - z\|_2 < \varepsilon$. Then $x - z \in 2B_{\rho_0} \subset \varepsilon\tau_0 B_{\rho_0}$ (since $2 \le \varepsilon/M(E_{\rho_0}^*)(n/k)^{1/2} \le \varepsilon\tau_0$). Thus

$$(9) \qquad S_{\rho_0} \subset \cup_{z\in\Lambda}(z + \varepsilon W_{\tau_0}).$$

Using Lemmas 1 and 2 we see that there exists an isometry U on ℓ_2^n such that

$$\|P_k Uz\|_2 \ge a(k/n)^{1/2}\rho_0 \quad \text{for every } z \in \Lambda,$$
$$\|P_k U : X_{\tau_0} \leftarrow \ell_2^n\| \le 2(2k/\eta n)^{1/2}.$$

Set $F = \ker P_k U$. Clearly, $\dim F = n - k$. Let $y \in S_{\rho_0}$ and, by (9), let $z \in \Lambda$ such that $y \in z + \varepsilon W_{\tau_0}$. Then

$$\begin{aligned}
\|P_k Uy\|_2 &\ge \|P_k Uz\|_2 - \|P_k U(z - y)\|_2 \\
&\ge a(k/n)_{\rho_0}^{1/2} - 2(2k/\eta n)^{1/2}|z - y|_{\tau_0} \\
&\ge a(k/n)_{\rho_0}^{1/2} - \varepsilon 2(2k/\eta n)^{1/2} = a(k/n)_{\rho_0}^{1/2}/2 > 0.
\end{aligned}$$

So $S_{\rho_0} \cap F = \emptyset$ and hence $F \cap B_E \subset \rho_0 B_2^n$, which completes the proof of the theorem. ∎

Remark: An operator version of the Sudakov minoration theorem involves so-called entropy numbers of compact operators (cf. [P-T]). Then (7) is equivalent to the following inequality valid for arbitrary compact operator $u: X \to \ell_2$

$$\text{(10)} \qquad\qquad \sup_{k \geq 1} k^{1/2} e_k(u) \leq 3\ell(u^*).$$

Since, by [C], $\sup_{k \geq 1} k^{1/2} e_k(u) \leq K \sup_{k \geq 1} k^{1/2} c_k(u)$, where K is a universal constant, Theorem 2 improves (10). For the formal identity operator $i: \ell_p^n \to \ell_2^n$, $1 < p < 2$, the order of growth of $e_k(i)$, as $k \to \infty$, is known (cf. [S]) and it can be easily checked that in this case the lower estimate given by Theorem 2 is essentially stronger than (10).

Now let us give applications of Theorem 3 which improve some estimates for Gelfand numbers. The key of the method is to make an optimal choice for ρ. In fact, it is again the technique from [M-P], in a different context.

Let us recall that for an operator $u: X \to \ell_2^n$, the dual norm ℓ^* is defined by

$$\ell^*(u) = \sup\{|\text{trace } uv| \mid v: \ell_2^n \to X, \ell(v) \leq 1\}.$$

Corollary 1. *Let $E = (\mathbf{R}^n, \|\cdot\|)$ and let $u: E \to \ell_2^n$ be an isomorphism. For every integer k, $1 \leq k \leq n$, one has*

$$k^{1/2} c_k(u) \leq K\ell^*(u) \log(2 + \|u^{-1}\|\ell^*(u)/\sqrt{k}),$$

where K is a universal constant.

Proof: Choosing an appropriate Euclidean norm on $\mathbf{R}^n$ we may assume that u is the formal identity. By Theorem 3 we have

$$c_k(u) \leq \rho_0 = \inf\{\rho > 0 | \ell(u^*: \ell_2^n \to E_\rho^*)/\rho \leq K\sqrt{k}\}.$$

Clearly,

$$\|u : E_\rho \to \ell_2^n\| \le \rho \quad \text{and} \quad \|u^{-1} : \ell_2^n \to E_\rho\| \le \max(1/\rho, \|u^{-1}\|),$$

hence

$$d(E_\rho, \ell_2^n) \le \max(1, \rho\|u^{-1}\|).$$

By the well-known K-convexity argument (cf. e.g. [T.2] II.5)

(11)
$$\ell(u^* : \ell_2^n \to E_\rho^*) \le K(E_\rho)\ell^*(u : E_\rho \to \ell_2^n)$$
$$\le \log(1 + d(E_\rho, \ell_2^n))\ell^*(u : E_\rho \to \ell_2^n).$$

Finally observe that $\ell^*(u : E_\rho \to \ell_2^n) \le \ell^*(u : E \to \ell_2^n) = \ell^*(u)$, so that we get

$$\rho_0 \le \inf\{\rho > 0 \mid \log(1 + \max(1, \rho\|u^{-1}\|))/\rho \le Ck^{1/2}/\ell^*(u)\},$$

where C is a universal constant. A simple calculation completes the proof. ∎

Consider now the case when E has a 1-unconditional basis, say $(e_j)_{j \le n}$, and an operator $u : E \to \ell_2^n$ is diagonal (i.e. (ue_j) is an orthogonal basis). Clearly, $(e_j)_{j \le n}$ is a 1-unconditional basis in E_ρ too. In this case the general estimate for the K-convexity constant used in (11) can be replaced by a stronger upper estimate by $(\log(1 + d(E_\rho, \ell_2^n)))^{1/2}$ (cf. [P]). The same proof as in Corollary 1 gives then

Corollary 2. *Let $E = (\mathbf{R}^n, \|\cdot\|)$ have a 1-unconditional basis and let $u : E \to \ell_2^n$ be a diagonal one-to-one operator. For every integer k, $1 \le k \le n$, one has*

$$k^{1/2}c_k(u) \le K\ell^*(u)(\log(2 + \|u^{-1}\|\ell^*(u)/\sqrt{k}))^{1/2},$$

where K is a universal constant.

Remark: It is important to observe that the latter result cannot be obtained by the iteration method (cf. e.g. [M]), because this procedure does not preserve the unconditional structure.

In particular one has $\ell^*(i\colon \ell_1^n \to \ell_2^n) \le c$, where c is a universal constant. So as an immediate consequence of Corollary 2 we get the following estimate ([G-G], [T.1]) which improves the result of Kashin [K].

Corollary 3.   *For every integer k, $1 \le k \le n$, one has*

$$c_k(i\colon \ell_1^n \to \ell_2^n) \le K(\log(1 + n/k)/k)^{1/2},$$

where K is a universal constant.

Remarks: 1) Observe that as $\ell(i^*\colon \ell_2^n \to \ell_\infty^n) \sim (\log n)^{1/2}$, the estimate from Corollary 3 cannot be obtained from Theorem 1. Garnaev and Gluskin showed in [G-G] that this estimate is optimal.

2) Let us sketch an alternative proof of Corollary 3 which does not use the K-convexity argument. Fix an orthonormal basis $(e_j)_{j \le n}$ in ℓ_2^n and let $v\colon \ell_2^n \to F$ be an isomorphism. Set $r(v) = \mathbf{E}\|\sum_{j=1}^n \varepsilon_j v e_j\|$, where ε_j's are Rademacher functions. Using relations between Gaussian and Rademacher averages one can show

$$(12) \qquad \ell(v) \le K r(v)(\log(1 + \|v\|\,\|v^{-1}\|))^{1/2},$$

where K is a universal constant.

Now, let $u\colon E \to \ell_2^n$ be an isomorphism. An analogous argument as in Corollary 1 combined with (12) shows the inequality

$$(13) \qquad k^{1/2} c_k(u) \le c r(u^*)(\log(2 + \|u^{-1}\| r(u^*)/\sqrt{k}))^{1/2}.$$

The interest of this estimate is that it uses Rademacher averages rather than Gaussian. In particular for $i\colon \ell_1^n \to \ell_2^n$ and $(e_j)_{j \le n}$ the standard unit vector basis in ℓ_2^n one has $r(i^*) = 1$, which gives the estimate of Corollary 3.

Finally, we have results on Euclidean subspaces of spaces of cotype 2.

Corollary 4. *Let* $E = (\mathbf{R}^n, \| \cdot \|)$ *have a cotype 2 constant* $C_2(E)$. *For every integer* k, $1 \leq k \leq n$, *there exists a subspace* F *of* E *with* $\dim F = n - k$ *such that*

$$d(F, \ell_2^{n-k}) \leq KC_2(E)(n/k)^{1/2} \log(1 + C_2(E)n/k).$$

Moreover, if E *has a 1-unconditional basis, then*

$$d(F, \ell_2^{n-k}) \leq KC_2(E)(n/k)^{1/2}(\log(1 + C_2(E)n/k))^{1/2}.$$

Proof: It is well-known (cf. e.g. [**T.2**] III.2) that there exists $u \colon E \to \ell_2^n$ such that $\|u^{-1}\| = 1$ and $\pi_2(u) = \sqrt{n}$ (it is so-called John's embedding). A dual version of a known estimate (cf. e.g. [**T.2**] V.3) says

$$\ell^*(u) \leq C_2(E)\pi_2(u) = C_2(E)\sqrt{n}.$$

Now the first conclusion follows from Corollary 1 and the second from Corollary 2. ∎

References

[C] B. Carl, *Entropy numbers, s-numbers and eigenvalue problems*, J. Funct. Anal. **41** (1981), 290–360.

[G-G] A. Yu. Garnaev and E.D. Gluskin, *On widths of the euclidean ball*, Soviet Math. Dokl. **30** No. 1 (1984), 200–204.

[G1] E.D. Gluskin, *On some finite-dimensional problems in width theory*, Vestnik Lening. Univ. **13** (1981), 5–10.

[G2] ——————, *Norms of random matrices and widths of finite-dimensional sets*, Math. USSR. Sbornik **48** No. 1 (1984), 173–182.

[K] B.S. Kashin, *Diameters of some finite dimensional sets and some classes of smooth functions*, Math. USSR. Izv. **11** (1977).

[M.1] V.D. Milman, *Volume approach and iteration procedures in local theory of normed spaces*, pp. 99–105 in "Banach Spaces", Proceedings of Missouri Conf. 1984, Lecture Notes in Math. **1166**. Springer Verlag 1985.

[M.2] —————————, *Random subspaces of proportional dimension of finite-dimensional normed spaces: approach through the isoperimetric inequality*, pp. 106–115 in "Banach Spaces", Proceedings of Missouri Conf. 1984, Lecture Notes in Math. **1166**. Springer Verlag 1985.

[M-P] V.D. Milman and G. Pisier, *Banach spaces with a weak cotype 2 property*, Israel J. of Math. **54** (1986), 139–158.

[M-Sch] V.D. Milman and G. Schechtman, "Asymptotic theory of finite dimensional normed spaces," Lecture Notes in Math., Springer-Verlag, 1986.

[P-T] A. Pajor and N. Tomczak-Jaegermann, *Subspaces of small codimension of finite-dimensional Banach spaces*, Proc. AMS **97** (1986), 637–642.

[P] G. Pisier, *Remarques sur un résultat non publié de B. Maurey*, Séminaire d'Analyse Fonctionnelle 1980/81. Ecole Polytechnique-Palaiseau exposé, No. 5.

[S] C. Schütt, *Entropy numbers of diagonal operators between symmetric Banach spaces*, J. App. Theory **40** (1983), 121–128.

[Su] V.N. Sudakov, *Gaussian random processes and measures of solide angles in Hilbert spaces*, Soviet Math. Dokl. **12** (1971), 412–415.

[T1] N. Tomczak-Jaegermann, *On widths of finite-dimensional spaces*. Unpublished.

[T2] —————————, "Banach-Mazur distances and finite-dimensional operator ideals." To be published by Pitman.

Université des Sciences et Techniques de Lille, U.E.R. de Mathematiques, 59644 Villaneuve d'Ascq

Department of Mathematics, University of Alberta, Edmonton, Alberta, Canada T6G 2G1

The Law of the Iterated Logarithm
for Empirical Processes

J.E. YUKICH

Let $(A, \mathcal{A}, P)$ be a probability space and $\mathcal{F} \subset L^2(A, \mathcal{A}, P)$. Let X_i, $i \geq 1$, be a sequence of i.i.d. random variables with distribution P and let $P_n = n^{-1}(\delta_{X_1} + \cdots + \delta_{X_n})$ be the n'th empirical measure for P. Using the methods employed to describe functional Donsker classes, we characterize when the normalized empirical process

$$\nu_n(f) := n^{1/2} \int f(dP_n - dP), \quad f \in \mathcal{F}$$

satisfies the compact and bounded law of the iterated logarithm (LIL) uniformly over $\mathcal{F}$. Sufficient conditions implying the bounded LIL are obtained. In particular, we obtain two new metric entropy integral conditions implying the bounded LIL. Moreover, the integral condition is essentially the best possible.

1. Introduction

Let $(A, \mathcal{A}, P)$ be a probability space and let X_i, $i \geq 1$, be independent random variables with distribution P. We shall consider the X_i to be the coordinates for a countable product $(A^\infty, \mathcal{A}^\infty, P^\infty)$ of copies of $(A, \mathcal{A}, P)$. Let $P_n := n^{-1}(\delta_{X_1} + \cdots + \delta_{X_n})$, where δ_x is the unit mass at x, be the n'th empirical measure for P. Let $\mathcal{F} \subset L^2(A, \mathcal{A}, P)$ be a class of real-valued functions with envelope $F = F_{\mathcal{F}}(x) := \sup\{f(x) : f \in \mathcal{F}\}$. Let $\|f\|$ be the $L^2(A, \mathcal{A}, P)$ norm of $f \in \mathcal{F}$. Let $\ell^\infty(\mathcal{F})$ be the Banach space of real-valued functions h on $\mathcal{F}$; equip $\ell^\infty(\mathcal{F})$ with the sup norm $\|h\|_{\mathcal{F}} = \sup_{f \in \mathcal{F}} |h(f)|$. When $\mathcal{F}$ is infinite, we note that $\ell^\infty(\mathcal{F})$ is non-separable.

In this article we shall be concerned with the processes

$$\nu_n(f)(\omega) = n^{1/2} \int f(dP_n - dP)(\omega), \quad f \in \mathcal{F}, \quad \omega \in A^\infty,$$

and

$$I_n(f)(\omega) = \left(\frac{n}{2\log\log n}\right)^{1/2} \int f(dP_n - dP)(\omega), \quad f \in \mathcal{F}, \quad \omega \in A^\infty,$$

and law of the iterated logarithm properties for $\mathcal{F}$. More precisely, as in [15] say that $\mathcal{F}$ is a Strassen log log class for P, or equivalently, that $\mathcal{F}$ satisfies the compact law of the iterated logarithm (CLIL), if and only if there is a set $A_0 \subset A^\infty$ with $P^\infty(A_0) = 1$ such that for all $\omega \in A_0$, the sequence $\{I_n(f)(\omega), f \in \mathcal{F}, \omega \in A_0\}$ is relatively compact in $(\ell^\infty(\mathcal{F}), \|\cdot\|_{\mathcal{F}})$. Say also that $\mathcal{F}$ is a log log class for P, or equivalently, that $\mathcal{F}$ satisfies a bounded law of the iterated logarithm (BLIL), if and only if

$$\limsup_{n\to\infty} \sup_{f \in \mathcal{F}} |I_n(f)| < \infty \quad \text{a.s. } (P^\infty).$$

Clearly, every Strassen log log class is a log log class; the converse, of course, is false.

The Dudley-Philipp LIL results for empirical processes (especially Theorems 1.3 and 1.5 of [7]) strengthen the related Kuelbs-Dudley results [15] and perhaps provide the most streamlined methods for obtaining LIL properties; unfortunately, their results assume that $\mathcal{F}$ is a Donsker class, which, in general, need not be the case. Our approach, especially suited for both the CLIL and BLIL, does not require a weak convergence hypothesis, and in this way provides results which more satisfactorily describe LIL properties. For more on the LIL reader is referred to the recent work of Anderson et. al. [2], Ledoux [16], and Ledoux and Talagrand [17].

In this article, we first use tightness considerations to help characterize Strassen log log classes and log log classes. Using the resulting description of the CLIL and BLIL, we indicate how to obtain LIL properties through the use of majorizing measures and an Orlicz norm condition; such a result suggests how to obtain BLIL properties through the use of a new metric entropy integral condition. We produce the BLIL analogs of the Pollard [19] and Ossiander [18] central limit theorems and also show that the new metric entropy integral condition is essentially the best possible. In the process we answer a conjecture raised by the author [26].

Before discussing the main results we need a little notation. We first give two definitions for the metric entropy of $\mathcal{F} \subset \mathcal{L}^2(A, \mathcal{A}, P)$, corresponding to those introduced by Dudley [4] and Pollard [19], respectively.

Definition 1.1.   *Given $\mathcal{F} \subset \mathcal{L}^2(A, \mathcal{A}, P)$ and $\varepsilon > 0$, let $N_{[\,]}^{(2)}(\varepsilon, \mathcal{F}, P)$ be the minimum number of functions $f_1, \ldots, f_m \in \mathcal{L}^2(A, \mathcal{A}, P)$ such that $\forall f \in \mathcal{F} \exists i, j \leq m$ with $f_i \leq f \leq f_j$ and $\int (f_j - f_i)^2 dP < \varepsilon^2$. $H_{[\,]}(\varepsilon) :=$ $\log N_{[\,]}^{(2)}(\varepsilon, \mathcal{F}, P)$ is called the metric entropy of $\mathcal{F}$ with $\mathcal{L}^2$ bracketing.*

For each finite subset S of A and each positive ε define $N(\varepsilon, S, \mathcal{F})$ as the smallest value of m for which there are functions $f_1, \ldots, f_m \in \mathcal{F}$ such that

$$\min_{i \leq m} \sum_{x \in S} (f(x) - f_i(x))^2 < \varepsilon^2 \sum_{x \in S} F^2(x)$$

for every $f \in \mathcal{F}$.

Definition 1.2.   *Define $N(\varepsilon, \mathcal{F}) := \sup_S N(\varepsilon, S, \mathcal{F})$, where the supremum runs over all finite subsets of A. Let $H(\varepsilon, \mathcal{F}) := \log N(\varepsilon, \mathcal{F})$.*

Finally, given $\mathcal{F} \subset \mathcal{L}^2(A, \mathcal{A}, P)$ define the pseudo-metrics e_P and ρ_P by

$$e_P^2(f, g) := \int (f - g)^2 dP \quad \text{and}$$

$$\rho_P^2(f, g) := \int (f - g)^2 dP - \left(\int (f - g) dP \right)^2.$$

Using ρ_P define for all $\delta > 0$ the classes

$$\mathcal{F}_\delta := \{ f - g : f, \, g \in \mathcal{F} \ \text{ and } \rho_P(f, g) < \delta \}.$$

For definiteness we take as our underlying probability space

$$(\Omega, S, Pr): = (A^{\infty}, \mathcal{A}^{\infty}, P^{\infty}) \times ([0,1], \mathcal{B}, \lambda),$$

where $\mathcal{B}$ is the usual Borel σ-algebra. Recall that for a function $h: \Omega \to \mathbf{R}$ the upper integral of h is defined as

$$E^*h: = \inf \left\{ \int g(\omega)dPr(\omega): g \text{ is } S \text{ measurable, } g \geq h \right\}.$$

Also, let $Pr^*(B): = E^*(1_B)$. Throughout we write $Lx: = 1 \vee \log x$ and $L_k x$ for the k'th iteration of Lx.

2. Statement of main results

We take as our starting point a fundamental result of Dudley [4,5] concerning the weak convergence of $\nu_n(f)$, $f \in \mathcal{F}$. Recall [5] that if $\nu_n(f)$ converges weakly to a Gaussian process uniformly over $\mathcal{F} \subset \mathcal{L}^2(A, \mathcal{A}, P)$ then $\mathcal{F}$ is called a P-Donsker class or, equivalently [6], a functional Donsker class. See [6] for a precise statement of the Donsker class property. For ease of reference we recall Dudley's result.

Theorem A. *(cf. Theorem 4.1.1 of [5]). If $\mathcal{F} \subset \mathcal{L}^2(A, \mathcal{A}, P)$, then $\mathcal{F}$ is a functional Donsker class if and only if both*

(a) *$(\mathcal{F}, \rho_P)$ is totally bounded in $\mathcal{L}^2(A, \mathcal{A}, P)$, and*

(b) *for every $\varepsilon > 0$ there is a $\delta > 0$ and an N such that for $n \geq N$,*

$$Pr^*\{ \sup_{f \in \mathcal{F}_\delta} | \int f \, d\nu_n| > \varepsilon\} \leq \varepsilon.$$

If $\|F/L_2 F\| < \infty$ and (a) and (b) hold, then $\mathcal{F}$ is a Strassen log log class [7]; the converse is not generally true. Our first result shows that a slight modification of the asymptotic equicontinuity condition (b) provides an elegant characterization of Strassen log log classes.

Theorem 1.   *Suppose that $\mathcal{F} \subset \mathcal{L}^2(A, \mathcal{A}, P)$ and $F \in \mathcal{L}^2(A, \mathcal{A}, P)$; then $\mathcal{F}$ is a Strassen log log class (for P) if and only if both*

   (a) *$(\mathcal{F}, \rho_P)$ is totally bounded in $\mathcal{L}^2(A, \mathcal{A}, P)$, and*
   (b') *for every $\varepsilon > 0$ there is a $\delta > 0$ such that*

$$Pr^* \{\limsup_{n \to \infty} \sup_{f \in \mathcal{F}_\delta} | \int f \, dI_n| > \varepsilon\} \le \varepsilon.$$

Remark: Theorem 1 should be compared with Theorem 4.1 of Kuelbs [14], which treats the separable case. Kuelbs' theorem cannot be extended to the non-separable case; on the other hand, Kuelbs' characterization of the BLIL (cf. Theorem 4.2 of [14]) does extend to the non-separable case:

Theorem 2.   *Suppose that $\mathcal{F} \subset \mathcal{L}^2(A, \mathcal{A}, P)$ and that $F \in \mathcal{L}^2(A, \mathcal{A}, P)$. Then $\mathcal{F}$ is a log log class (for P) if and only if $\{I_n\}_{n \in \mathbb{N}}$ is stochastically bounded in $(\ell^\infty(\mathcal{F}), \| \cdot \|_\mathcal{F})$; i.e., for all $\varepsilon > 0$ there exists $M > 0$ such that*

$$\sup_{n \ge 1} Pr^* \{\sup_{f \in \mathcal{F}} | \int f \, dI_n| \ge M\} \le \varepsilon.$$

Our next result is essentially a corollary to Theorem 2, and provides conditions on $\mathcal{F}$ and P insuring that $\{I_n\}_{n \in \mathbb{N}}$ is stochastically bounded.

Theorem 3.   *Suppose that $\mathcal{F} \subset \mathcal{L}^2(A, \mathcal{A}, P)$ and that $F \in \mathcal{L}^2(A, \mathcal{A}, P)$. Then $\mathcal{F}$ is a log log class for P if both*

   (a) *$(\mathcal{F}, \rho_P)$ is totally bounded in $\mathcal{L}^2(A, \mathcal{A}, P)$ and*
   (b'') *for every $\varepsilon > 0$ there is a $\delta > 0$ and an N such that for $n \ge N$*

$$Pr^* \{\sup_{f \in \mathcal{F}_\delta} | \int f \, dI_n| > \varepsilon\} \le \varepsilon.$$

As is well known, there is a variety of sufficient conditions on $\mathcal{F}$ and P implying hypotheses (a) and (b) of Theorem A. One of the main ideas of this article is that such sufficient conditions can be weakened and still yield the conclusions of either Theorems 1, 2 or 3.

The following result provides one approach to the verification of hypotheses (a) and (b); one may show that it is actually a consequence of a more general result of Talagrand [21]. The significance of this result lies in the fact that with slight modifications it becomes an LIL statement.

Theorem 4. *Let* $\mathcal{F} \subset \mathcal{L}^2(A, \mathcal{A}, P)$ *be countable and let* $\phi: x \to \exp(x^2)$. *Then* $\mathcal{F}$ *is a functional Donsker class whenever the following conditions are both satisfied:*

(i) $\forall \varepsilon > 0 \exists n_0 := n_0(\varepsilon)$, $U := U(\varepsilon)$, *and* $D := D(\varepsilon)$ *such that for all* $n \geq n_0$:

$$\sup_{f,g \in \mathcal{F}} \mathbf{E}\phi\left(\frac{\nu_n(f) - \nu_n(g)}{D\rho_P(f,g)}\right) \leq U < \infty,$$

(ii) *there is a probability measure* μ *on* $(\mathcal{F}, \rho_P)$ *such that*

$$\limsup_{\delta \downarrow 0} \sup_{f \in \mathcal{F}} \int_0^\delta \left(\log \frac{1}{\mu(g \in \mathcal{F}, \rho_P(f,g) < u)}\right)^{1/2} du = 0.$$

Let us make a few comments about this theorem. Condition (i), which is rather strong, implies stochastic boundedness for the sequence $(\nu_n(f) - \nu_n(g))/\rho_P(f, g)$ in an appropriate Orlicz space. The measure μ of condition (ii) is called a majorizing measure; see [9, 22, 2] for more on the general importance of majorizing measures. Condition (ii) is equivalent to the pregaussian ($G_P BUC$) property for $\mathcal{F}$ [22] and is thus necessary for the Donsker class property [4], but clearly not necessary for the LIL properties. Actually, it is not difficult to see that if we replace the function ϕ in condition (i) by $\Phi: x \to \exp(x^2 L_2 x - 1)$ and the log function in (ii) by $\psi: x \to \left(\frac{\log x}{L_2 \log x}\right)^{1/2}$, then we obtain the log log property for $\mathcal{F}$. For more on majorizing measures and their use in functional limit theorems, see the work of Anderson et.al. [2].

While these remarks show that majorizing measures may be used to obtain a functional BLIL, they also indicate the existence of weak metric entropy integral statements assuring the BLIL property of $\mathcal{F}$. In other words, when $F_{\mathcal{F}} \in L^2$, the above observations suggest that we ought to be able to weaken the metric entropy integral statements

$$\int_0 (\log N(\varepsilon, \mathcal{F}))^{1/2} d\varepsilon < \infty \Rightarrow \mathcal{F} \quad \text{is Donsker,}$$

and

$$(2.1) \qquad \int_0 (\log N_{[\,]}^{(2)}(\varepsilon, \mathcal{F}, P))^{1/2} d\varepsilon < \infty \Rightarrow \mathcal{F} \quad \text{is Donsker}$$

of Pollard [19] and Ossiander [18], respectively, to statements assuring the log log property to $\mathcal{F}$. This is indeed the case, as shown by the following results, which are consequences of Theorem 2, and which answer in the affirmative a conjecture raised by the author [26].

Theorem 5. *Let $\mathcal{F} \subset \mathcal{L}^2(A, \mathcal{A}, P)$ and $F \in \mathcal{L}^2(A, \mathcal{A}, P)$. If*

$$(2.2) \qquad \int_0 \left(\frac{\log N(\varepsilon, \mathcal{F})}{L_2 \log N(\varepsilon, \mathcal{F})} \right)^{1/2} d\varepsilon < \infty,$$

then $\mathcal{F}$ is a log log class for P.

Theorem 6. *Let $\mathcal{F} \subset \mathcal{L}^2(A, \mathcal{A}, P)$ and $F \in \mathcal{L}^2(A, \mathcal{A}, P)$. If*

$$(2.3) \qquad \int_0 \left(\frac{\log N_{[\,]}^{(2)}(\varepsilon, \mathcal{F}, P)}{L_2 \log N_{[\,]}^{(2)}(\varepsilon, \mathcal{F}, P)} \right)^{1/2} d\varepsilon < \infty,$$

then $\mathcal{F}$ is a log log class for P.

It is not clear whether conditions (2.2) and (2.3) also imply that $\mathcal{F}$ is a Strassen log log class. Using an entropy condition which is stronger than (2.3), Kolcinskii [13] provides sufficient conditions assuring the Strassen log log property to uniformly bounded $\mathcal{F}$. We note that statement (2.3), which involves a deterministic entropy condition, is relatively easy to verify.

It is well known to specialists that the Ossiander central limit theorem (2.1) is sharp (e.g., the theorems of Borisov-Dudley-Durst [3,8] and Yukich [23] may be trivially modified to show that (2.1) is sharp). Analogously, our last result shows that the BLIL statement (2.3) is essentially the best possible. As in [23], this may be seen by examining the class of functions

$$\mathcal{G}: = \{x \rightarrow e^{itx} : |t| \leq 1\}$$

and probability measures P on $\mathbf{R}$ satisfying

(2.4) P has a density $f(x)$ with $(f(x) + f(-x))$ decreasing for x large.

We shall also need to consider the function $H_1(x): = (x(\log \frac{1}{x})^2 L_2 \frac{1}{x})^{-1}$, $0 < x \leq e^{-1}$.

Theorem 7.   *Let $\mathcal{G}$ be as above and let P satisfy (2.4); assume that $N_{[\,]}^{(2)}$ satisfies one of the regularity conditions: $\log N_{[\,]}^{(2)}(\varepsilon, \mathcal{G}, P) = O(H_1(\varepsilon))$ or $H_1(\varepsilon) = O(\log N_{[\,]}^{(2)}(\varepsilon, \mathcal{G}, P))$. If $\mathcal{G}$ is a log log class then $\log N_{[\,]}^{(2)}(\varepsilon, \mathcal{G}, P) = O(H_1(\varepsilon))$ and therefore for all $\tau > 0$*

$$(2.5) \qquad \int_0 \left(\frac{\log N_{[\,]}^{(2)}(\varepsilon, \mathcal{G}, P)}{(L_2 \log N_{[\,]}^{(2)}(\varepsilon, \mathcal{G}, P))^{1+\tau}} \right)^{1/2} d\varepsilon < \infty.$$

The proof of (2.5), detailed in [27], rests upon Gaussian randomization [11], Fernique's minorization for stationary Gaussian processes [9], and a delicate comparison between random and non-random L^2 distances for elements of $\mathcal{G}$.

In the remainder of this paper we will provide proofs of the above results; only Theorems 1, 5, and 6 need explicit proofs.

3. Proof of Theorem 1

This section is devoted to the proof of Theorem 1. We begin with two small preliminary remarks.

First, let B denote the unit ball of the Hilbert space

$$H_0 := \{f \in L^2(A, \mathcal{A}, P) \quad \text{such that} \int f dP = 0\}.$$

As in [15], B defines a collection $B_{\mathcal{F}}$ of functions on $\mathcal{F}$ by

$$B_{\mathcal{F}} := \{f \to \int gf \, dP : f \in \mathcal{F}, \ g \in B\}.$$

The set $B_{\mathcal{F}}$ will be used in the proof of our first result.

Also, since $F_{\mathcal{F}} \in L^1(A, \mathcal{A}, P)$, $\mathcal{F}$ is e_P-totally bounded iff it is ρ_P-totally bounded (cf. p. 942 of [11]). Throughout we assume without loss of generality that $\int (F_{\mathcal{F}}(x))^2 dP \leq 1$. Using the above terminology we turn to the

Proof of Theorem 1: We first prove sufficiency. The proof will consist of showing that $B_{\mathcal{F}}$ is compact for $\|\cdot\|_{\mathcal{F}}$, that $(I_n(\cdot))_{n \in \mathbb{N}}$ is a.s. relatively compact, and that the set $L := L(\omega)$ of limit points of the sequence $(I_n(\cdot))_{n \in \mathbb{N}}$ is a.s. equal to $B_{\mathcal{F}}$. We follow the proof of Theorem 2.1 of [15].

Note that for any finite collection $\mathcal{G} = \{g_j\}_{j=1}^r \subset \mathcal{F}$, the set of restrictions of functions in $B_{\mathcal{F}}$ to $\mathcal{G}$ is exactly B. Now for $g \in B$ and $f_1, f_2 \in \mathcal{F}$ we have

$$(3.1) \qquad |\int g(f_1 - f_2)dP| \leq (\int g^2 dP)^{1/2} e_P(f_1, f_2) \leq e_P(f_1, f_2)$$

and also $|\int gf dP| \leq 1$. Thus $B_{\mathcal{F}}$ is a uniformly bounded uniformly equicontinuous collection of functions on $(\mathcal{F}, e_P)$. By the Arzelà-Ascoli theorem, $B_{\mathcal{F}}$ is totally bounded for $\|\cdot\|_{\mathcal{F}}$. Since $B \subset H_0$ is weakly sequentially compact, $B_{\mathcal{F}}$ is also compact (cf. [15]).

Now by hypothesis (ii), for all $\varepsilon > 0$ there exists an integer $k: = k(\varepsilon)$ and functions $f_1, \ldots, f_k \in \mathcal{F}$ such that $\forall f \in \mathcal{F}$ $\exists j$, $1 \le j \le k$, with $\rho_P(f, f_j) \le e_P(f, f_j) < \inf(\delta, \varepsilon)$. Let $\mathcal{F}_j: = \{f \in \mathcal{F}: e_P(f, f_j) < \inf(\delta, \varepsilon)\}$. Clearly, by (ii) we have

$$P^{\infty *}\{\limsup_{n \to \infty} \sup_{1 \le j \le k} \sup_{f \in \mathcal{F}_j} |I_n(f) - I_n(f_j)| > \varepsilon\} \le \varepsilon.$$

Now $\{f_j\}_{j=1}^k$ is a finite collection and thus satisfies the compact LIL a.s. (P^∞); see Lemma 2 of [10]. Thus, for n large enough, there therefore exists a function g belonging to a countable dense subset of $B_\mathcal{F}$ such that

$$\sup_{1 \le j \le k} |I_n(f_j) - g(f_j)| < \varepsilon \quad \text{a.s. } (P^\infty).$$

Thus, on a set with probability greater than $1 - \varepsilon$ we have

$$\begin{aligned}
\|I_n - g\|_\mathcal{F} &\le \sup_{1 \le j \le k} [\sup_{f \in \mathcal{F}_j} |I_n(f) - I_n(f_j)| \\
&\quad + |I_n(f_j) - g(f_j)| + \sup_{f \in \mathcal{F}_j} |g(f) - g(f_j)|] \\
&\le 3\varepsilon,
\end{aligned}$$

where the last inequality follows from (3.1). Letting $\varepsilon \downarrow 0$ we see that the set L of limit points is almost surely a subset of $B_\mathcal{F}$.

To show inclusion in the other direction, note that there exists a countable dense subset $\mathcal{F}'$ of $\mathcal{F}$ for ρ_P. Let $(\mathcal{F}_m)_{m \in \mathbb{N}}$ be an increasing sequence of classes of functions converging towards $\mathcal{F}'$. For each m, the set of functions of L restricted to $\mathcal{F}_m$ is $B_{\mathcal{F}_m}$ (Lemma 2 of [10]). For any $g \in B_\mathcal{F}$ there exist $f_m \in L$ almost surely such that $f_m = g$ on $\mathcal{F}_m$ for all m. Now we can find a subsequence $(f_{m(j)})_{j \in \mathbb{N}}$ which converges towards an element f of L such that $f = g$ on $\mathcal{F}_{m(j)}$ for each integer j. Thus $f = g$ on $\mathcal{F}'$ and by continuity we also have $f = g$ on $\mathcal{F}$. Taking a countable dense set of $B_\mathcal{F}$ gives $B_\mathcal{F} = L$ a.s., completing the proof of the sufficiency.

Conversely, if $\mathcal{F}$ satisfies the compact LIL then (ii) is satisfied since $B_\mathcal{F}$ is a class of functions uniformly continuous for ρ_P. The necessity of (i) is also easily verified, for if $\mathcal{F}$ satisfies the compact LIL then the set

$$\{f \to \int gf\,dP \colon f \in \mathcal{F},\ g \in \mathcal{F}\} \subset B_\mathcal{F}$$

is relatively compact and thus totally bounded in $(\ell^\infty(\mathcal{F}), \|\cdot\|_\mathcal{F})$. Thus $\forall \varepsilon > 0,\ \exists g_1, \ldots, g_k \in \mathcal{F}$ such that $\forall g \in \mathcal{F} \exists j,\ 1 \le j \le k$ with

$$\sup_{f \in \mathcal{F}} |\int (g - Eg)f\,dP - \int (g_j - Eg_j)f\,dP| < \varepsilon^2/2.$$

In particular, for $f = g$ and $f = g_j$ this gives

$$(3.2) \quad |\int g^2\,dP - (\int g\,dP)^2 - \int gg_j\,dP + \int g\,dP \int g_j\,dP| < \varepsilon^2/2$$

and

$$(3.3) \quad |\int gg_j\,dP - \int g\,dP \int g_j\,dP - \int g_j^2\,dP + (\int g_j\,dP)^2| < \varepsilon^2/2.$$

A simple application of the triangle inequality shows that $\rho_P^2(g, g_j)$ is bounded by the sum of (3.2) and (3.3), giving the desired result. ∎

4. Proof of Theorem 5

For the proof of Theorem 5 we shall slightly modify Pollard's approach to the CLT [19]; as in [24], the key idea consists of chaining up to a fixed level depending upon specified n and ε. By Theorem 3 it is enough to show that $\forall \varepsilon > 0 \exists \delta > 0$ and $n_0 := n_0(\varepsilon)$ such that

$$Pr^*\{\sup_{f \in \mathcal{F}_\delta} |I_n(f)| > \varepsilon\} \le \varepsilon \ \forall n \ge n_0(\varepsilon).$$

As in [19], [20] define and choose independently of the X_i, $i \ge 1$, a sequence of r.v. σ_i, $i \ge 1$, on $[0,1]$ with $\lambda(\sigma_i = 2i - 1) = \lambda(\sigma_i =$

$2i) = \frac{1}{2}$, and from this construct the symmetrized processes ν_n^0 and $I_n^0: = (L_2 n)^{-1/2} \nu_n^0$. Writing $\|\cdot\|_{2n}$ for the $L^2(P_{2n})$ semi-norm, it suffices to show $\forall \varepsilon > 0 \,\exists\, \delta > 0$ and $n_0: = n_0(\varepsilon)$ such that

$$Pr^* \{ \sup_{\|f-g\|_{2n} < \delta} |I_n^0(f-g)| > \varepsilon \} \leq \varepsilon \quad \forall n \geq n_0.$$

To this end, put $\delta_i: = 2^{-i}$. Choose finite subclasses $\mathcal{F}_i$ of $\mathcal{F}$ such that $\min_{\mathcal{F}_i} \|f - \phi\|_{2n} \leq \delta_i \|F\|_{2n}$ for each fixed f; $|\mathcal{F}_i| \leq H_i: = H(\delta_i, \mathcal{F})$. Replacing H by a larger more regular function, if necessary, we may assume WLOG that $\sum_{j=1}^{\infty} \exp(-2H_j) < \infty$ and that for all j large

$$(4.1) \qquad\qquad H_j: = H(\delta_j) = o(2^{2j}).$$

Fix $\varepsilon > 0$. The hypotheses imply the existence of a fixed integer $r: = r(\varepsilon)$ simultaneously satisfying the following three inequalities:

$$(4.2) \qquad\qquad \sum_{j=r}^{\infty} 72\delta_j \|F\| (H_j / L_2 H_j)^{1/2} < \varepsilon/8,$$

$$(4.3) \qquad\qquad \sum_{j=r}^{\infty} \exp(-2H_j) < \varepsilon/8, \quad \text{and}$$

$$(4.4) \qquad\qquad 2\exp\{2H_r - \frac{\varepsilon^2 2^{2r}}{200\|F\|^2}\} < \varepsilon/2.$$

Also, let $n_0: = n_0(\varepsilon)$ be the smallest integer such that for all $n \geq n_0$,

$$1 + [\log(16\|F\| n^{1/2}(L_2 n)^{-1/2}/\varepsilon)] > r(\varepsilon)$$

and

$$(4.5) \qquad\qquad \frac{\log\log n}{\log\log(16\|F\|/\varepsilon)^2 n/\log\log n} \geq 4.$$

For all $n \geq n_0$ let $s := s(n) := 1 + [\log(16\|F\| n^{1/2}(L_2 n)^{-1/2}/\varepsilon)]$. Then $s > r$ by definition.

For the remainder of the proof let $n \geq n_0(\varepsilon)$ be fixed. Let S be the realization of $X_1, \ldots, X_{2n}$.

Now given $f \in \mathcal{F}, S$ and r, let $f_r \in \mathcal{F}_r$ be the $2^{-r} \|\cdot\|_{2n}$-approximating function to f. We note, as in [19], that

$$\sup_{\|f-g\|_{2n}<\delta} |I_n^0(f-g)| \leq 2 \sup_{h \in \mathcal{F}} |I_n^0(h-h_r)| + \sup_{\|f-g\|_{2n}<\delta} |I_n^0(f_r-g_r)|.$$

It thus suffices to show that

(4.6) $$Pr^* \{\sup_{h \in \mathcal{F}} |I_n^0(h-h_r)| > \varepsilon/4\} < \varepsilon/2$$

on the set $A: = \{\|F\|_{2n} \leq 2\|F\|\}$ and, for $\delta > 0$ small enough, that

(4.7) $$Pr^* \{\sup_{\|f-g\|_{2n}<\delta} |I_n^0(f_r-g_r)| > \varepsilon/2\} < \varepsilon/2$$

on the same set. As in [19], (4.6) and (4.7) may be shown through the use of a chaining technique via the deterministic sequence $\eta_j, j \geq 1$, defined by

$$\eta_j^2: = 72\|F\|^2 \delta_j^2 H_j / L_2 H_j, \quad j \geq 1,$$

where, by hypothesis, $\sum_{j=1}^{\infty} \eta_j < \infty$. The remainder of the proof shows how to prove (4.6) and (4.7).

Keeping in mind (4.2) and (4.3), holding S fixed, and working on the set A we obtain (cf. [19], eq. (14)):

$$\lambda^\infty \{\sup_{h \in \mathcal{F}} |I_n^0(h-h_r)| > \varepsilon/4\}$$

$$\leq \lambda^\infty \{\sup_{h \in \mathcal{F}} |I_n^0(h-h_r)| > \sum_{j=r}^{s} \eta_j + \varepsilon/8\}$$

$$\leq \sum_{j=r+1}^{s} \lambda^\infty \{\sup_{\mathcal{F}} |I_n^0(f_j-f_{j-1})| > \eta_j\}$$

$$+ \lambda^\infty \{\sup_{\mathcal{F}} |I_n^0(f_s-f)| > \varepsilon/8\}$$

(4.8) $$\leq \sum_{j=r+1}^{s} |\mathcal{F}_j||\mathcal{F}_{j-1}| \sup_{\mathcal{F}} \lambda^\infty \{|I_n^0(f_j-f_{j-1})| > \eta_j\},$$

by definition of s. A straightforward application of Hoeffding's sub-gaussian inequality and the definition of η_j show that (4.8) is bounded by

$$
\begin{aligned}
(4.9) \quad &\leq 2 \sum_{j=r+1}^{s} \exp(2H_j)\exp(-\eta_j^2 L_2 n/72\delta_j^2\|F\|^2) \\
&\leq 2 \sum_{j=r+1}^{s} \exp(2H_j)\exp(-H_j L_2 n/L_2 H_j).
\end{aligned}
$$

Using (4.1) for $j \geq r$ and the definition of s, we deduce for all $r < j \leq s$ that

$$
L_2 H_j \leq L_2 H_s \leq L_2 2^{2s} \leq L_2((16\|F\|/\varepsilon)^2 n/\log\log n);
$$

using this estimate, (4.5) and (4.3) in that order, we see that (4.9) is bounded by $\varepsilon/4$. Integrating over P^∞ yields (4.6).

Next we show (4.7). For this, let $\delta := \delta_r$. If $\|f - g\|_{2n} < \delta$ then on the set A we have

$$
\|f_r - g_r\|_{2n} < \delta + 2\delta_r\|F\|_{2n} \leq 5\delta_r\|F\|.
$$

Hoeffding's inequality shows that (4.7) is thus bounded by

$$
\begin{aligned}
2|\mathcal{F}_r|^2 &\sup_{\|f-g\|_{2n}<\delta} \exp\{-\varepsilon^2/8\|f_r - g_r\|_{2n}^2\} \\
&\leq 2\exp\{2H_r - (\varepsilon^2 2^{2r}/200\|F\|^2)\} \leq \varepsilon/2,
\end{aligned}
$$

where the last inequality follows by (4.4); this completes the proof. ∎

5. Proof of Theorem 6

For the proof of Theorem 6 we will modify the truncation and stratification methods used to obtain Ossiander's CLT [18]. In view of Theorem 2 it will clearly suffice to show that hypothesis (2.3) implies the exponential bound of Theorem 3.4 of [18] with $\phi_n = (\log\log n)^{1/2}$ and $\delta = \operatorname{diam} \mathcal{F}$ there. This exponential bound, together with Theorem 3.4

of [18], give the probabilistic bound required by Theorem 2 and thus complete the proof.

Thus it only remains to show the stated exponential bound of Theorem 3.4 of [18]. Throughout let the $H(\varepsilon)$ of Ossiander [18] be our $H_{[\ ]}(\varepsilon)$.

Using all of the notation and terminology of [18], follow the proof of Theorem 3.3 verbatim until (3.33), where the values of a_k, $\eta_k^{(1)}$ and $\eta_k^{(2)}$ are specified. The key to the proof centers around letting $\phi_n = (\log\log n)^{1/2}$ and choosing a_k, $\eta_k^{(1)}$ and $\eta_k^{(2)}$ in such a way so as to preserve the inequalities (3.38) and (3.39) of [18] and to also preserve, through the new entropy hypothesis (2.3), the small upper bound on the sum

$$\sum_{k\geq 0}\eta_k^{(1)} + \sum_{k\geq 1}\eta_k^{(2)}.$$

As in [18], fix n, $\delta > 0$. For $k \geq 0$ let $\delta_k = \lambda\delta\beta^k$ and $\gamma_k = \sum_{j\leq k} H_{[\ ]}(\delta_j)$, where $0 < \lambda \leq 1$ and $0 < \beta \leq 1$ are constants with values as in Ossiander [18]. Let

$$a_k: = \delta_k(3/8)^{1/2}(\gamma_k/L_2 H(\delta_k) + 2Lk + \eta^2/2\delta)^{-1/2},$$

and let $\eta_k^{(1)}: = 2\delta_k^2/a_{k+1}$, $\eta_k^{(2)}: = \delta_{k-1}^2/a_k$. As in [18] define the sequence k_n in such a way that

$$\frac{na_{k_n+1}}{\log\log n} < C: = \int_0^{\delta_0}\left(\frac{\log N_{[\ ]}(\varepsilon)}{L_2 \log N_{[\ ]}(\varepsilon)}\right)^{1/2} d\varepsilon + \eta\delta_0 \leq \frac{na_{k_n}}{\log\log n}.$$

Using the notation of [18], we first show that for all $k \leq k_n$

(5.1) $$\gamma_k - 3\eta_k^{(1)}\phi_n^2/16a_k \leq -2\phi_n^2 Lk - \eta^2\phi_n^2/2\delta$$

and

(5.2) $$\gamma_k - (3\eta_k^{(2)})^2\phi_n^2/2(3\delta_{k-1}^2 + a_k\eta_k^{(2)}) \leq -2\phi_n^2 Lk - \eta^2\phi_n^2/2\delta.$$

We will only show (5.1) since (5.2) is obtained using similar methods. To show (5.1), we note that its left-hand side is bounded by

$$\gamma_k - \phi_n^2 \gamma_k / L_2 H(\delta_k) - \phi_n^2 (2Lk + (\eta^2/2\delta)),$$

and thus it suffices to show for $k \leq k_n$ that

$$\gamma_k (1 - \phi_n^2 / L_2 H(\delta_k)) \leq 0,$$

or, equivalently,

(5.3) $$\log \log n / \log \log H(\delta_{k_n}) \geq 1.$$

To this end, notice that $(\delta_k/C)(\gamma_k/L_2 H(\delta_k))^{1/2}$ is the k'th term of a series which converges and so it is bounded by one for k large. Therefore, for n large we obtain via the definition of a_n:

$$a_{k_n} \leq \delta_{k_n} \left(\frac{L_2 H(\delta_{k_n})}{\gamma_{k_n}} \right)^{1/2} \leq \frac{CL_2 H(\delta_{k_n})}{\gamma_{k_n}} < \frac{CL_2 H(\delta_{k_n})}{H(\delta_{k_n})}.$$

Using this estimate we obtain via the definition of the sequence k_n:

$$0 < C \leq \frac{na_{k_n}}{\log \log n} \leq \frac{CnL_2 H(\delta_{k_n})}{H(\delta_{k_n})L_2 n}.$$

This last inequality will be satisfied only when $H(\delta_{k_n}) \leq n$, showing (5.3) and thus (5.1). The companion inequality (5.2) is proved similarly.

Having shown that (3.38) and (3.39) of [18] hold, now continue with the proof of Theorem 3.4 there. A close examination of this proof reveals that it only remains to bound $\sum_{k \geq 0} \eta_k^{(1)} + \sum_{k \geq 1} \eta_k^{(2)}$ by $K \left(\int_0^{\delta_0} \left(\frac{H_{\lfloor \rfloor}(u)}{L_2 H_{\lceil \rceil}(u)} \right)^{1/2} du + \eta \delta_0 \right)$ for some fixed constant K. This bound, however, follows easily from the estimates in [18] as well as the inequality

$$\sum_{k \geq 1} \delta_k \left(\frac{\gamma_k}{\log \log H(\delta_k)} \right)^{1/2} \leq (1-\beta)^{-2} \int_0^{\delta} \left(\frac{H(u)}{\log \log H(u)} \right)^{1/2} du.$$

∎

References

[1] K.S. Alexander, *Probability inequalities for empirical processes and a law of the iterated logarithm*, Ann. Prob. 12 (1984), 1041–1067.

[2] N.T. Anderson, E. Giné, M. Ossiander and J. Zinn, *The central limit theorem and the law of the iterated logarithm for empirical processes under local conditions*, Prob. Theory Relat. Fields 77 (1988), 271–305.

[3] I.S. Borisov, *Problem of accuracy of approximation in the central limit theorem for empirical measures*, Siberskij Matematicheskig Zhurnal 24, No. 6 (1983), 14–25. and Siberian Mathematical Journal, July issue, 1984, pp. 833–843.

[4] R.M. Dudley, *Central limit theorems for empirical measures*, Ann. Prob. 6 (1978), 899–929. Correction 7 (1979), pp. 909–911.

[5] —————————, *A course on empirical processes*, pp. 1–142 in "École d'Été de Probabilités Saint-Flour XII-1982", Lecture Notes in Math. 1097 (1984). Springer Verlag 1984.

[6] —————————, *An extended Wichura theorem, definition of Donsker class, and weighted empirical distributions*, pp. 141–178 in "Probability in Banach Spaces V", Lecture Notes in Math. 1153 (1984). Springer Verlag 1985.

[7] R.M. Dudley and W. Philipp, *Invariance principles for sums of Banach space valued random elements and empirical processes*, Z. Wahr. v. Geb. 62 (1983), 509–552.

[8] M. Durst and R.M. Dudley, *Empirical processes, Vapnik-Chervonenkis classes and Poisson processes*, Prob. math. Statist. (Wrocław) 1, No. 2 (1981), 109–115.

[9] X. Fernique, *Regularité des trajectoires des fonctions aléatoires gaussiennes*, pp. 1–96 in "École d'Été de Probabilités Saint-Flour IV-1974", Lecture Notes in Math. 480 (1974). Springer Verlag 1975.

[10] H. Finkelstein, *The law of the iterated logarithm for empirical distributions*, Ann. Math. Statist. 42 (1971), 607–615.

[11] E. Giné and J. Zinn, *Some limit theorems for empirical processes*, Ann. Prob. 12 (1984), 929–989.

[12] —————————, *Lectures on the central limit theorem for empirical processes*, pp. 50–113 in "Probability and Banach Spaces", Proceedings Zaragoza 1985, Lecture Notes in Math.. Springer Verlag 1986.

[13] V.I. Kolcinskii, *On the law of the iterated logarithm in the Strassen form for empirical measures*, Theor. Prob. and Math. Stat. 25 (1982), 43–49.

[14] J. Kuelbs, *Kolmogorov's law of the iterated logarithm for Banach space valued random variables*, Ill. J. Math. 21 (1977), 784–800.

[15] J. Kuelbs and R.M. Dudley, *Log log laws for empirical measures*, Ann. Prob. 8 (1980), 405–418.

[16] M. Ledoux, *Loi du logarithme itéré dans C(S) et fonction caracteristique empirique*, Z. Wahr. v. Geb. 60 (1982), 425–435.

[17] M. Ledoux and M. Talagrand, *Characterization of the law of the iterated logarithm in Banach spaces*, Ann. Prob. 16 (1988), 1242–1264.

[18] M. Ossiander, *A central limit theorem under metric entropy with L^2 bracketing*, Ann. Prob. 15 (1987), 897–919.

[19] D. Pollard, *A central limit theorem for empirical processes*, J. Australian Math. Soc., Ser. A 33 (1982), 235–248.

[20] ————, *Limit theorems for empirical processes*, Z. Wahr. v. Geb. 57 (1981), 181–195.

[21] M. Talagrand, *Donsker classes and random geometry*, Ann. Prob. 15 (1987), 897–919.

[22] ————, *Regularité des processus gaussiens*, C.R. Acad.Sc. Paris, t. 301, Serie I, No. 7 (1985), 379–381.

[23] J.E. Yukich, *Weak convergence of the empirical characteristic function*, Proc. Amer. Math. Soc. 95 (1985), 470–473.

[24] ————, *Uniform exponential bounds for the normalized empirical process*, Studia Mathematica 84 (1986), 71–78.

[25] ————, *Théorème limite centrale et l'entropie metrique dans les espaces de Banach*, C.R. Acad. Sci., Paris, t. 301, Serie I, no. 6 (1985), 333–335.

[26] ————, *Metric entropy and the central limit theorem in Banach spaces*, pp. 113–128 in "Geometrical and Statistical Aspects of Probability in Banach Spaces", Lecture Notes in Math. 1193. Springer Verlag 1986.

[27] ————, *Convergence rates for function classes with applications to the empirical characteristic function*, Illinios Journal Math. 32 (1988), 81–97.

Department of Mathematics, Lehigh University, Bethlehem, PA 18015, USA

Universal Donsker Classes and Type 2

Joel Zinn[*]

In this note we give a partial analogue of a theorem of Pisier [6], which relates the universal Donsker property for classes of sets to a type 2 condition. Actually what we give here is less delicate, since it does not require the useful Lemma 7.13 of Dudley [1]. We start with some

1. Notation and definitions

Let $(S, \mathcal{S})$ be a measurable space and let $X_i : S^{\mathbb{N}} \to S$ be the coordinate maps which will be independent with respect to $P^{\mathbb{N}}$. $\mathcal{F}$ will denote a collection of real-valued measurable functions on S. Further, let $\mathcal{M}$ be the collection of finite signed measures on $\mathcal{S}$. For $\nu \in \mathcal{M}$ and $f \in \mathcal{F}$ define

$$\nu(f) = \int f d\nu$$

$$\|\nu\|_\infty = \sup_{A \in \mathcal{S}} |\nu(A)|$$

$$\|\nu\|_{\mathcal{F}} = \sup_{f \in \mathcal{F}} \left| \int f d\nu \right|$$

and if $\mathcal{F} = \{I_C : C \in \mathcal{C}\}$

$$\|\nu\|_{\mathcal{C}} = \sup_{c \in \mathcal{C}} |\nu(c)|.$$

More generally if Φ is a real-valued function on $\mathcal{F}$,

$$\|\Phi\|_{\mathcal{F}} = \sup_{f \in \mathcal{F}} |\Phi(f)|.$$

To state the results we will need the following definitions.

---

*Partially supported by an NSF grant.

1.1 Definition. *$C \subseteq S$ is a Vapnik-Červonenkis collection (abbreviated $C \in VC$) if there exists $n > 1$ such that*

$$\sup_{\substack{F \subseteq S \\ \text{card } F = n}} card\{C \cap F : C \in C\} < 2^n.$$

As usual, $\{\varepsilon_i\}_{i=1}^{\infty}$, $\{g_i\}_{i=1}^{\infty}$ will denote an independent Rademacher sequence and an independent sequence of $N(0,1)$'s, respectively.

1.2 Definition. *For Banach spaces B_1 and B_2 a linear map $v \colon B_1 \rightarrow B_2$ is said to be a type 2 map if there exists a constant c such that for all $n \geq 1$ and all $x_1, \ldots, x_n \in B_1$,*

$$E\Big\| \sum_{i=1}^{n} g_i v(x_i) \Big\|_{B_2} \leq c \left(\sum_{i=1}^{n} \|x_i\|_{B_1}^2 \right)^{1/2}.$$

We denote by $T_2(v)$ the smallest such constant (if one exists).

To avoid measurability problems we will assume from here on that $\mathcal{F}$ is countable.

1.3 Definition. *$\mathcal{F}$ is a universal Donsker class (abbreviated $\mathcal{F} \in UD$) if for every probability P on S*

$$\left\{ n^{-1/2} \sum_{j=1}^{n} (\delta_{X_i} - P)(f) \right\}_{f \in \mathcal{F}}$$

converges in distribution in $\ell_\infty(\mathcal{F})$ to a Gaussian Radon measure. (See [2],[5] for more details.)

The limiting Gaussian process is denoted G_P and satisfies

$$E\, G_P^2(f) = \int f^2 dP - \Big(\int f dP \Big)^2 \quad \text{for all } f \in \mathcal{F}.$$

If $\Phi \colon (S^{\mathbf{N}}, S^{\mathbf{N}}) \rightarrow \mathbf{R}_+$ we'll write

$$E_P \Phi(X_1, X_2, \ldots) = \int \Phi(X_1, X_2, \ldots) dP^{\mathbf{N}}.$$

1.4 Definition. $\mathcal{F}$ *is a universal bounded Donsker class (abbreviated* $\mathcal{F} \in UBD$*); if for every probability P on S*

$$\sup_n E_P \| n^{-1/2} \sum_{j=1}^{n} (\delta_{X_j} - P) \|_{\mathcal{F}} < \infty.$$

A simple consequence of definition 1.4 is the

1.5 Lemma. $\mathcal{F} \in UBD$ *implies $E\|G_P\|_{\mathcal{F}} < \infty$ for all P.*

Proof: If $\mathcal{F}_0 \subseteq \mathcal{F}$ is finite, then by the finite dimensional central limit theorem

$$E\|G_P\|_{\mathcal{F}_0} = \lim_{n \to \infty} n^{-1/2} E_P \| \sum_{j=1}^{n} (\delta_{X_j} - P) \|_{\mathcal{F}_0}$$

$$\leq \sup_n n^{-1/2} E_P \| \sum_{j=1}^{n} (\delta_{X_j} - P) \|_{\mathcal{F}}.$$

Hence, $E\|G_P\|_{\mathcal{F}} < \infty$.

2. The Theorems

In the case $\mathcal{F} = \{I_C : C \in \mathcal{C}\}$ there are rather complete results. Combining some of the results in Dudley [1] and Durst and Dudley [4] we have the

2.1 Theorem. $\mathcal{C} \in VC$ *if and only if $\mathcal{C} \in UD$.*

In [6] Pisier further clarified the situation by relating these concepts to that of type 2. To be more specific first fix $x_0 \in S$ and let $\mathcal{F}_{x_0} = \{f - f(x_0) : f \in \mathcal{F}\}$. Now for $\nu \in \mathcal{M}$ define $j(\nu) = \{\nu(k)\}_{k \in \mathcal{F}_{x_0}}$. In Theorem 2.3 we'll show j is a type 2 map into $\ell_\infty(\mathcal{F}_{x_0})$ if and only if $\mathcal{F} \in UBD$. A Corollary of some of the results of [1], [4] and [6] is the

2.2 Theorem. *If $\mathcal{F} = \{I_c : C \in \mathcal{C}\}$, then the following are equivalent.*

   (i) $\mathcal{C} \in UD$.

   (ii) $\mathcal{C} \in VC$.

   (iii) j *is type 2.*

The above suggests asking whether an analogous result holds for general $\mathcal{F}$. While we do not answer this completely we can give an analogue of (i) if and only if (iii).

2.3 Theorem. *The following are equivalent.*

(i) $\mathcal{F} \in UBD$.

(ii) *For all probabilities P on S $E\|G_P\|_{\mathcal{F}} < \infty$.*

(iii) $\sup_P E\|G_P\|_{\mathcal{F}} < \infty$.

(iv) *j is type 2.*

Proof: (i) $\Rightarrow$ (ii): This is Lemma 1.5. (ii) $\Rightarrow$ (iii): Assume (ii). By Remark 1.6 we have $E\|G_P\|_{\mathcal{F}} < \infty$ for all P. If $\sup_P E\|G_P\|_{\mathcal{F}} = \infty$, then there exists Q_n with

$$E\|G_{Q_n}\|_{\mathcal{F}} \geq 2^n.$$

Further we may assume $\{G_{Q_n}\}$ are independent. Now, let

$$Q = \sum_{n=1}^{\infty} \frac{1}{2^n} Q_n \quad \text{and} \quad G = \sum_{n=1}^{\infty} \frac{1}{2^{n/2}} G_{Q_n}.$$

Then, by Hölder's inequality for f, $f' \in \mathcal{F}$

$$E|G(f) - G(f')|^2 \leq E|G_Q(f) - G_Q(f')|^2.$$

Hence by the Slepian-Fernique inequality ([5], Theorem 2.17) and Jensen's inequality

$$\frac{1}{2^{k/2}} E\|G_{Q_k}\|_{\mathcal{F}} \leq E\|G\|_{\mathcal{F}} \leq 2E\|G_Q\|_{\mathcal{F}} < \infty,$$

which yields a contradiction. Hence (iii) holds. (iii) $\Rightarrow$ (iv): We first note that, if for $x, y \in S$ we let $P_{x,y} := \frac{1}{2}(\delta_x + \delta_y)$, then we have by (iii) $\sqrt{\frac{2}{\pi}} \sup_{f \in \mathcal{F}} \sup_{x,y \in S} (EG_{P_{x,y}}^2(f))^{1/2} \leq \sup_P E\|G_P\|_{\mathcal{F}} < \infty$. But

$EG^2_{P_{x,y}}(f) = \left(\frac{f(x)-f(y)}{2}\right)^2$. Hence, $\sup_{f \in \mathcal{F}} \sup_{x,y \in S} |f(x)-f(y)| =: \alpha < \infty$. In particular, for $\nu \in \mathcal{M}$, $j(\nu) \in \ell_\infty(\mathcal{F}_{x_0})$. Now, for $\nu_1, \ldots, \nu_n \in \mathcal{M}$, $h = f - f'$, $f, f' \in \mathcal{F}_{x_0}$,

$$E\left|\left(\sum_{j=1}^n g_j \nu_j\right)(h)\right|^2 = \sum_{j=1}^n \left|\int h \, d\nu_j\right|^2$$

$$\leq \sum_{j=1}^n \left(\int |f|^2 d|\nu_j|\right) \|\nu_j\|_\infty.$$

Now let $Q = \left(\sum_{j=1}^n \|\nu_j\|_\infty^2\right)^{-1/2} \sum_{j=1}^n |\nu_j| \|\nu_j\|_\infty$, and $\bar{G}_Q(f) = G_Q(f) + gQ(f)$ where g is $N(0,1)$ and independent of G_Q. Then the Slepian-Fernique inequality gives

$$E\left\|\sum_{j=1}^n g_j \nu_j\right\|_{\mathcal{F}_{x_0}} \leq 2E\|\bar{G}_Q\|_{\mathcal{F}_{x_0}} \left(\sum_{j=1}^n \|\nu_j\|_\infty^2\right)^{1/2}.$$

Hence j is type 2 with $T_2(j) \leq 2[\sup_P E\|G_P\|_{\mathcal{F}} + \sqrt{2/\pi}\alpha]$.

(iv) $\Rightarrow$ (i): This is well known. It follows by symmetrization and comparison of Rademacher coefficients with Gaussian coefficients.

2.4 Remark: One consequence of Theorem 2.3 is that the UBD property depends only on the existence of almost surely bounded versions of certain Gaussian processes, namely G_p. This makes certain results easier to see.

For one example, notice that in proving "(iii) $\Rightarrow$ (iv)" we've shown that $\mathcal{F} \in UBD$ implies $\sup_{f \in \mathcal{F}} \text{diam}(f) < \infty$ (Proposition 1.1 in Dudley [3]).

As another example, note that Proposition 6.2 in Dudley [3] follows from Theorem 2.3 (iii) and Sudakov's minoration ([5], Theorem 2.16).

References

[1] R.M. Dudley, *Central limit theorems for empirical measures*, Ann. Probab. 6 (1978), 899–929.

[2] ——————, *A course on empirical processes*, pp. 1–142 in "École d'Été de Probabilités de Saint-Flour XII-1982", Lecture Notes in Math. 1097. Springer Verlag 1984.

[3] ——————, *Universal Donsker classes and metric entropy*. Preprint.

[4] M Durst and R.M. Dudley, *Empirical processes, Vapnik-Červonenkis classes and Poisson processes*, Probab. Math. Statist. 1 (1981), 109–115.

[5] E. Giné and J. Zinn, *Some limit theorems for empirical processes*, Ann. Probab. 12 (1984), 929–989.

[6] G. Pisier, *Remarques sur les classes de Vapnik-Červonenkis*, Ann. Inst. H. Poincaré 20 (1984), 287–298.

Texas A & M University, Department of Mathematics, College Station, TX 77843.

Progress in Probability

Edited by:

Professor Thomas M. Liggett
Department of Mathematics
University of California
Los Angeles, CA 90024-1555

Professor Charles Newman
Department of Mathematics
University of Arizona
Tucson, AZ 85721

Professor Loren Pitt
Department of Mathematics
University of Virginia
Charlottesville, VA 22903-3199

Progress in Probability includes all aspects of probability theory and stochastic processes, as well as their connections with and applications to other areas such as mathematical statistics and statistical physics. Each volume presents an in-depth look at a specific subject, concentrating on recent research developments. Some volumes are research monographs, while others will consist of collections of papers on a particular topic.

Proposals should be sent directly to the series editors or to Birkhäuser Boston, 675 Massachusetts Avenue, Suite 601, Cambridge, MA 02139.